# Invisible Hands

# Invisible Hands

## SELF-ORGANIZATION AND THE EIGHTEENTH CENTURY

Jonathan Sheehan & Dror Wahrman

The University of Chicago Press CHICAGO & LONDON

The University of Chicago Press, Chicago 60637
The University of Chicago Press, Ltd., London

Published 2015
Paperback edition 2022
Printed in the United States of America

31 30 29 28 27 26 25 24 23 22     1 2 3 4 5

ISBN-13: 978-0-226-75205-1 (cloth)
ISBN-13: 978-0-226-82404-8 (paper)
ISBN-13: 978-0-226-23374-1 (e-book)
DOI: https://doi.org/10.7208/chicago/9780226233741.001.0001

Library of Congress Cataloging-in-Publication Data

Sheehan, Jonathan, 1969– author.
Invisible hands : self-organization and the eighteenth century / Jonathan Sheehan & Dror Wahrman.
pages cm
Includes bibliographical references and index.
ISBN 978-0-226-75205-1 (cloth : alk. paper)
ISBN 978-0-226-23374-1 (e-book)
1. Philosophy, Modern—18th century. 2. Order—Religious aspects. 3. Social sciences—Philosophy. 4. Enlightenment. I. Wahrman, Dror, author. II. Title.
B802.S47 2015
117—dc23
2014044792

♾ This paper meets the requirements of ANSI/NISO Z39.48-1992 (Permanence of Paper).

*This book is dedicated to Shani, Maya,*
*Alice, and Jacob, self-organizers all.*

# CONTENTS

# PREFACE

It is unusual to know who introduced a common word into the English language. It is even more unusual to have a record of its creation as a deliberate act at a particular moment. On January 28, 1754, Horace Walpole, cultural connoisseur and correspondent extraordinaire, announced in a letter to a friend his invention of "*Serendipity*, a very expressive word," in order to denote unintended consequences and accidental discoveries. Walpole's neologism derived from his idiosyncratic reading of "a silly fairy tale, called *the three Princes of Serendip*." It may well have also owed something to Voltaire's *Zadig; or, The Book of Fate* (the title of the English edition that Walpole read upon its publication in 1749), which also drew on an episode based on the Serendip princes to illustrate the play of chance and providence in human life.[1]

Walpole's invention was far from serendipitous, however. The renowned sociologist Robert K. Merton asked a long time ago why it was Walpole, in the middle of the eighteenth century, to whom it fell "to fill a minute space in the English language by creating this strange new word." Merton and his collaborator Elinor Barber suggested two answers: the contemporary interest in the Orient, and "Walpole's idiosyncratic propensities" and eccentric interests.[2] *Invisible Hands* proposes a third answer. Viewed in the broad sweep of the period, we suggest, the significance of the space filled by "serendipity" was anything but minute. It was, rather, a telling clue to a sea change in the eighteenth-century West, a revolution in notions of chance and order, accidents and causality, agency and aggregation. This revolution generated a host of innovations in domains ranging from religion and philosophy through science and economy to law and politics, innovations that became signature features of our modern era. It is this revolution that *Invisible Hands* proposes to unearth, chart, and evaluate.

Where does order come from? And how are seemingly random moments of disorder accounted for—be they comets or cattle plagues or the fall of kings? Once upon a time, the usual story goes, Europeans had a clear answer for both questions. God planned the world, his benevolent oversight maintained its order, and he made his active will known in unusual events that amaze or bewilder mere human beings. But thanks to the "disenchantment of the world" diagnosed by scholars ranging from the historian Keith Thomas to the sociologist Max Weber to the philosopher Marcel Gauchet, the early modern belief in the supernatural involvement of God in daily affairs grew weak and eventually disappeared. In or about 1700, early modern Europeans discovered Newton: enter the mechanical realm of the Enlightenment. Thereafter, God receded from mundane things, perhaps setting the gears of the world in motion, but above all content to let the natural order unfold in precise and orchestrated ways.

Even Newton would have been dissatisfied with this story. Like many of his contemporaries, the great champion of natural law doubted that metronomic regularities could really account for the chance, disorder, and dynamism that are such cornerstones of both natural systems and human experience. Compared to a rigid mechanical order, after all, Christian divine providence was a reassuringly supple tool for explaining the vagaries of life and nature. As organizer of both the cosmic and the quotidian, it provided the means for understanding both heavenly regularity *and* those disruptive moments of chaos. The world is no clockwork, sneered the conservative critic Joseph de Maistre in the 1790s, dismissing mechanical explanations of the world. Only an ordering device that could combine the infinitely various actions of free beings with the regularity of general processes stood a chance of competing with the work of Christian providence: "If one imagines a watch all of whose springs continually vary in power, weight, dimension, form, and position, and which nevertheless invariably shows the right time, one can get some idea of the action of free beings in relation to the plans of the Creator."[3]

*Invisible Hands* charts how eighteenth-century Europeans learned to imagine an order that moved beyond both the mechanical worldview and traditional providence. In place of a universe designed by a divine architect, this alternative conjured up a world vulnerable to the operations of chance yet organized by unseen and powerful forces. Neither a clock nor an engine, this imagined world generated its orders in unexpected ways, shaped by aggregate systems that worked in apparent independence of the parts that made them up. Imagining the world in this way required new appreciations for complexity, new understandings of causality, and new functions for the divine hand. In response to the question of what *did* guarantee harmony and order in a world replete with randomness, contemporaries came up with what we propose was one of the most distinctive and far-reaching innovations in this period: the notion that complex systems, left to their own devices, generated order immanently, without external direction, through *self-organization.*

"Self-organization" is a term used today most commonly by scientists, often in conjunction with the term "emergence." Here is one definition, from a primer on self-organization in biological systems:

> Self-organization is a process in which pattern at the global level of a system emerges solely from numerous interactions among the lower-level components of the system. Moreover, the rules specifying interactions among the system's components are executed using only local information, without reference to the global pattern. . . . Emergent properties are features of a system that arise unexpectedly from interactions among the system's components.[4]

A classic example is a school of fish, which behaves as a superorganism whose movements are unintended by any individual fish. Instead, the behavior of the school emerges from the local interactions between every fish and its close neighbors, relations themselves determined by a few simple rules (one fish cannot get too close to another fish, and the like). Scientists distinguish different modes of interaction in such systems. Negative feedback mechanisms, for instance, restore or maintain equilibrium in a system, damping down the effects of change. Positive mechanisms, by contrast, amplify effects in a system, so that over time the system changes its fundamental nature. In either case, though, the overall order of the system is produced immanently, by the operations of the system itself in dialogue with, but not determined by, the world around it.

The previous paragraph is all you will find in this book about modern scientific uses of the term "self-organization," uses that extend into realms as diverse as cybernetics, systems theory, cognitive science, chemistry, physics, law, and sociology.[5] Although this modern conversation obviously shapes our conceptual toolbox, the notion of self-organization in *Invisible Hands* is more flexible and capacious, and not as rigorous. This book does not chart the genealogy of an idea nor the straightforward prehistory of a twentieth-century scientific theory (though it does endow this theory with a long history largely unacknowledged by its champions). Self-organization does not actually represent for us a single idea, concept, or model. Rather, taking the innovative insights of eighteenth-century Europeans on their own terms, we will speak of the *language* of self-organization and the family of ideas that this language coordinates.

Let us explain. The language of "language" that we use in this book draws on Ludwig Wittgenstein.[6] Whatever its status now, self-organization in the eighteenth century was not a coherent theory with distinct rules governing the relations of the parts, let alone the logical or consistent set of relations one hopes to find in a philosophy. It was something far more nebulous and mobile than that, a "complicated network of similarities overlapping and criss-crossing," in Wittgenstein's terms, with different spheres and outcomes of its application. This language cohered analogically, or in his terms via family resemblance, in which things belong together despite evident differences and even contradictions. Put five siblings in a room, and they all look sufficiently alike to be recognizable as members of one family, even if they do not sport a particular feature common to all. Similarly, our mapping of the variations on eighteenth-century self-organization establishes their recognizable family resemblance even if they differed from each other in significant ways. Key characteristics of this family included the following: operations that distinguished between rules applying to the aggregate and those applying to the parts; causal links that were discontinuous and nonlinear; appreciation for the constructive possibilities of chance and randomness; understandings of providence with latitude for indeterminate events; self-propelling nodes or agents within a system; unintended consequences of actions cumulatively leading to beneficial outcomes; and the mysterious emergence of order out of the chaos of discrete particulars.

As this list makes clear, self-organization tends toward contradiction, brazenly assuming what it needs to prove, relying on endlessly circular accounts of order and its generation. Adam Smith's "invisible hand"—a rather belated example of the many in this book—explains much even as it explains little: Whose hand? Where does it come from? To what end? How do we know it is working? Self-organization also ranges in the *kinds* of worlds it can explain. In politics, for example, it can underwrite the most heterogeneous of projects, from laissez-faire liberalism through constitutional republicanism to unabashed conservatism. To the question posed to Wittgenstein—but is this amorphous thing really a *language*?—we can only reply in his terms, that a language of this sort is understood only by describing examples of its use. The language of self-organization defied easy definition, but it was exactly because of this that it flourished in such different contexts and had such powerful historical effects.

Self-organizing thinking emerged piecemeal, haltingly, sometimes incompletely, and often in the face of considerable resistance to its seemingly heretical ramifications. Yet it gradually moved into the intellectual, cultural, and political mainstream. It took a variety of forms and shapes that we will note as we proceed. In some instances, the emphasis is on the randomness that precedes the harmonious order; in others, on the separate rules that apply to individuals and to their aggregates. In some instances a self-organizing system gravitates toward a stable and durable equilibrium, so that time is not a meaningful variable. In other instances, time is the key variable, as multitudes of events aggregate into orders that themselves shift in dynamic and unpredictable ways. In some cases, the individual components of a system would be found to have a real internal "swerve," some kind of force or activity immanent in their nature. In others, this was not so at all. In some cases, the language of self-organization referred to the realm of ontology: things actually *do* organize themselves. Other times, it operated in the realm of epistemology: however things actually are, we humans understand them *as if* they organized themselves. In exploring these variations, our point is not that some are more authentic examples than others. Rather, we want to show how an elastic language of self-organization emerged during this period, and how different contemporaries availed themselves of its resources and effectively adapted them to different, and sometimes incompatible, needs. We do not intend to lump together apples and oranges, but we are interested in their qualities as fruits.

*Invisible Hands* is the fruit of collaboration between two authors with a deep, shared passion for the question of order in the origins of modernity. At the same time, unlike those instances of collaborative authorship that emerge from a unanimity of position, opinion, and approach, in our case what drew us together into this project was precisely that we are not alike. We differ in our intellectual commitments, one tending more toward intellectual history, the other toward cultural history. We differ in how we read our materials, one interested more in the self-consciousness of textual expression, the other at least as often in a textual unconscious. We differ in our historical sensibilities, one drawn more to moments of historical rupture, the other more to strands of continuity. We differ in our styles of historical argument, one more cumulative and extensive, one more exegetical and intensive.

We have not tried to paper over these differences, or to resolve them into some consensual middle ground, or to hide the seams of our moments of divergence. To the contrary, we are convinced that it is precisely these divergences that turn the whole into something greater than its parts. To echo the themes of our book, we have experienced the benefits of the *concordia discors*, and the reader may in turn experience our text as itself partly emergent and self-organized. Indeed, we have increasingly come to see this as in the very nature of our investigation. Why has the history of the language of self-organization not been written before? Is it not, at least in part, because it is an amphibian kind of creature, partly a set of ideas or constellation of theoretical insights about the world, partly an attitude or strategy or sensibility toward the world? It may well be the case that neither intellectual history nor cultural history alone could have produced a history of self-organization: that it could only be excavated—and presented—through an intertwined mixture of both.

Our jointly told story, therefore, depends equally on the high intellectual analysis (how certain ideas were developed, how their potential and their limits were explored by the most creative thinkers of the era) and on a broader, medium-level analysis (how these ideas caught on, what happened to them as they were transmitted more broadly, what the limits were beyond which they could not reach, and what the social and material circumstances were that contributed to their contemporary resonance). The model of historical causality underlying this book gives neither the sociocultural nor the intellectual side primacy in driving a complex historical story. Although the writing of a book requires chapters to be laid out in a linear sequence, the order of presentation implies no explanatory precedence. Rather, we see them as braided together and reciprocally constitutive, a causal complexity that is often lost when each approach is practiced apart.

It is thus no coincidence that this book moves back and forth between different perspectives, sometimes focusing in depth on certain key thinkers, other times taking a broader look at a whole swathe of contemporary conversation; zeroing in sometimes on the Enlightenment, other times on a more loosely demarcated eighteenth-century culture. We reserve the term "Enlightenment" for that realm of "high" intellectual production, deliberately avoiding the expansion of the term, common in recent scholarship, to include every manifestation of literate eighteenth-century culture. Such a move, it seems to us, results in the loss of the analytical coherence of "Enlightenment" and its reduction to a mere descriptor of a historical period.

The different disciplines and domains of knowledge among which our chapters travel—theology, biology, cognition, economy, society, law, meteorology, international relations, politics—are likewise not a coincidental mix but the result of a series of choices about where we find the language of self-organization to be especially meaningful. At the same time, we recognize that this mixture of emphases, types of analysis, genres of sources, and disparate bodies of knowledge can be taxing. In particular, since our respective intellectual concerns aim at audiences that overlap but are not identical, our respective chapters have a different feel, and make rather different demands on the reader. Sheehan's chapters (1, 4, 5) take up the intellectual history more intensively, while Wahrman's chapters (3, 6, 7) put a greater emphasis on social and cultural developments. Chapter 2 was an experiment in joint writing, with both of our approaches arranged side by side in easy-to-identify sections. Consequently, we anticipate that readers more sympathetic to one or another approach may find some parts easier to get through than others, and for this we ask you, the reader of whatever stripe, for your patience in advance.

Our map of the eighteenth-century intellectual landscape is extensive but not exhaustive, and others may well find parallel conceptual developments elsewhere. As regards the specific materials we present, here the role of chance and circumstances—we might say serendipity—is naturally greater, and the coverage far from complete. These choices also depended on our own expertise, which means that British and German materials are more heavily represented, though we often draw on French and American ones, and less often on examples from other places (Italy, Spain, the Netherlands). As in the case of the historiographical approaches, so in these other aspects of our collaboration we chose to leave visible the seams between our respective contributions.

The main seam in the book, however, is the shift between its first and second halves. The first part of the book answers simple historical questions: When did the language of self-organization emerge in the West as a robust cultural resource? What questions could it help to resolve, and why at this particular juncture? Our story does not offer a full-scale intellectual genealogy of self-organization, tracing its roots back to their origins in the ancient world. In our view, the modern history of self-organization really begins in the late seventeenth century and culminates in an explosive moment in the 1720s. Chapter 1 sets the scene with new departures in the European understanding of God's place in the material and moral world, amounting to a profound reconceptualization of the functions of providence without which self-organization would have been unthinkable. Chapter 2 plots the simultaneous emergence from the late seventeenth century onward of multiple new experiences of complexity, be it in new social structures that introduced aggregation, chance, and choice into the lives of a growing number of people, or in increasingly sophisticated tools of probabilistic analysis, or in new understandings of human agency. Chapter 3 then focuses on the sudden eruption of experimentation with the language of self-organization in the 1720s in the wake of Europe-wide financial crises. It explains why this particular moment witnessed, in multiple contexts, the public emergence of new understandings of complexity, randomness, causality, providence, and order that had been building up since the late seventeenth century.

The second half of *Invisible Hands* maps the application and elaboration of the language of self-organization later in the century, across multiple disciplines and domains of knowledge, and examines its implications and consequences. In part 2, chapters 4 and 5 take up, respectively, the domains of biology and cognition. Living systems were objects of constant eighteenth-century interest, we argue, because they made concrete the limits of mechanical explanation and thus were fertile ground for self-organization. The transformations necessary to generate organized beings from inanimate matter perplexed and astounded scientists, philosophers, and laypeople of all sorts. The result was a ferment of speculation that brought the language of self-organization to bear on everything from the smallest elements of the living world to the very organization of nature. Cognition was an equally dynamic arena for self-organizing approaches. If we stipulate that all thought is a product of experience, then how can we possibly account for all the complexities of the human mind? How do we get from simple inputs to complex outputs? From brain research through the philosophy of causality itself, notions of immanent order were both developed and applied in efforts to produce new genealogies of human cognition.

Part 3 turns from the single human being to larger human systems. Their understanding and description in self-organizing terms proved to be especially productive, given that they are inhabited by unreliable individuals with minds of their own. So whereas the mechanical order was least applicable to the human realm, the whole point of self-organizing systems was that they are full of precisely such erratic, internally moved elements that nonetheless come together to form one ordered whole. Chapter 6 examines varieties of self-organization in social and economic thinking in the second half of the eighteenth century, while also trying to gauge the spread of the language of self-organization as it became more readily available for the analysis of different human phenomena in different contexts. Chapter 7 shifts to the realm of politics, where it turns out that by late in the century self-organization spread far enough and proved pliable enough to be mobilized in support of different and even incompatible political positions, some more surprising than others. By the end of this book, in short, it becomes evident that Adam Smith's famous self-organizing "invisible hand" was but one product, and not a particularly remarkable one, of this much wider and varied cultural-intellectual environment. It was the advent of this environment, ultimately, that was signaled by Horace Walpole's felicitous discovery of "serendipity" as that seemingly random unfolding of events that can indeed bring about, the obliviousness of the actors notwithstanding, those unexpected beneficial outcomes.

As we look back from the end of this project on what we have done, we believe that *Invisible Hands*, beyond its own story, contributes to a revision of current views of the long eighteenth century in two ways. First, it counters the effect of the last generation of scholarship that has splintered the Western eighteenth century between rationality and irrationality, between order and chaos. Instead it opens up a new way of thinking about the Enlightenment in its eighteenth-century context, one that integrates well-known urges for systemic thinking together with the interests in chance, complexity, sentiment, passions, and disorder that have often been posited as their diametric opposites. Self-organization was an important tie that bound these supposedly incompatible strands together, allowing eighteenth-century Europeans to imagine an orderly world built on disorderly foundations.

Second, this book throws new light on the place of the eighteenth century in the transition from the premodern to the modern. It shows how a variety of different developments from the late seventeenth century onward—social developments like urbanization and commercialization, together with intellectual and cultural developments in religion and science—came together and opened up new possibilities for human thought and experience. In the process, we argue, stock narratives that accompany familiar accounts of modernity require rethinking. The rise of individual agency, for example, is a notion challenged by self-organizing systems, in which unfettered agency can be combined with deterministic outcomes. More broadly, *Invisible Hands* shifts our sense of the eighteenth century as the threshold to our disenchanted age. If the dwindling place of God in the modern world gave narrative urgency to Keith Thomas's pathbreaking *Religion and the Decline of Magic* (1971), we see our book as a sequel, as it were, looking forward to a more complicated future. It is a future where we have learned to live without God, and to discover the meanings of the quotidian and the cosmic on our own, but where, nonetheless, the divine can remain a mysterious guarantor of the order of things. It is also a future marked by what J. G. A. Pocock once called the "alienation of man from his history," where one lives without the full knowledge of the designs in which one participates and with only a hazy sense of the moral consequences of one's actions; a world in which we can explain both our own motivations and larger forces of change but cannot integrate the two in a fully satisfactory way. Insofar as the language of self-organization insisted on irreducible differences between individuals (their intentions, their desires, their engagement with the world) and the orders toward which they contribute, it both registered and gave shape to the experience of complexity that was so fundamental to European life with the coming of the eighteenth century.[7]

# * 1 *

PROLOGUE

# Europeans at the Threshold

We begin with a Frenchman, an Italian, a Dutchman, and an Englishman. Not as a bar joke set in an early-eighteenth-century context, but in order to raise the possibility of a perhaps surprising common ground among four individuals who were not very likely to have walked together into a bar.

Charles-Louis de Secondat, baron de Montesquieu, the Frenchman, hatched in 1721 a literary conundrum that three centuries later still has critics scratching their heads. The *Lettres persanes* is a book that conspicuously lacks any kind of order. It moves erratically and unpredictably between letters, between two plots (in Europe and in Persia), and especially between themes, a disorder mirrored in various moments in the contents of the letters themselves. What transformed this absence of an identifiable organizational principle from a puzzling writerly deficiency into a renowned literary parlor game was a self-conscious and tantalizing retrospective from Montesquieu himself, a year before his death and more than thirty years after the fact. He had chosen to write an epistolary work, the older Montesquieu explained, because

> in using the letter form, in which neither the choice of characters, nor the subjects discussed, depend on any design or preconceived plan (*ne sont dépendants d'aucun dessein ou d'aucun plan déjà formé*), the author allowed himself the advantage of being able to unite philosophy, politics, and moral discourse into a novel, and to tie it all together by a secret chain which is in a sense unknown (*une chaîne secrète et, en quelque façon, inconnue*).[1]

It is easy to see why so many readers have tried their skill at unraveling Montesquieu's last riddle, tacked on like Fermat's last theorem in the margin of the *Persian Letters*. But another possibility—suggested, inter alia, by the mounting heap of unsatisfactory solutions for the "secret chain"—is that Montesquieu was not really challenging his readers with a brainteaser. Perhaps, rather, he was reporting to them on a different *mode* of order that secretly underlay this particular work. "Nothing was more pleasing in the *Persian Letters*," he began his rather self-congratulatory retrospective, "than finding there, unexpectedly, a sort of novel." What was surprising was not the disorder but its opposite: here was a collection of letters, each of which was allowed to proceed seemingly more or less at random, without a preconceived plan or deliberate design, that nonetheless came together through an invisible and somehow mysterious mechanism into a unified, meaningful aggregate. Read thus, "*secrète*" and "*inconnue*" may have referred not to a trick played on the reader but to a perhaps "unexpected" principle of aggregate order in the whole emerging from the disorder at the lower level of the constituent parts.[2]

Almost at the same time, in Naples, an Italian scholar was completing his original and quirky masterpiece on the nature of nations and the laws of human history, one he would continue to rework and rewrite until his death some twenty years later. In what again could be read as a retrospect, in the third revised edition of *Scienza nuova* of 1744, Giambattista Vico characterized the "New Science" he had been seeking for a quarter of a century as

> a demonstration, so to speak, of what providence has wrought in history, for it must be a history of the institutions by which, without human discernment or counsel, and often against the designs of men, providence has ordered this great city of the human race.

"Without human discernment or counsel" is uncannily reminiscent of Montesquieu's "without a preconceived plan or deliberate design," each referring in its own context to spontaneously emerging order. In Vico's case, though, the mechanism that transforms random or misguided individual actions into a beneficial aggregate outcome was perhaps invisible but not mysterious: this was divine providence. What confirmed the involvement of providence was "with what ease the institutions are brought into being, by occasions arising often far apart and sometimes quite contrary to the proposals of men, yet fitting together of themselves."[3] Still a bit mysterious, then: somehow, of themselves, those contrary pulls in the history of human social institutions come together to produce an aggregate order that bears little direct relationship to the disorder of its constituent elements.

And while this aspect of the operation of divine providence became more pronounced between the different editions of Vico's work, it was there already in its first incarnation in 1725, that which following Vico himself is now often called *The First New Science*. Let us reflect, Vico had stated then, how,

> through her infinite bounty, Providence disposes the things that particular men or peoples order for their own particular ends, things that would lead them principally to their own ruin, towards a universal end, beyond, and very often contrary to, their every intention."[4]

Providence here was already conjuring a higher-rank social order and purpose from the lower-level chaos of multiple individuals following their particularized purposes.

Set against this formulation, the identity of our third *dramatis persona* becomes rather overdetermined. The Dutchman—now based in London—is of course Bernard Mandeville, probably the most incisive early-eighteenth-century champion of the collective benefits of disorder, as immortalized in the subtitle of his *Fable of the Bees*: "Private Vices, Publick Benefits." As he is also the best known, we need here only remind ourselves of the basics of his social vision. Individuals in Mandeville's world are erratic, driven by "the Passions within, that, unknown to themselves, govern their Will and direct their Behaviour." "The Frailties of Men often work by Contraries": some "are too stingy," others "spen[d] their Money freely," and so on. Their particular choices, however, do not really matter in the greater scheme of things: it is the very multiplicity of these contradictory pulls—and their social effects, most famously on employment and commerce—that guarantees overall social well-being. "The various Ups and Downs compose a Wheel that always turning round gives motion to the whole Machine."[5]

The key, then, to Mandeville's theorizing of the benefits of disorder is the leap between the level of individual persons and that of the aggregate of the actions of the many.

> The short-sighted Vulgar in the Chain of Causes seldom can see further than one Link; but those who can enlarge their View, and will give themselves the Leisure of gazing on the Prospect of concatenated Events, may, in a hundred Places, see *Good* spring up and pullulate from *Evil*, as naturally as Chickens do from Eggs.

Good springs from evil when the effects of "concatenated events" are considered as parts of a greater whole—or when a great number of people are considered as parts of a greater whole: "the Fitness of Man for Society," Mandeville liked to insist, "is hardly perceptible in Individuals, before great Numbers of them are joyn'd together." Here, then, is Mandeville's theory of social order in a single phrase. Put a great number of erratically and selfishly driven people together, shake vigorously, and a unity of purpose and order would emerge spontaneously, independent of any human design. "Men by degrees, and great Length of Time, fall as it were into these Things spontaneously." Again, these notions were uncannily mirrored in Vico's work: "out of the passions of men each bent on his private advantage," Vico asserted, "[divine providence] has made the civil institutions by which they may live in human society."[6]

Finally, the Englishman. In 1733–34 Alexander Pope, England's first professional writer and the greatest English poet of the age, published his renowned *Essay on Man* in the form of epistles to his friend Lord Bolingbroke. It begins with an invitation to Bolingbroke to stroll with Pope in his Twickenham garden:

> A mighty maze! But not without a plan;
> A Wild, where weeds and flow'rs promiscuous shoot,
> Or Garden, tempting with forbidden fruit. (1.6–8)

The significance of this image of the wild, seemingly randomly disordered garden that nonetheless has a plan was emphasized by Pope himself, in notes that he left in manuscript: "The 6th, 7th and 8th lines allude to the Subjects of This Book, the General Order and Design of Providence." And indeed, the *Essay on Man* returns time and again to the general design of providence, a design that secures beneficial order in the unified whole from the promiscuous jumble of natural, or human, contradictory pulls. The first epistle, which begins with the garden's mighty maze, gestures back at this framing device in its final verses:

> All Nature is but Art, unknown to thee;
> All Chance, Direction, which thou canst not see;
> All Discord, Harmony, not understood;
> All partial Evil, universal Good. (1.289–92)

The second epistle, dedicated to the nature of man, makes the point more explicitly with regard to the capricious, unpredictable actions of self-motivated individuals:

> 'Tis but by parts we follow good or ill,
> For, Vice or Virtue, Self directs it still;
> Each individual seeks a sev'ral goal;
> But HEAV'N's great view is One, and that the Whole:
> That counter-works each folly and caprice;
> That disappoints th' effect of ev'ry vice. (2.235–40)

And thus, despite the seeming chaos created by natural forces (Pope's example is the weather) or by human capricious follies,

> The gen'ral ORDER, since the whole began,
> Is kept in Nature, and is kept in Man. (1.171–72)

Once again, randomness, chance, caprice, vice—that is, all manner of chaotic and contradictory disorder on the level of individuals—are transformed into a beneficial order at a higher aggregate level. Small wonder, therefore, that in an earlier, unpublished version of the lines just quoted (2.238–40) Pope had actually included a knowing nod to Bernard Mandeville:

> But HEAV'N's great view is One, and that the Whole:
> That counter-works each folly and caprice;
> *And public good extracts from private vice.*[7]

Like Mandeville's, Pope's vision of aggregate order also presupposes erratic individuals led by their passions in multiple contradictory directions (e.g., 2.161–66). The result—and thus the behavior of the individual—remains unpredictable:

> Extremes in Nature equal ends produce,
> In Man they join to some mysterious use. (2.205–6)

Note again the mystery in these workings of providence.

It is elsewhere, however, in Pope's *Epistle to Bathurst*, published almost simultaneously with the *Essay on Man*, that the poet draws out most clearly the implications of this vision of what propels the individual from the point of view of the greater social whole. Indeed, Pope draws attention to this shift of emphasis by self-quoting and reusing his own lines from the *Essay on Man*:

Hear then the truth: " 'Tis Heav'n each Passion sends,
"And diff'rent Men directs to diff'rent Ends.
"Extremes in Nature equal Good produce,
"Extremes in Man concur to general Use." (*To Bathurst*, lines 161–64)

While these four lines rely closely on the previously quoted passages regarding the directive power of passions, their ending shifts focus from the motivation and drive of the individual to the general social benefit of the concurrence—the mysterious concurrence—of these extremes in individual behavior. Somehow, again, disorder and discordance at the level of multiple individuals are transformed into order and concurrence in the aggregate. And this alchemy of disorder, as it were, seems to happen spontaneously. This point is never more apparent than when Pope sketches the analogous situation in the political realm, when the opposite pulls are those of different interests, and the desired aggregate order is a harmonious polity:

Till jarring int'rests of themselves create
Th' according music of a well-mix'd State. (*Essay on Man* 3.293–94)

The key phrase here, of course—echoing Vico's characterization of the emergence of social institutions out of chaos—is "of themselves."

In 1735 Pope collected the epistles of the *Essay on Man* and four "Moral Essays," of which the *Epistle to Bathurst* was one, into the second volume of his self-edited *Works*, adding copious notes and explanations. Rarely, very rarely, he also inserted cross-references between the different poems. The whole of the *Epistle to Bathurst* boasts one single cross-reference: Pope attached it to the verses just quoted from this poem, directing the reader to the two verses of the *Essay on Man* that we saw two paragraphs ago, likewise dealing with the formation of order from disorder through the mysterious operations of providence.

Furthermore, the same passage in *To Bathurst* is itself cross-referenced in another of the "Moral Essays," the *Epistle to Burlington*. *To Burlington* raises the Mandevillian question of the social value of avaricious man, and comes up with a Mandevillian answer:

Yet hence the *Poor* are cloath'd, the *Hungry* fed;
Health to himself, and to his Infants bread
The Lab'rer bears: What his hard Heart denies,
His Charitable Vanity supplies. (*To Burlington*, lines 169–72)

Avarice and vanity end up supporting the laborer, as Pope spells out in his note: "A bad Taste employs more hands, and diffuses Expence, more than a good one." Here, then, was one specific providential mechanism through which disorder and vice produced that vaunted social benefit. This note is accompanied by two cross-references, to passages we have just discussed in the *Epistle to Bathurst* and the *Essay on Man*—the only two cross-references to be found in the whole of the *Epistle to Burlington*.[8]

These observations are not pedantic. Pope painstakingly referenced and cross-referenced these verses, all addressing the providential emergence of order from the contradictory and erratic actions of individuals, across three different poems that he had originally published within about two years in the early 1730s, and these were the *only* cross-references he chose to add to those poems. Surely this tells us that this particular point was of especial importance to Pope at that juncture, or one that he found to be new and surprising. It was certainly an insight to which Pope very much wanted to draw our attention.

Montesquieu, Vico, Mandeville, Pope. Four European writers who presented visions of order in the aggregate, order emerging—mysteriously, and perhaps by itself—from chaotic disorder at the level of the constituent elements of the greater whole. All four thus elaborated key components of the family of conceptual moves that we group together in this book under the umbrella of "self-organization." To repeat again what we mean by this, the fundamental insight that was common to them all was the notion that even if God was no longer the active hands-on guarantor of order, complex systems, left to their own devices, still generated order immanently, without external direction, through self-organization.

These four writers who never walked together into a bar are not a random group, however. All four resorted to the language of self-organization within just over a decade, in the 1720s and early 1730s. In fact, as we shall soon see, they were far from alone: the same decade saw an unprecedented concentration of similar visions. Some of them, while penned by lesser-known figures, presented the self-organizing insight even more explicitly and in greater detail. Why thus, and why now?

In the 1720s, we want to suggest, European imaginings of order crossed a threshold. For more than half a century previously, key elements in early modern thinking about order and disorder had been expanding in new directions with great transformative potential; be it in understandings of God and providence, or of complexity and causality, or of aggregation and large numbers, or of human agency and choice. Together, these late-seventeenth- and early-eighteenth-century threads, sketched out in the next two chapters, gradually entwined to form the fertile vocabulary of the self-organizing idiom.

And then came a dramatic and unsettling constellation of events that drew this emergent idiom into the open. Around 1720, Europe witnessed an unprecedented international financial crisis that posed a radical challenge to prevailing understandings of causality. The crisis generated a sometimes hysterical preoccupation with the apparent dangers of man-made disorder, of multitudes acting erratically, and of events driven by what contemporary observers experienced as randomness run amok. Chapter 3 anatomizes the particular pressures generated by these events, and how they led some people—including, arguably, the foursome with which we introduced this prologue—to experiment with self-organization as a conceptual move that could provide an effective if revolutionary response to these pressures. The crossing of the threshold in the wake of these unusual events was twofold. First, European self-organizing thinking, albeit still tentative, was now articulated with a heightened clarity, concreteness, and comprehensiveness. Second, self-organization burst into the scene in the 1720s on an unprecedented scale, with many more people than ever before testing its potential and ramifications. It is for this reason that we propose this particular moment in European history as the first public appearance of self-organization as a meaningful (Western) cultural phenomenon.

CHAPTER 1

# Providence and the Orders of the World

## INTRODUCTION: 0 + 1 = EVERYTHING

In 1718, some eighteen months after the death of Germany's great polymath Gottfried Wilhelm Leibniz, the mathematician Johann Wiedeburg published a dissertation praising the virtues of binary arithmetic. On its title page was an imprint of a medal celebrating the strings of zeros and ones that would, he hoped, replace the ten ordinary digits (fig. 1). As Leibniz had already discovered fifty years earlier, simple rules could coordinate these strings, producing an entire system of mathematics consisting only of 0s and 1s. The architecture of modern computing had its origins here, but for Leibniz and Wiedeburg, the excitement of binary lay as much in metaphysics as in mathematics. "*Unus ex nihilo omnia*," "*imago creationis*," and "*unum autem necessarium*" were the truths concealed inside binary: the truths that "the one is necessary," that "the one from nothing [makes] all things," and that this very emergence of something from nothing is "the image of the creation." Leibniz himself had drawn these metaphysical conclusions circa 1696 in conversation with his patron, Duke Rudolf August of Wolfenbüttel-Brunswick, who was apparently dumbstruck by such an elegant version of divine creation. In a 1701 letter to the mathematician Johann Bernoulli, Leibniz commented that Rudolf August had "had 0 and 1 engraved on a medal, which he uses to seal the letters he sends to me." Leibniz sketched at least six versions of his own medal, some more complicated than others, but all featuring some rough equivalent of the epigraph "*unus ex nihilo omnia fecit*," "the one made all from nothing." "In all of mathematics, it is hard to find anything more beautiful" than this

FIGURE 1. *Unus ex nihilo omnia*. From Johann Wiedeburg, *Dissertatio mathematica* (1718). Courtesy of The Bancroft Library, University of California, Berkeley.

marvelous system of arithmetic, he commented in an unpublished fragment, since it shows "the divine footsteps of the creator in the creation by means of the miraculous order and harmony of things."[1]

Two notes resounded in these exclamations. First, a delight that the "one"—God, more or less—should be the only element needed to turn zero into everything. Second, a hope that the binary might answer a challenge that had puzzled thinkers since antiquity. Posed by pre-Socratics like Parmenides, the challenge was essentially this: why is there something rather than nothing? Since "what is, cannot come to be (because it *is* already), and from what is not, nothing could have come to be (because something must be underlying)," then how did the stuff of things come to be? The one question carried a host of others in its wake. What is a thing? How does it change? What is change? The greatest of Greek metaphysicians, Aristotle, certainly felt the force of these. "Nothing can be said without qualification to come from what is not," he agreed, but insisted that "a thing may come to be from what is not in a qualified sense, i.e. accidentally." Although nothing can come from nothing, new properties can attach to extant substances. So when two balls collide in space and bound off in new directions, he argued, their substances stay intact, but their accidents—direction, speed, and so on—change. Although it did not precisely answer Parmenides's question, this powerful description of the order of things served as a foundation of ancient natural science and metaphysics.[2]

From a Christian point of view, however, Aristotle's account of things was perplexing. The Greek denied, after all, that anything new was really possible. His creation was no more than the ringing of changes on an eternal substratum, an a priori matter "receptive of coming-to-be and passing-away," as he put it in *On Generation*. This was a hard pill to swallow for anyone who looked to Genesis for the authoritative account of God's formation of the world. Swallowing the pill—and conforming Aristotle to Christian theology—was one of the major projects of medieval science and philosophy.[3]

This project lost traction in the early modern period, as Aristotle became the bête noire of Europe's new humanistic, philosophical, and scientific movements. No longer did Aristotelian metaphysics answer in a satisfying way the Parmenidean challenge. But the *questions* that Aristotle felt compelled to address were not withdrawn just because his *answers* became increasingly unpersuasive in the face of the new science. How, after all, did something come from nothing? God did this in the book of Genesis, but the mechanism was, to put it mildly, obscure. Leibniz's discovery of the binary seemed an almost magical solution, a formula for the emergence of the world's things from an absolutely minimal substrate. Nothing can come from just nothing, Leibniz admitted. Yet given only the "one," all is possible. The immense variety of things, the world's great diversity of plants, animals, and people, the stars, the mountains: all from nothing and one, coordinated by simple rules but arranged in strings of labyrinthine complexity.

This marvelous binary was just one element of a bigger intellectual pattern that shaped the later seventeenth century. At this moment, and for the next hundred years, the worlds of science, theology, and moral theory converged on the problem of order, its origins, and its dynamics. People from across a huge spectrum of intellectual inquiry enthusiastically generated new answers to old questions: How is the world organized? Where does this organization come from? Is it just the unfolding of divine design? And if so, are human actions free? How do they fit into this design? Their answers were, no surprise, heterogeneous. Like Leibniz's binary, however, they were saturated with a deep appreciation of the world's complexity and an intense curiosity about its origins and its dynamics.

At their heart, this chapter argues, was a renewed and energized vision of providence. In the late seventeenth century, providence was the most dynamic account of how and why order exists in the natural and human worlds. It played an absolutely central role in the development of the new natural and moral sciences. A building block of new materialisms, providential thought activated the things of the world and animated a mechanical universe, giving it purpose, variety, and motion. Yet this was hardly an age of orthodoxy. On the contrary, the older tradition of Christian providence would not emerge from this period intact. The late-seventeenth-century fascination with order and complexity generated a simultaneous expansion and reorientation of the power of providence—indeed, the power of God himself.

In this, the binary was exemplary. Leibniz was surely pleased to find in it new support for God's providential design and activity in the world. Nevertheless, this was balanced by a remarkable revision of what God and providence might be. If God is the "one" in the dyadic system, after all, he need be no more than a vertical scratch on a page. Nor is it the "one" that gives this system its dynamic flexibility, but rather the arithmetical rules that govern the combinations of binary strings. And these rules that generate the complex variety of things—perhaps the equivalent of providence?—apply no less to the "one" than to the zero, apparent organizers of God himself. This was not your grandfather's providence.

This chapter explores both the late-seventeenth-century curiosity about order—its origins, its dynamics, and its complexity—and the workings of a new providentialism. It tours a wide landscape of European intellectual, scientific, and religious production. Across this landscape, it maps a set of common concerns and projects. It describes the widespread dissatisfaction with the so-called mechanical philosophy of Descartes and others, with its stress on regularity, simplicity, and predictability. Mechanism not only left human beings subject to fate and caprice, many felt, but also failed in front of the reality of the world's complexity and variety. The chapter also describes the innovative responses to this. Older Christian verities, it shows, could not simply be reasserted. As the precise symmetry between heaven and earth guaranteed in traditional Christian metaphysics became itself a threat to human freedom and dignity, natural and moral orders began to part ways. The felt alignment between the quotidian chaos that besets us and the regularities that scientists and theologians assure us are the reality of the world's operations weakened. Here at the dawn of the eighteenth century, the experience of things came unhinged from the order that subtends them. New concepts of providence were not so much the means of their ultimate reconciliation as efforts to allow us to live with this dissonance. In short, the chapter maps the new intellectual matrix of the late seventeenth century. It was out of this creative matrix—with its embrace of immanence, its attention to complexity and variety, its impatience with strict rules of causal explanation—that the language of self-organization developed.

## NATURE, FREEDOM, AND THE POWER OF GOD

Interest in complexity took shape against the background of great simplifications. Most notoriously, the simplifications were those of René Descartes, who by the 1640s was a dominant figure in both natural science and philosophy. His approach to the order of things—mechanism, as it has come to be called—was severe and parsimonious. Matter, he insisted, "consists simply in its being something which is extended in length, breadth, and depth." It is divisible and extended; it moves according to the laws of inertia; and it can always be reduced to uniformity once motion is subtracted:

> The matter existing in the universe is . . . one and the same, and it is always recognized as matter simply in virtue of its being extended. All the properties which we clearly perceive in it are reducible to its divisibility and consequent mobility. . . . Any variation in matter or diversity in its many forms depends on motion.

Instead of being filled with Aristotelian properties, let alone divine ones, matter is uniform and its motion predictable. These radical simplifications were correlates of Descartes's stark epistemology. Doubt was his method, as is well known, and given this doubt, that "there exists something extended in length, breadth, and depth" is the most that can be said about the world and its things. A spare model of human capacities demanded a spare model of the world's processes, one whose virtue lay above all in its availability to mathematical formalization.[4]

These models were just that, however: models, useful ways to think about things. For a guarantee that they are *right*, a guarantee that they are correct descriptions of how things actually are, Descartes had to look beyond the models. Since neither the world nor the human experience of it could confirm the truth of the models, in other words, Descartes reached beyond the world altogether, to the nature and qualities of a transcendent God. Although he always insisted that it was "quite against the grain" for him to "mingle religion with philosophy," the Christian divinity surrounded his thought with powerful arms. God's power sustained the very existence of the universe. His goodness guaranteed our ability to understand it. And his wisdom ensured that this universe would behave in regular and orderly ways. God alone, Descartes wrote, "is the author of the motions in the world in so far as they exist and in so far as they are rectilinear." Since he "always acts in the same way," he creates only simple motions, ruled by his laws of nature. Complexity grows from simplicity, then, in a deductive way. Given "a chaos as confused and muddled as any the poets could describe," Descartes maintained, "the laws of nature are sufficient to cause the parts of this chaos to disentangle themselves and arrange themselves . . . [into] a quite perfect world."[5]

In a sense, it is because God is God that we humans can have a science at all. This science is, however, limited. What Amos Funkenstein has called Descartes's "radical voluntarism" left the ultimate truths of the world, its telos and ontology, in the hands of the deity. The world is the way it is because God wanted it so. As a result, we humans work within serious restrictions. We can describe natural phenomena through the mathematics of matter and motion; we can be secure that we are accurate because God is no deceiver. But the *why* of the matter—why there is something rather than nothing, why processes unfold the way they do—is hidden in God's bosom. The search for final causes was "totally useless in physics," Descartes declared. Even the *how* of the matter was obscure to the mortal mind. "In the material world, at least, God is the only genuine causal agent," one recent Descartes scholar has written, "but it is not at all clear how he does it." How one action causes another—what causality *is*, in other words—is a mystery. Our job as knowers is to work inside our limits, to make the models as best we can.[6]

This limitation does not matter if you are trying to model, say, billiard balls. But sometimes the *how* and *why* matter. Sometimes the motivations of the pool players are exactly the point at issue. Are we agents, for example, of our own thoughts? Can we exercise our judgment and behave accordingly? Descartes thought so, but he was far from sure how this might be possible.[7] At stake was nothing less than human freedom, the ability of the human will to orient itself meaningfully to the world.

The very piety (such as it was) of Descartes's project exacerbated rather than relieved the force of this dilemma. Divine omnipotence solved one set of natural and epistemological quandaries, but it re-created others. Making God the only causal agent, most awkwardly, makes him responsible not just for our suffering but also for the malice that motivates so many human actions. When someone piously insists, as Descartes's later acolyte Nicolas Malebranche did in 1674, that "only God . . . is the true cause," then humans and matter alike both get caught in the net of divine necessity. All connections between our will and action in the world—any affirmation of our ability to *do* anything, to orient ourselves meaningfully in the world—are imaginary. "You cannot yourself move your arm . . . or effect the least change in the universe," this Oratorian priest went on. "Here you are in the world without a single power, immobile as a rock, as stupid, as it were, as a stump." You take a step and imagine that *you* lifted your leg. But without God, you are rooted to the spot. For it is God who moves your leg, and at just the right moment to produce the illusion that you did it yourself. The order of things became a "continuous creation, a single volition subsisting and operating simultaneously."[8] All motion, all action, all causation, all from God.

As Malebranche admitted, this piety carried heavy consequences for human liberty, but he also had no principled objection to it. The more Malebranche exalted God's power, the more he shrank the space of moral autonomy. What resulted was a curious convergence of total piety and utter impiety, a commitment to God's omnipotence so complete that the moral frame demanded by the Christian story of sin and salvation—where Christ's judgment and justice usually depended on an element of free moral action—seemed to vanish.

The priest Malebranche could never have abandoned this moral frame entirely, nor abandoned the moral order to the mechanical one. But others in the mechanist camp were less cautious. The natural philosophy of Thomas Hobbes, for example, had no use for *any* moral domain distinct from that of the mechanical. For Hobbes, as he put it in 1664, "there is nothing, that can give a beginning of motion to itself. . . . All . . . determinations must proceed from some other move[me]nt." There is "but one universal cause, which is motion." To imagine a framework of moral action independent of natural necessity would thus be an exercise in futility. "Moral motion is a meer Word," he argued in a great debate with the Anglican bishop John Bramhall, a circumlocution that we use to avoid the real consequences of a world where "God [is] the cause of all motion and of all actions." Hobbes's thought thus teetered between pious and (for a Christian) wicked understandings of the world. God stands at the beginning of all things: "The hand of God [is] the first of all causes," he would write. But the consequences of this were dire for the human ability to orient ethically and morally, to *choose* responsibly between good and evil. "*Liberty*, and *Necessity* are consistent," Hobbes famously argued in his *Leviathan*. Human actions may be free to the extent that they express a human will. But because "every act of mans will, and every desire, . . . proceedeth from some cause, and that from another cause, in a continuall chaine . . . they proceed from *necessity*."[9] You "choose" to sin or not just as much as a dropped stone "chooses" to fall or not.

The result was striking, both for us and for God. For us, we may *feel* as if we are choosing, but the order of the world shows this choice to be an illusion. God hardly fares better. Though present in all things, he is no longer manifest in any particular way. In Malebranche's terms:

> If . . . one drops a rock on the head of passers-by, the rock will always fall at an equal speed, without discerning the piety, or the condition, or the good or bad disposition of those who pass by. If one examines some other effect, one will see the same constancy in the action of the cause that produces it: but *no effect* proves that God acts by particular wills.

Given this reduction in God's felt presence in the world, it is hardly surprising that fellow Cartesians and Catholics alike excoriated this mechanistic piety. From Antoine Arnauld to Bishop Jacques-Bénigne Bossuet, French critics found it far too close to atheism (Hobbesian or Spinozistic) for comfort.[10] If God is responsible for everything, we are responsible for nothing.

## THE MORAL THEOLOGY OF PROVIDENCE

This early modern hybrid of piety and natural philosophy selected one theological aspect of God, his power, and used it like an axe to clear the ground for a mechanist epistemology. It is important to note what a peculiar use this was of the Christian theological archive. Go back, for example, to John Calvin. His keen sense of divine omnipotence bears more than a passing resemblance to our natural philosophers'. Recognition of God's power and his superintendence of all things around us was, for Calvin, a sine qua non of Christianity. But, *pace* Descartes and Hobbes, this recognition was not part of a structure of knowledge. It was instead the cornerstone of *faith*. It did not so much generate positive insights about the world—as it did in Cartesian metaphysics—as orient us in our relationship to God.

For this reason, Calvin's doctrine of providence was at heart a moral doctrine. It was not the " 'how' and 'why' of providence" that was important for Calvin, as Michael Witmore has argued, but the conviction " 'that' God is at work in the world." When faith learns that God made and governs the world, it should immediately conclude that

> he is also everlasting Governor and Preserver—not only in that he drives the celestial frame (*orbis machinam*) as well as its several parts by a universal motion, but also in that he sustains, nourishes, and cares for, everything he has made.

Living with this explicit rebuke to the human illusion of autonomy is, for Calvin, an exhortation to a life of obedience to God. Future events are uncertain for us, for example, and so "we hold them in suspense, as though they might incline to one side or another." Nonetheless, in the hearts of the truly pious, "it nonetheless remains fixed that nothing will take place that the Lord has not previously foreseen." In this sense, Michael Winship has commented, "Calvin discovered and interpreted God's power in the world's contingencies." These contingencies spur the Christian to pious living, in which the conviction of God's absolute power—not the cognitive dissonances this conviction might produce—governs the Christian's moral compass.[11]

This negative theology powerfully controlled the Calvinist imagination, not least its terrifying embrace of predestination. The idea that "before their birth, [God] has laid up for [men] individually the grace that He willed to grant them" was, first and foremost, a moral doctrine. God's power provides solace and lets us "rest in the protection of him . . . whose authority curbs Satan with all his furies." It also confirms man's humiliation in front of God and inspires him to repentance. Similarly, doctrines of fated election and reprobation crush the pride of mankind and put a stop to all of our positive knowledge of the world. "If we annex Providence to order, or motion," the English moralist Sir Thomas Culpeper later argued, "we bring our thoughts into a Labarinth, and trace a circle to no other purpose, but to make our understandings giddy."[12] Negative theology, in short, made a virtue of its own incompletion. In the face of our ignorance, we make an existential decision and cast ourselves onto God's mercy. In this sense, the inscrutability of divine power was a virtue, since it served as an instrument of moral and religious regeneration. Election kills knowledge but births faith.

The Biblical locus classicus for this insight was the book of Job. Job's life, remember, is effectively destroyed on a whimsical bet between God and the Devil. Sons and daughters dead, left only to sit among the ashes and scrape his diseased skin with potsherds, Job refuses to accept any easy explanation for the terrors that afflict him. His friends insist with "windy words" that his trials must be punishments for his sins (16:3). "If you are pure and upright, surely then he will rouse himself for you," they argue, since the world's order must reflect God's justice (8:6). But Job has only scorn for them, knowing his own innocence and enraged at the injustice of his pain. So he demands to hear from God himself why he must suffer. "I would lay my case before him and fill my mouth with arguments," Job angrily cries (23:4). At last, his reproaches—"Behold, I cry out 'Violence!' but I am not answered" (19:7)—are indeed answered. The only answer he gets, however, is more violence. Stamping on the earth and unveiling his tremendous might, God forces Job to cower for having dared demand account for "things too wonderful" (42:3). In this final moment, Job (like us) learns that there *is* an order underlying the universe, albeit an order that he cannot ultimately know.

As with Calvin, the message is not just humiliation. Job *is* humiliated by God, but in this humiliation, he discovers that he matters to God, and that, by extension, the world and its vagaries have a human significance. This discovery was as important in ancient times as in the seventeenth century. "No event could disprove God's care," as Barbara Donagan wrote about Puritan providentialism, however trivial, disastrous, or apparently contradictory.[13] Ancient and modern, providence wove man and cosmos together into a seamless web of order: even if we cannot always understand why things are the way they are, we know that there is a meaning, and that the meaning has something to do with us.

This anthropomorphized understanding of order made providence an immensely powerful tool for explaining even the most bewildering events. From the quotidian to the cosmic, from the human to the natural, all things found their place in it. It explained why "great and epidemical diseases" spread across the land, why it "rain[s] upon one City and not upon another," why some "are in Authority to rule and others in awe and subjection to be ruled." It explained regularities and irregularities alike. Humans are, for example, "not of one disposition: some are hot and fiery suddain, rash and violent . . . others are more solid, of riper wits, and more deliberate." And this is a good thing, since "if all men were hot and heady, the world would soon be in a flame; if all were dull and heavy, it would lie flat and dead." Regular disorders—weather patterns, seasonal changes, human diversity—fall into patterns governed by so-called secondary causes, the "means" by which God proximately enforces his will. But even the truly irregular event—the sudden accident, the comet, and, of course, the miracle—had its place, subject to the rule of God's omnipotent decision, his special providence that operated "without" or even "against means."[14]

Apparent contradictions, moreover, only increased the wonders of providence. "God by his Providence makes contrary things contribute to his glory," the nonconformist English minister Stephen Charnock wrote in 1680, ". . . as in some Engines you shall see wheels have contrary motions, and yet all in order to one and the same end." Providence moves by a "cross and perplex motion," mused the Independent minister and Westminster divine William Strong; "one Wheel acts another by a . . . quite contrary motion." A wheel in the midst of a wheel: the metaphor had both biblical and mechanical resonances. An age fascinated by engines and gears took inspiration from Ezekiel's vision of God's wheels, "rims . . . full of eyes round about" (1:18). "Though matters might seem to run upon wheels . . . to follow their own courses, without any speciall guidance, to go at random," John Wilkins, a Puritan clergyman and later cofounder of the Royal Society, insisted in 1649, "yet *these wheels have eyes in them*."[15]

Relentless and alert, Ezekiel's wheels of providence were powerful enough to resolve disorder into harmony:

> In the naturall body, the variety and dissimilitude of parts, is required to the beauty of the whole; . . . though they may seem to argue some opposition . . . yet look upon them as they stand in relation to the whole frame, and . . . they do each of them conduce to its comlinesse and order. If this lower world had in it no changes and varieties, but were in all respects alike, it would not then be so properly a beautiful world, as a lump or masse.

An older tradition echoed here: the *concordia discors*, the harmony of discordant opposites that found a home in antiquity and the Middle Ages alike. Human beings were a favorite example of this kind of discordant production. The twelfth-century poet Alain of Lille, for example, described the body as a series of resolved opposites, the product of "concordant discord," "dissonant consonance," "similar unsimilarity," "unequal equality," and so on. In the seventeenth century, Thomas Browne's *Religio medici* (1642) was only one text among many that sought to reconcile the contradictions of man, that "masse of Antipathies," as he put it. We are all "at variance" with ourselves, and as such, "we doe but imitate our great selves, the world, whose divided antipathies and contrary faces doe yet carry a charitable regard to the whole by their particular discords, preserving the common harmony."[16] The symmetry between humans and the world at large—a symmetry more generally between microcosm and macrocosm—secured human beings as authoritative readers of the moral and physical qualities of the world. Indeed, it put human beings at the center of this world. Even if we cannot exhaust their meaning, terrible or tumultuous events mean *something*. They happen for our pleasure, our punishment, our correction, and our pedagogy. And even, finally, when our understanding reaches its last limit, and we throw ourselves down in frustration, then, too, there is a Christian lesson for us: that God's ways are not those of men. Faith, not knowledge, is the object of Christian providence.

## ANCIENT HERESIES REBORN

When mechanists discarded this moral frame—when they collapsed moral and natural motions, in Hobbes's terms, and made God's power and governance do scientific rather than religious work—they did not simply threaten to abandon the world to necessity. They also left us with a terrifying question: What if the world is nothing for human beings at all? What if it is a world "*void of God*," where only "blind fortune [does] sit at the helm" of things?[17] If science and philosophy cannot tell us why things are the way they are, if there is no moral meaning to the physical world, then perhaps the world was not made for us at all. Perhaps there is no reason at all why things happen the way they do; perhaps chance is all there is.

The demon of fatal necessity was not the only one summoned by mechanism, in short. It also summoned the laughing demon of chance, the specter of a world hurried pell-mell by chaos alone. This was the impish laughter of Lucretius and Epicurus—one must "laugh and philosophize at the same time," as one Epicurean fragment has it—and its echoes grew deafening in the early modern period.[18] "Never any man denied the Divine providence more hardily" than these ancient materialists, the French philosopher Pierre Bayle commented in his *Historical and Critical Dictionary* (1st ed. 1697).[19] In place of providence, they offered a picture of a world moving according to its own unpredictable logic, an order of things with no order at all:

> For surely not by planning did prime bodies
> find rank and place, nor by intelligence,
> nor did they regulate movement by sworn pact;
> no, changing by myriads myriad ways, they sped
> through the All forever, pounded, pushed, propelled,
> till, trying all kinds of movements and arrangements,
> they came at last into such patterned shapes
> as have created and formed this Sum of Things.[20]

The Lucretian universe emerged in a flickering of atoms, which dance in unpredictable ways, resolving and dissolving the patterned shapes of things. Complicating the dance was what the Roman poet called the *clinamen*, the inbuilt diminutive "swerve" that "at uncertain times and at uncertain points" pushed atoms out of their paths and set off the chain of events that, eventually, resulted in our universe (2.217–18). The result was a swirl of unpredictable change. In Lucretius's terms:

> . . . our world was made by nature, when
> atoms, meeting by chance, spontaneously,
> and joined in myriad useless, fruitless ways,
> at last found patterns, which when thrown together

> became at once the origin of great things—
> earth, sea, and sky, and life in all its forms. (2.1058–63)

All things living and dead were spontaneous and accidental products, the fortuitous joinings of uncomprehending atoms. When he summarized the teachings of Epicurus in 1660, the English scholar Thomas Stanley put it this way: "Atoms concurre, cohere, and are co-apted, not by any designes, but as chance led them." Indeed, atoms do not just move by chance: "The very Atoms themselves are called chance."[21] At the very bottom of things, right at their root, is chaos.

Needless to say, providence had no role to play here. No benevolent or caring God created the world that we live in. Surely, Lucretius chuckled, the "gods most certainly never made the world / for you and me: it stands too full of flaws" (2.180–81). Anticipating the medieval king Alfonso X of Portugal, who apparently commented that had God consulted *him* when making the world, "many things would have been ordered better," Lucretius could not imagine how a perfect God could make a world so messy, so unpleasant, and so chaotic. As Thomas Stanley later put it in his paraphrase of Epicurus:

> I think it may not be ill argued thus: Either God would take away ills and cannot, or he can and will not, or he neither will nor can, or he both will and can. If he would and cannot, he is impotent, and consequently not God; if he can and will not, envious, which is equally contrary to God's nature; if he neither will nor can, he is both envious and impotent, and consequently not God; if he both will and can, which only agrees with God, whence then are the ills? Or, why does he not take them away?[22]

In an important way, Lucretius and Epicurus not only thought that providence was a wrong idea. They also thought it was a *bad* idea. It would be positively immoral, in Lucretius's mind, to imagine that the world came about according to some plan. What kind of God—now laughing through tears—would plan the disaster that is our universe?

Chance might be seen, then, as a way to protect the gods from the reality of the world's evils. The feeling that the gods (or God) might need protection like this also energized one of the great heretical challenges to Christianity in the ancient world, Manichaeism. It was Manichaeism that, in the third and fourth centuries, endlessly pressed the issue of the moral significance of things. How can it be that a good God created evil things? The answer, for the

Manichaeans, was that he *did not*. "We assign," Faustus the Manichaean declared, "all the power to do evil to Hyle [matter], and all the power to do good to God, as is proper."[23] To preserve the eternal goodness of God, Faustus and others elevated evil to an independent nature, a thing uncreated by God and in eternal struggle with him.

This was a serious historical challenge to ancient Christianity: St. Augustine was a Manichaean, for example, before his conversion. Afterward, he dedicated much of his polemical labor to its refutation. Evil was, for the bishop of Hippo, not a *real* thing in the world but only the privation of good, the "desertion of a better [nature]." It was, literally, nothing. "All things in proportion as they are better measured, formed, and ordered, are assuredly good in a higher degree."[24] As they fall away from order, they move toward evil, toward chaos and nothingness.

Augustine's was a neat scheme. But it did not explain why God created a world as fallen as this one. Clarifying this—protecting the goodness of God, in essence—was always the central virtue of the Manichaean worldview. In the late seventeenth century, Pierre Bayle saw the problem clearly. It is one thing to believe that "a Being which exists by itself, which is Necessary and Eternal, must be one," he pointed out. It is quite another to make this Being responsible for the real events of the world. "History, properly speaking, is nothing but a Collection of the Crimes and Misfortunes of Mankind" was Bayle's famous dictum in his article on the Manichaeans. We see "how difficult it would be to refute this false System" when we lay these crimes at the feet of our omnipotent creator.[25]

Augustine did have an answer, but it was not an answer that very many people liked. "The human race's present condition of misery," the Church Father wrote in his *City of God*, "is a punishment for which we can only praise God's justice." The world, he cheerfully argued, is a penal colony for humans subject to an original sin so vile that it makes them only "worthy of an eternal evil." Bayle doubted this would persuade any Manichaean, however, ancient or modern. After all, he asked, how could an "infinitely Good" God not foresee that man would incline to sin? And why would an infinitely powerful God create a man who would so incline?[26] These were, of course, exactly the questions that Epicurus put to the providential system.

The shades of Epicurus and Lucretius thus not only invited chance into the world but also challenged the benevolence of God. No surprise, then, that "Lucretian" would become a stock late-seventeenth-century epithet used to smear anyone seen as threatening to a providential order. For Edward Stillingfleet—the pillar of England's apologetic church establishment in the later seventeenth century—the Lucretian atheist crumbles "the whole *Universe* into *little particles* . . . [and] *grind*[s] the *Sun* to *powder*" and then stands back "to behold the *friskings* and *dancings* about of these little *particles* of *matter*." The world *must* be designed, insisted the Dutch political philosopher Hugo Grotius in a text that was translated and reprinted scores of times in the century. Anyone who thought it a product of chance

> might as well believe that Stones and Timber got casually together, and put themselves into the form of a House. . . . A thing so unlikely, that even a few *Geometrical figures* espied on the *Sea-shore*, gave the beholder just ground to argue, that some *man* had been there; it being evident enough that such things could not proceed from *meer chance*.

Just as the footsteps on Robinson Crusoe's beach marked it with the signs of intelligent life, so did the world, for Grotius, reveal the unmistakable marks of God's footsteps. On the edge of the ocean, the primal site of chaos, design stands out against the immoral "nothing" of chance. To admit this chaos into the world's order would be to crumble the integrity of God, his design, and our confidence about our place in it.[27]

The late seventeenth century saw, in short, a real dismay about the natural and moral order of things. Old questions—Where does order come from and how do we fit into it? What is the relationship between our experience of contingency and the structures that organize the natural and moral worlds?—were made suddenly urgent by the new science (and by more than this, as we will discuss in the next chapter). Yet the older Christian answers could in no way satisfy them. Reapplications of the theology of omnipotence raised more dilemmas for human freedom than they solved. Calvin's logic of humiliation did not have a place in a positive science of order. Nor was antiquity much help. Aristotle was sidelined as retrograde, and ancient materialism was both morally and philosophically indigestible. How then to solve the problem of order, both its origins and its dynamics, both its natural and its moral forms? These were, we suggest, *the* big questions of the age, ones that mechanism and Lucretianism sharpened rather than answered.

## HETEROSIS: A NEW PROVIDENTIAL MATERIALISM

In the natural world, there is a phenomenon called heterosis, in which hybrid offspring surpass the parents in vigor and productivity. Elements foreign to each parent find a home in the child, whose newfound fitness gives it powerful abilities for survival and propagation. In the intellectual world of the late seventeenth century, there was a moment of striking heterosis, an emergence of a new hybrid that quickly took over the intellectual environment of its parents. Its vigor came through a singular act of synthesis, a wedding of the most unlikely bedfellows. The result was materialism beyond mechanism, one that reconnected natural and moral orders in novel ways. It was a form of materialism that fused elements apparently inimical to one another into a new intellectual organism.[28]

If there was one person who can be credited with creating this hybrid, it was Pierre Gassendi, the seventeenth-century French priest and scientist who, in a lifetime of work, dedicated himself to crossing the most unlikely traditions. Gassendi had the remarkable insight that Lucretian and Epicurean metaphysics could provide new scientific energy and moral purpose to materialism, if it was crossed with its most ancient enemy, Christian providence.[29]

When he first turned to Epicurus and Lucretius, Gassendi was looking for alternative natural philosophies. On the one hand, like many others in this period, he was eager to escape what felt like the straitjacket of Aristotle. Gassendi's publishing career in fact began with the 1622 *Exercitationes paradoxicae adversus Aristoteleos*, a defense of "philosophical freedom" against the dominance of the Stagirite texts, and an attack on their physics, metaphysics, and logic. On the other hand, however, he also came quickly to chafe under the bridle of Cartesian mechanism. Gassendi was close to the world of Descartes, an integral part of his social and intellectual networks, which loosely centered on the Minim monk Marin Mersenne. From the 1620s until his death in 1648, Mersenne was famed for the virtual academy of natural philosophers that he gathered in Paris, which included Pierre de Fermat, the young Blaise Pascal, Jan Baptist van Helmont, and Descartes.[30] Gassendi was deep in these circles but was also an early opponent of Descartes's mechanical philosophy. "O Mind," Gassendi addressed Descartes in their correspondence. "O Flesh" was Descartes's rejoinder.

The epithets were accurate. To the epistemological severity of Descartes, Gassendi opposed a form of materialism rooted in the ancient Greek atomists, whose company he never left between 1626 and his death in 1655.[31] The great virtue of Epicurus, for Gassendi, was his refusal to discard the material phenomena of nature, whether in favor of Platonic ideas or Aristotelian substances or Cartesian rationalism. That is, rather than reducing nature's variety to some underlying constants, Epicurean atomism *began* with variety as a fundamental feature of our world. In Gassendi's words:

> Atoms are the elements of things from which first the most attenuated concretions, or molecules, are formed, and after that, larger and larger ones, and [then] small bodies, and bigger ones, and finally the largest ones. . . . [So also] atoms with their innumerable shapes variously composed may produce their [own] diversity of qualities, or appearances, innumerable beyond any proportion.

From the ground up, the world is a heterogeneous place. The elements of things are not undifferentiated points but "innumerable" in their shapes and compositions. From this basic variety grows the infinite diversity of things in the world. Even this growth does not follow from uniform natural laws, moreover, but rather from the various powers inherent in these diverse atoms. Atoms are not, Gassendi happily reported in a reference to Lucretius, "inert or immobile, but most active and mobile." In fact, "*all* motive power compounded in things is from the atoms," he went on. Without mobility, "how could it be that some things move themselves perpetually and boundlessly, that some things are roused from their torpor, and after rest, renew their motion?"[32] From the various qualities of atoms and their various interactions, active elements made the world come alive.

Dynamic immanence entailed a radically different vision of how the world changes, how one state of nature becomes another, and how causality operates. A cause, the Aristotelians had maintained for centuries, had material, formal, efficient, and final components, all of which needed to be understood to describe a natural process. No, said the Neoplatonist, a cause is something incorporeal acting through matter. Foolishness, Descartes cried; we cannot understand the mechanisms of causality at all, but must attribute all action to God's will. Gassendi was persuaded neither by Aristotle's causal categories nor by Plato's "world soul" as the "cause of all things" nor by Descartes's causal transcendentalism. Only Lucretius, Gassendi figured, offered a causal formulation flexible enough to account for the diversity and variety of the world's phenomena. Atoms cause events in themselves and others via their internal *vis motrix*, a principle of change inherent in the material itself. "When it creates itself," Gassendi declared, this *vis motrix* "compels the body in which it resides to move itself, and often even to move another external body."[33]

In a sense, for Gassendi as for the ancient atomists, matter *is* causality. To say that one event causes another is to say that physical matter is at work. In nature, it is possible to "join matter and its agent into one entity that would be therefore capable of making itself . . . and consequently would be both the artisan and the artifact." What would be absurd when considering an "artificial thing"—when the builder and the built are distinct entities—made sense in nature, where "the agent is hidden inside [the material]." All natural substances—all the things in the world that we see with our eyes, touch with our hands, hear with our ears, all the order and all the diversity—are self-creating, formed by their own internal principles, moving each other by the coordination of internal physical mobilities. Atoms "by themselves . . . supply motion to all things," Gassendi wrote, in praise of his pagan poet Lucretius.[34]

Whatever the advantages of the Lucretian genome, it nonetheless bore substantial defects. Most obvious was its utter impiety. But Gassendi cleverly saw that he did not have to treat Lucretius the way a Renaissance antiquary might, as a venerable authority of ancient wisdom. That is, he did not have to accept Lucretius as a total package—atomic physics and irreligion—but could instead treat him as stock, as material for making a new materialism.

The key was a proper crossing, a hybrid form of materialism and Christianity, a materialism in which providence could find a home. After all, Gassendi reasoned, there is no reason to assume that self-movement and divine direction are incommensurable. "Certainly there is no motion that cannot be considered natural seeing that there is none that does not come from the fundamental [parts] of things," Gassendi wrote in an explicitly materialist vein. But, he continues, it was their Author who "willed that their nature should always include a force by which they can be moved." The things of this world "would return to nothing, or to the ancient Chaos, if [God] were to withdraw his hand in the tiniest way."[35] Self-creation—the emergence of something from nothing—then becomes just the means by which God implements his creative plan. The world did not *have* to be made of mobile atoms, but God's decree made it so.

This moment of divine oversight doubtless made Epicurus more digestible for the pious. Even more importantly, it also *corrected* philosophical weaknesses in the Epicurean and Lucretian systems, and especially their recourse to the *clinamen*, that arbitrary swerve that saved the world from fatalistic necessity. The *clinamen* had never made much sense in a metaphysics where every effect has a prior cause. Given an omnipotent God, however, this causal conundrum was no problem. Like all things, Gassendi insisted, the *clinamen* was installed "by God the author, when he created [the atoms]," and thus the swerves find a home in the bosom of his wise decree.[36] From this decree in turn grows the world that we see around us: its causalities, its organisms, its order and its disorder.

This heterosis explains why, despite the impiety of Lucretius and Epicurus, their materialist physical theory was "the classical theory most on the rise in scientific circles" circa 1650, as Richard Kroll has observed.[37] Although hardly without opponents, Lucretianism gathered strength in the mid-seventeenth century, especially in England. A partial list of people interested in Epicurus, in different ways, would include Henry More, Ralph Cudworth, Thomas Hobbes, Robert Hooke, Thomas Browne, Robert Boyle, and many others. Thomas Stanley's account in the third volume of his *History of Philosophy* (1660) helped to introduce Epicurus to a wide audience of interested readers. Although Stanley was no explicit Epicurean, his section on the ancient atomist was twice as long as any other section on ancient philosophy.[38] And its neutral tone gave Epicurus as much authority as any other ancient philosopher. Beginning in the midcentury, too, Lucretius's poem left its Latin confines and entered the vernacular idiom. In the 1650s, for example, the Puritan Lucy Hutchinson completed an English translation of the entire poem—the first known, even if it did remain in manuscript for 350 years.[39] And in 1656, John Evelyn authored the first published (if partial) English Lucretius. When his translation of the first book of *De rerum natura* appeared, "little of the Epicurean philosophy was then known among us," this famous diarist, botanist, and later cofounder of the Royal Society commented. His exposition and notes on Lucretius followed Gassendi (at times quite explicitly), aiming to sanitize the work for a Christian audience, just as the "captive woman was in the old law to have . . . her excrescences pared off, before she was brought as a bride to the bed of her lord," as he colorfully put it.[40]

"Be this our Faith, and farewell all Religion," Evelyn's commentary on Lucretius began, rebuking the poet for his rejection of divine oversight. But by its end, the commentary had moved Lucretius into the Christian fold, precisely by laving his wickedness with the waters of providence. The idea that bodies "continually justling [*sic*], urging and crowding one another by so incessant an inquietude . . . for so many *myriads* . . . of ages . . . chanced (O wondrous chance!) at last, *once*, every one of them, to . . . fall into that goodly Fabrick and admirable *Architecture* of the *Universe*" seems both irreligious and nonsensical, Evelyn admitted. Still, he went on, even this hypothesis "is not of so vast difficulty for a rational, pious, and practical *Philosopher* to believe," as long as one concedes that God "afforded his concurrence" to it, by setting in motion a world so composed.[41]

A Lucretian body but a Christian soul: this was the heterosis that grew wildly in the later seventeenth century. Take, for example, the project of Walter Charleton, a doctor who translated the chemist Jan van Helmont and was deeply engaged with the works of both mechanists like Hobbes and materialists like Gassendi. Over the course of the 1650s, Charleton unfolded a baroque natural philosophy that refuted "Modern *Anti-Epicureans*," conserving pagan materialism inside a Christian frame. Nature is made of "insensible particles," he argued in his 1654 *Physiologia Epicuro-Gassendo-Charletoniana*, but since "*Nature can produce Nothing out of Nothing*," then "there must be some *Common Stock*, or an Universal Something, *Ingenerable*, and *Incorruptible*, of which being praeexistent, all things are Generated." The immediate question—where does this "universal something" come from?—had an equally immediate answer: "The fruitful *Fiat* of God, out of the *Tohu*, or infinite space of *Nothing*, called up a sufficient stock of the First Matter, for the fabrication of the World." It is, he conceded, "but one degree removed from the monstrous absurdities of Lunacy," to imagine that "all Atoms were from all Eternity endowed . . . of their uncreate [*sic*] and independent Essence . . . whereby they are variously agitated in the infinite space." For both pious Christian and scientific reasons, atoms cannot per se move and interact with others, "like Bees variously interweaving in a swarm." But it only took one intellectual shift to make this nonsense sensible: "*that, at their Creation, God invigorated or impraegnated them with an Internal Energy* . . . [and] *that their internal Motive virtue necessitates their perpetual Commotion among themselves.*"[42]

The larger point is this. Providence was not a kind of theological residue preserved for propriety's sake in late-century materialism.[43] Rather, it supplied this materialism with essential resources for its own growth and productivity. From the body of Lucretius came a materialism that paid attention to complexity and variety, and made the diversity of things an inescapable attribute of the universe. But from the providential soul, this materialism discovered not only the origin of things but also their telos, their operating system, their dynamic structure of change. It offered a set of descriptions that enclosed the bony skeleton of things in living flesh, giving both a *moral* and a *material* form to the world and our study of it.

## A THIRD WAY: EMERGENT ORDER AND THE ETHICS OF SCIENCE

"We have a mean conception of the System of the World if we think it is no more than a Clock or a Watch," the clergyman John Edwards declared in 1697. Nor should we believe our untutored eyes, which show us a world "in a Continual Hurly-burly, a Pell-mell, a Confusion."[44] Edwards did not have to choose between metronomic clock and random swirls, since revelation supplied him his security. But for those more philosophically inclined, providential materialism represented a new third way, a materialism beyond necessity and beyond chance. This providential materialism offered a theory of the world's construction. It also entailed a specific *ethic* of investigation, a disposition toward knowledge that would be essential in the ordering projects of the eighteenth century. In this final section, we want to take readers on a trip through the landscape of providential materialism, to show the power of its vision and the sensibility it brought to natural philosophy.

For there are clocks and there are clocks. Modern clocks are so exact that it seems inevitable that they should serve, as with Edwards, as metaphors for mechanical regularity. Early modern clocks were a rather different story, though, needing frequent resetting according to venerable instruments like the sundial. They could, moreover, be fiendishly complex in their structure.[45] The English experimental philosopher Robert Boyle took a particular interest in one of the most complicated clocks, the astrological timepiece in the Strasbourg cathedral built in the fourteenth century. There he discovered other metaphors, beyond regularity and predictability, for approaching the natural world. In the Strasbourg clock, he commented, "the numerous Wheels . . . move several ways, and that without anything either of Knowledge or Design," and yet each moves "as regularly and uniformly as if it . . . were concern'd to do its Duty." From their own perspective, the gears of the clock act freely. From the outside, however, they harmonize into a complicated and formal dance. These views, inside and outside, cannot be collapsed into one. The gears work as if they were dutiful, just as creatures act "as if . . . [they] had a Design of Self-Preservation" and bodies move "as if there were diffus'd through the Universe an intelligent Being."[46] But their own purposes are, at the end of the day, their own. Not even the clockmaker can command complete knowledge, since the mechanism "knows" something he does not, namely, the time. Different orders of knowledge, Boyle hinted, were distributed across a single mechanism.

The ethic of providential materialism began here. Right knowledge began, in its view, in medias res. *Pace* Descartes and others, it did not propose genetic explanations for phenomena or reduce experience to metaphysical first principles. Instead, it took things up only *after* "God gave Motion to Matter" and "so guided the various Motions . . . as to contrive them into the World he design'd they should compose."[47] We must, it insisted, observe and describe a world already in action, diverse, and revealing in its subtle organization the design of its creator. Things act *as if* they had both freedom and purpose. The goal of philosophy is not to filter out the "as if" but to embrace it as a necessary element in all descriptions of the phenomenal world.

Above all, this ethic of science in the later seventeenth century proposed to reconnect, if in novel ways, natural and moral orders. One way it did so was by retooling an older theological idiom, one common in the Middle Ages (and beyond), the idiom of design. Since the age of the scholastics, design theory had insisted firmly on the concinnity of natural and moral things. Thomas Aquinas's *Summa theologiae*, for example, offered the ordered behavior of the world not only as proof of God's existence but also as clue to his nature. The design of the world reflected, Aquinas argued, God's wisdom, his symmetrical ordering of things human and divine, natural and moral. Because the regularities in one domain mirror those in the other, we can therefore *use* the world as a mirror of God's purposes. The late seventeenth century saw an informal Thomist renaissance of design theory, one that insisted on the harmony of orders and the visible purposes in things. When Bishop Samuel Parker demanded that natural philosophers "take the Divine Wisedom into their Mechanicks, and make their several ways of mechanism the effects of his contrivance," it was to the end, in his view, that the "great lines of Morality" might be uncovered in natural things.[48]

As with Gassendi earlier, apologetics were at work here. The early modern version of the Discovery Institute—Robert Boyle's endowed lecture series on Christian natural philosophy—opened in 1692 with an elaborate defense of the world's design against atheism. In it, the textual scholar Richard Bentley showed how natural orders—the movement of the stars, the flow of the tides, and so on—evidenced the hand of providence. Natural *disorders* were his real quarry, however, and he systematically converted each apparent defect of nature into a virtue. The weakness of the human eye was actually a good thing, for example, since with perfect vision, the "smoothest Skin would be beset all over with ragged Scales, and bristly Hairs." The ocean, full of "Chance and Confusion," is commodious to "the true Uses of Life and the Designs of Man's Being on the Earth." Jagged cliffs, raging storms, deserts: Bentley's text relentlessly excluded any threat to the regularity of God's plan. He just as relentlessly attacked anyone who would disagree, those "so abandon'd to sottish credulity as to think," for example, "[t]hat a clod of Earth in a Sack may ever by eternal shaking receive the Fabrick of Man's Body?"[49] Apologetics at its clearest.

But, again as with Gassendi, apologetics were not the only game in town. Design theorists were hardly a coterie of the orthodox. They were not Thomists in any strict sense, nor were they, like Calvin above, interested in reproducing those dynamics of humiliation and regeneration that were so crucial to providence's older theological function. This was, instead, a new form of positive providentialism, in which a theological structure was reoriented to secure new intellectual objectives. First, it confronted pure mechanism with its own explanatory weaknesses, asking how complexity and diversity can be reconciled with simple natural law. Second, it insisted that teleology was imperative for any human understanding of order, both for cognitive reasons and moral ones. And third, it affirmed this new ethic of natural philosophy, a *posture* toward the world that was rooted in a new sense of how to relate the experience of the world to our knowledge of it. Together, these three goals were—as this book will show—immensely productive. They did not so much preserve theological orthodoxy as enable new modes of scientific, moral, and ultimately social thought.

The complex creativity of design theory was already evident in its most famous seventeenth-century advocate, the botanist John Ray, who explicated it in a series of "physico-theological" works published in the 1690s. Boyle had used the term in a 1675 treatise on resurrection, but Ray made it common currency for the century to follow. Ray positioned his *Wisdom of God Manifested in the Works of the Creation* (1691) exactly between chance and mechanism. The *clinamen* of Epicurus, he argued, was an idea that beggared the philosophical imagination. Why do atoms swerve at all? "Do they cast Lots among themselves which shall decline, and which not?" he asked incredulously. Even worse, however, were those mechanists who made the world into a pure automaton, clicking along with no need for God at all. For Ray, this offense "quite outstripped" that of Lucretius: at least the ancient atomists were sincere in their rejection of providence. The "*mechanical Atheists*" simply made God wholly irrelevant.[50]

Given the choice between automata and chaos, Ray preferred the latter. Here he stood in sympathy with earlier poets of Creation like the Frenchman Guillaume Du Bartas, whose 1578 *Divine Weekes* eulogized the "most formelesse *Forme*," the "confus'd Heape, [and] *Chaos* most disforme," with which God began making the world.[51] Or, closer to home, with John Milton, whose *Paradise Lost* conjured a chaos where

> Chance governs all. Into this wild abyss
> (The womb of Nature and perhaps her grave)
> of neither sea, nor shore, nor air, nor fire,
> But all these in their pregnant causes mixed
> Confused'ly, and which thus must ever fight
> Unless th' Almighty Maker them ordain
> His dark materials to create more worlds.[52]

Milton's chaos was hellish, but God still found in it "the seeds of all subsequent good."[53] For Ray, like Milton, chaos was far more susceptible to divine alchemy than soulless automata, chaos understood "not as self-existent and improduced [*sic*]" but created by God at the beginning of things. This *tohu vabohu*, Ray commented, was consonant with the temporality of God's creativity. He did not create "all things in an instant" but "proceeded gradually," seeing positive virtue in the movement from disorder to order.[54] The slow move from dark materials to ordered world confirmed the majesty of God.

In consequence, Ray's book was a hymn to complexity and variety. Nature's chaotic surfaces only amplified the marvel of its hidden organization. Craggy mountains are far better than a "perfectly level Country," Ray insisted, since their bulges and valleys afford pleasant views and healthful sites for human dwelling. Their different terrains engender the "greatest variety" of plants, nurture more animals than the plains, provide rivers that power human mills, and store up snow for the summer cultivating season. Peculiarities of natural organization were particularly fruitful to think with: the dog's ability to make itself vomit, the mole's lack of a tail, the woodpecker's backward toes, the prickly thorns of hedges, and so on, endlessly.[55] The more bizarre and unlikely a natural phenomenon, by this logic, the more decisive a witness to the ordering creativity of God.

Clearly design arguments are curiosities for most modern readers, since they seem to get the order of things so completely backward. Thorns, tails, and toes: these, we would say, are outcomes, not causes, of how organisms fit into the world. But we should not miss their innovations, less in terms of their answers than in their questions. The Platonist Henry More began with this one: How is *difference* introduced into the world? How can a "universall uniform Matter" be variously differentiated into the diverse forms around us? Bentley too wondered how the "diffused Chaos" of disunited matter might be interrupted, that it might not "reign . . . to all eternity." These were real questions, not just apologetic ones. Ray the botanist, who spent much of his early career documenting all the differences between various English plants, realized they were crucial. After all, he commented, the "Formation and Organization of the Bodies of Animals" consist of "such variety and curiosity" that they cannot be explained by a uniform law. Why does one seed become a tree and another a daffodil? Mechanists "prudently . . . break off their System there," when it comes to plants and animals, but any real understanding of nature must take account of these differences.[56]

To explain them, Ray looked for help to the theologian and philosopher Ralph Cudworth. Like More, Cudworth was a crucial intermediary between Cartesian thought and the English scientific scene. By the time that he published his 1678 *True Intellectual System of the Universe*, however, Cudworth had become thoroughly disenchanted with mechanism. Modern mechanists, he insisted, fell into the same atheism that their ancient forerunners had. People like Empedocles and Democritus had wondered, rightly, where things come from, if indeed "*Nothing can come from Nothing, nor go to Nothing*." What led them to "*Atomick Atheism*," however, was their belief that this principle demanded the eradication of all incorporeal agents. "Admitting no other *Causes* of things . . . save the *Material* and *Mechanical* only," they made "the whole World to be nothing else, but a mere *Heap of Dust*, Fortuitously agitated."[57]

What was needed was not mechanism but materialism, a way of bridging matter and a creative principle. Some of the ancients had figured this out, those that "joyned both *Active* and *Passive* Principles together, the *Corporeal* and *Incorporeal*, *Mechanism* and *Life*." A proper materialism, therefore, started here, with what Cudworth called "plastic nature," a force that acts "*for the sake of something*, and *in order to Ends*, Regularly, Artificially and Methodically." Plastic nature is a teleological force—it drives toward greater complexity and variety in an organized fashion—but it operates immanently in natural things. It is "*Incorporated* and *Imbodied in matter*" and operates as its "*Living Soul* or *Law*." It is not, however, a conscious designer. It can neither "properly *Intend* those *Ends* which it acts for" nor have any knowledge of what it does. And yet, unlearned and untaught—"*Magically*," Cudworth called it—it precipitates out the diverse forms that inhabit the natural world. It makes the eye different from the foot; it "designs and Organizes the Heart and Brain"; it coordinates the entire system of animated beings.[58]

Plastic nature, for Cudworth and Ray alike, made a natural philosophy of animate things possible. It guarded against atheistic mechanism. And it guarded against the overenthusiastic theism that demanded God's immediate intervention into every aspect of life. Without plastic nature, Cudwoth wrote,

> it seems that one or other these Two things must be concluded, That Either in the Efformation and Organization of the Bodies of Animals . . . every thing comes to pass *Fortuitously* . . . Or else, that God himself doth all *Immediately*, and as it were with his own Hands, Form the Body of every Gnat and Fly, Insect and Mite, as of other Animals in Generations.

Both of these positions led to impiety. The former denied divine providence, while the latter rendered it "Operose, Sollicitous, and Distractious" (in Cudworth's curious language) by shoving every tiny event into the zone of the miraculous.[59] A healthy materialism embraced teleology in nature without making a fetish of it.

This materialism also demanded what we called above an ethic, a posture toward knowledge. Plastic nature "doth not comprehend the *Reason* of its own action" Cudworth wrote, and we are in a similar position with regard to the world. We cannot approach it with anything but modesty, with a sense of our own limitations. Although Robert Boyle would have no truck with plastic nature, what he called his mechanical philosophy also stipulated our limited abilities to know the world. "Our knowledge is not very *deep*, not reaching with any certainty to the bottom of Things, nor penetrating to their intimate or innermost Natures," he wrote in his 1674 work, *The Excellency of Theology Compar'd with Natural Philosophy*. Historians like Steven Shapin and Simon Shaffer have described this method of restraint as "nescience," a principled refusal of arguments about causality in favor of the establishment of observational matters of fact.[60]

Nescience sounds metaphysically innocent, but it was not. It insisted on empirical restraint even as it routinely committed itself to positions unsupportable by empirical means. The treatise that Boyle attached to his *Excellency of Theology* made this clear. While we may not *know* nature in its all its particulars, he argued in it, we should still feel comfortable projecting this modest empiricism beyond the threshold of our senses. Mechanical, material, and corpuscular principles "can be apply'd to the hidden Transactions that pass among the minute Particles of Bodies," he wrote, not because we can observe these hidden transactions but because there is no reason *not* to think that nature operates uniformly even at the smallest scale. Most of the operations of nature, after all, employ "finer materials" and "more curious contrivances than Art."[61] Beyond the threshold of sight, then, the *real* work of nature takes shape.

Consistently in the late seventeenth century, providential materialism focused its attention downward, beyond the realm of sight. The proper level for emergent order—the place where nothing became something, as materials ordered themselves according to divinely implanted forces and in harmony with a divine plan for man and universe—was the microscopic. With great zeal, design theorists embraced preformation, that popular seventeenth-century doctrine that God implanted the future seeds of all creatures in the moment of creation (for a more expansive discussion, see chapter 4). This doctrine demanded enormous and obviously antiempirical investment in the possibilities of the very small. While it may seem impossible that "the Ovaries of one Female should actually . . . contain the innumerable myriads of Animals . . . as have been and shall be" as long the world lasts, John Ray commented, we cannot imagine the "unconceivable" tininess of things. Metaphysical speculation served as a prosthetic for senses unable to see, and a mind unable to imagine, the microlevels at which the orders of nature emerged.[62]

"All philosophy," declared the French Cartesian Bernard de Fontenelle in 1686, "is based on two things only: curiosity and poor eyesight," as if to say, if only we saw better, speculation could end. But when he further remarked that "beyond the mite an infinite multitude of animals begins for which the mite is an elephant," he raised a real question: Is there any end to this regress? The suspicion that there was not could, in the hands of someone like the mathematician and mystic Blaise Pascal, become the spur for a desperate faith:

> Let a mite be given [a man], with its minute body and parts incomparably minute. . . . Dividing these last things again, let him exhaust his powers of conception. . . . Perhaps he will think that here is the smallest point in nature. I will let him see therein a new abyss. I will paint for him not only the visible universe, but all that he can conceive of nature's immensity in the womb of this abridged atom. Let him see therein an infinity of universes . . . the same thing without end and without cessation.

Like the infinitely large, for Pascal, the infinitely small could ground no positive knowledge at all. Instead, like Calvin's providence, it gave us knowledge only of our own limits. It forced us to see that man is "equally incapable of seeing the Nothing from which he is made, and the Infinite in which he is swallowed up."[63]

Few followed Pascal into such abyssal speculations. When he looked at tiny objects, for example, the English scientist Robert Hooke still assumed that, no matter how far you go, *some* kind of mechanical structures are still at work. His colleague Joseph Glanville agreed: the real workings of things lay "not in the *greater* Masses, but in those little *Threds* and *Springs* which are too *subtile* for the *grossness* of our *unhelp'd Senses*." Of course, there was no way to say what lay beyond the range of the visible. As Richard Westfall pointed out long ago, "a form of matter, unobservable by definition, from which the imagination could construct fictitious mechanisms to generate intractable phenomena" was the "*sine qua non*" of the new science.[64] Little surprise, then, that for the pious materialists of the later seventeenth century, the subvisible would become the very zone where natural and moral orders converged.

You did not need a microscope to affirm this. As early as the 1650s, Gassendi had already woven the very small into the fabric of his pious Epicureanism. Drawing on an explicitly Lucretian idiom, he imagined every seed as "a little machine within which are enclosed in a way impossible to comprehend almost innumerable [other] little machines, each with its own little motions." Gassendi had never seen these machines, but he did not have to. For all his devotion to empirical fact, Gassendi *knew* that beyond the visible lurked the true springs of reality. His English follower Walter Charleton was equally convinced. Even if nature's "Instruments be invisible and imperceptible; yet are we not therefore to conclude, that there are none such at all." On the contrary, many of her "small Engines" exist "by incomputable excesses below the perception of our acutest sense."[65] There was a skeptical valence to these arguments. They clearly limited the abilities of human beings to know their world. But they also invested that space beyond empirical knowledge with substantial explanatory powers.

For third-way theorists, those powers were cast in theological terms. The subvisible was the very arena of providential action. As such, it was a crucial imaginative space for the new materialism. The "Stupendous Smallness of some Animals," in the physician Nehemiah Grew's words, was a clue to the microphysics of providential activity. The "Principles of Bodies . . . may be Infinitely Small, not only beyond all naked or assisted Sense; but beyond all Mathematical Operation and Conception," he argued in 1701. Take an object and divide it up as finely as you want, and still you will not find its animating principle. To imagine that "the Drops of a Mist, a Heap of Sand, or a Sack of Corn, will all have some Life" in them, simply because they are small, is "Subtile Nonsense." What is needed, insisted his colleague the physician George Cheyne, is an "active principle" that "animates, as it were, the dead Mass of Bodies" and that "owes its Origin to something different from Matter and Motion." There is no natural philosophy, he wrote, that "does not require some *Postulates* that are not to be accounted for *Mechanically*," since any mechanical system itself needs an ordering principle, an origin and telos, that it cannot encompass or describe. This was especially apparent in the micromachines of nature, whose hugely complicated organization testified to a principle transcending both Cartesian mechanism and Epicurean materialism: "when the *Complications* are infinite, the Machine [must be] altogether above the Power of *Mechanicks*."[66] Operating below the threshold of human vision, even cognition, are the divine structures that animate the world we inhabit.

Both Grew and Cheyne leaned on the authority of Isaac Newton. In the later seventeenth century, Newton tirelessly searched for a synthesis of divine order and natural philosophy. Like so many in this period, he too made the subvisible into a zone of their reconciliation. On the one hand, Newton's physics presumed uniformity in material things. "The foundation of all natural philosophy" lay in the belief that "every one of the least parts of all bodies is extended, hard, impenetrable, movable, and endowed with a force of inertia." Matter is dead, Newton's first law of motion declared, and his planetary mechanics resolved complex bodies into point-sized centers of gravity.[67] On the other hand, this uniform world was unstable. Motion is "always upon the Decay," Newton wrote in his *Opticks*, matter tends to "wear away, or break in pieces." His correspondent Richard Bentley spelled out the gloomy consequences, a future universe of "squander'd Atoms" separated by "vast intervals of empty Space," dead and cold. The "inequalities" of planetary motion, one popularizer paraphrased Newton, "render at length the present frame of nature unfit for the purposes it now serves." Until "this System wants a Reformation," Newton would have added to this—until God steps in either to keep things in motion or to end things altogether.[68]

The rejuvenation of things at God's behest was all that stood between our world and a sterile wasteland, but this rejuvenation could happen in various ways, not least through God's providence operating via the very mechanisms of nature itself.[69] Nature, Newton wrote to Henry Oldenburg, is "a perpetuall circulatory worker, generating . . . fixed things out of volatile & volatile out of fixed, subtile out of gross, & gross out of subtile." This was in 1675, but his quest to understand this dynamism only intensified as the century waned. When he wrote his queries on the *Opticks*, sometime after 1705, this search had reached something of an end, as he turned to "active principles" as the counterweight to the problem of the decaying world. Because the "variety of Motion . . . is always decreasing," he wrote,

> there is a necessity of conserving and recruiting it by active Principles, such as are the cause of Gravity . . . and the cause of Fermentation, by which the Heart and Blood of Animals are kept in perpetual Motion and Heat. . . . If it were not for these Principles, the Bodies of the Earth, Planets, Comets, Sun, and all things in them, would grow cold and freeze, . . . and all Putrefaction, Generation, Vegetation and Life would cease.

The *Opticks* did not investigate these active principles directly. But it did suggest that these would be at the center of any future natural philosophical enterprise. "Whence arises all that Order and Beauty which we see in the World?" he asked. The answer would come from beyond the range of the human senses, from the "Powers, Vertues, or Forces" that invisibly work their magic. No experience of the world could confirm either their nature or the providential ordering of things. God's miraculous work goes on unseen.[70]

The great Newton biographer, Richard Westfall, attributes Newton's turn to active principles to "the survival of animistic modes of thought" in seventeenth-century natural philosophy.[71] But this seems wrong. We would suggest, instead, that this apparent animism was less the detritus of some ancient natural theory than an integral component of late-seventeenth-century natural philosophy. This natural philosophy was structured in new and curious ways. On the one hand, it insisted on the utter priority of experience and empirical fact. On the other, it structured the space beyond the empirical in such a way that natural and moral orders stayed firmly knit together. Experience and theory were, in one sense, separated, but then reconnected in ways that can seem to modern readers utterly arbitrary. There *is* a natural and moral order to all things, the argument went, but experience can never completely confirm it. Quotidian life and cosmic order are connected, but the connections cannot be worked out in their deductive fullness. This will become a signature structure in the scientific, political, and social argumentation for the next century. And it will become *the* signature structure for that family of ideas that we are calling self-organization.

## LEIBNIZ'S SYNTHESIS: GOD AND NATURE, FREEDOM AND NECESSITY

> The relation of rest to motion is not that of a point to space, but that of nothing to one. . . . whatever moves is not in one place, not even in an infinitesimal instant.
>
> LEIBNIZ, letter to Antoine Arnauld, November 1671

Germany's great polymath—historian, mathematician, philosopher, librarian, diplomat—was the age's greatest confabulator of systems that no one could believe. He was also a tireless poser of questions, a barometer of his intellectual climate. The binary with which we started this chapter was only one of his repeated efforts to figure out how to understand order, complexity, and moral freedom in a modern age. He could be bleak. We are trapped in labyrinths, he worried, labyrinths where "our reason very often goes astray." Two in particular worried him:

> One concerns the great question of the Free and the Necessary, above all in the production and origin of Evil; the other consists in the discussion of continuity and of the indivisibles which appear to be the elements thereof, and where the consideration of the infinite must come in.[72]

His quest for an escape from these was the quest of an age. How to get from the zero to the one, from nothing to something, from chaos to order? Like so many others, Leibniz restlessly searched for Ariadne's thread, seeking to rediscover a morally meaningful world beyond the random motions of an Epicurus or Lucretius, beyond the older pieties of Christianity, and beyond the cold world of Descartes.

Rather than lose ourselves in Leibniz's writings, we will take as our thread an ass. Buridan's ass, that is, the ass famous for dying of indecision trapped between two equally delicious bags of oats. This casuist story was designed to produce an unpleasant dilemma. One might laugh at its absurdity and insist that, of course, the ass must make a choice rather than die. But this would affirm that an effect (the choice) might have no cause at all, no *reason* underlying it. Alternately, one might consign the hapless ass to death and, in doing so, deny his will all freedom. Leibniz's great adversary, the French skeptic Pierre Bayle, knew the story, if from nowhere other than Spinoza's *Ethics*. Here Spinoza had looked at Buridan's ass and chosen option two. "A man placed in such a state of equilibrium . . . will die of hunger and thirst," since his will is fully unable to make *any* free decision. "There is in the mind no volition," as he put it. Bayle's article on Buridan in the *Historical and Critical Dictionary* offered a solution to the predicament. A man could always "disengage himself from the Snare of Equilibrium," Bayle wrote, either by *pretending* to exert his will or by "Lot or Chance," just as a man with "a mind to divert himself" with two courtesans could decide the order of his pleasures by flipping a coin.[73]

A light-hearted story, perhaps. But it had serious consequences. Certainly Leibniz thought so, since it offered the two visions of the world that most plagued him and his age. On the one hand, pure necessity; on the other, the pure randomness of the rolled dice. "The case of Buridan's ass . . . is a fiction that cannot occur in the universe," he insisted. His way out of the labyrinth depended on explaining why, on explaining how it could be both that "absolute necessity . . . does not exist in free actions" and that "there is never an indifference of perfect equipoise," no pure act of random choice.[74]

Leibniz struggled with this and, in the end, resolved it by means of a distinction. On the one hand, he famously argued, there are the absolutely necessary, geometric truths that brook no contradiction: 2 + 2 = 4, for example. On the other, there are the *morally* and *physically* necessary truths, those things that might have been otherwise had things been organized differently. There is no *absolute* necessity that our earth should rotate around the sun, for example. One can imagine a world where different natural laws held the earth perfectly still. Similarly, there is no absolute necessity that Buridan's ass should choose this bag of oats. Since neither choice is impossible, he is "free" to choose either. The ass's choice will not be random, however. "If we were to choose without having anything to prompt us to the choice, chance would then be something actual, resembling what, according to Epicurus, took place in that little deviation of the atoms, occurring without cause or reason," Leibniz wrote in reference to Bayle. Buridan's ass makes his choice from necessity. But it is only a moral necessity, one incurred because he is an individual making *this* choice on *this* day, who must thus incline one way or another. It is necessary "only *ex hypothesi*, and so to speak by accident." Natural and moral orders thus stand shoulder to shoulder. They are linked together by God's wise organization of things. And the necessity that they impose on us—natural and moral creatures alike—is less than absolute. Order and spontaneity, causality and freedom, are perfectly harmonious.[75]

The precondition for this harmony is the world's enormous complexity. "There is no portion of matter, however tiny, in which there is not a world of creatures," he wrote in that 1679 essay, and "no created substance . . . whose complex concept . . . does not contain the whole universe." Only God, though, "can see everything that is in this series." As a result, we humans can never demonstrate the necessity of contingent truths, never reduce them "to an equation or an identity, [because] the analysis proceeds to infinity." Although "a perfect indifference is a chimerical . . . supposition"—that is, no choice is free from causal governance—still no finite amount of retrogression can tell us why the ass chooses *this* bag of oats and not another.[76]

Between theoretical order and the order *available to us*, as limited knowers of things, there is an unbridgeable space. But it is exactly this space between theory and experience that guarantees both human freedom and divine justice. After all, it would inspire only "horror," Leibniz wrote, if we imagined that God abandoned men "to the devil his enemy, who torments them eternally and makes them curse their Creator . . . [solely because he] exposed their parents to a temptation that he knew they would not resist."[77] The causes of Adam's choice, like those of Buridan's ass, are infinitely complex. The Fall occurred in a zone of freedom created by the inability of anyone besides God to predict the fortunes of the apple.

Throughout his life, then, Leibniz reveled in the complexity of the world's orders. "In any grain of sand whatever there is not just a world, but even an infinity of worlds," he wrote in an early essay on motion. Only at the submicroscopic level can anything like causality or motion be understood. And these causes themselves dwindle in size and predictability. Our gross choices arise from

> insensible stimuli, which, mingled with the actions of objects and of our bodily interiors, make us find one direction of movement more comfortable than the other. In German, the word for the balance of a clock is *Unruhe*—which also means disquiet; and one can take that for a model of how it is in our bodies, which can never be perfectly at their ease. . . . There is a perpetual conflict which makes up, so to speak, the disquiet of our clock.[78]

This perpetual conflict leaves us vulnerable to the smallest causal agents. Large effects in the world result, then, from the tiniest of circumstances, imperceptible nudges in one direction or another.

Significantly, Bayle spoke a similar language. In one of the great skeptical works on providence in the period, his *Various Thoughts on the Occasion of a Comet* (1682), the Frenchman scorned the notion of special providence, the idea that the world can offer a reliable guide either to God's will or the future more generally. The complexity of things, he would ever insist, belied any such predictive knowledge. The world is a place filled with uncertainty, where the "great events that overturn mankind" are products of the tiniest effects. "One blow of a horse's hoof, which in other circumstances would have been of no use, would have saved the lives of millions of men who perished because of Alexander and would have spared the world an infinite number of miseries," he wrote. And he was astonished by how even simple actions are "executed by the concurrence of an infinite number of occasional causes, the infinite diversity of which divides in a manner the general cause into an infinite number of particular causes that no longer appear to depend on one another."[79]

For Bayle, the only reasonable response to this mess was skepticism. Leibniz disagreed. Reveling in the "wonderful variety" of the natural and moral worlds, Leibniz made the infinite a subject of knowledge. He did this, of course, with the calculus. More importantly for our purposes, he did it via extreme metaphysical inventiveness. Underneath all things, he argued for nearly forty years, lies a single, infinitely various substance, the solution to the dilemmas of his age, what he called the monad. Here, in the language of Leibniz's binary, was the "one" out of which the zero became everything. The monad is the one thing that makes our world. And it is not a dead thing, mere matter, but alive and active. Each monad, Leibniz wrote,

> may be called a *living being* . . . . Every organic body of a living being is a kind of divine machine, or natural automaton, infinitely surpassing all artificial automatons. . . . The machines of nature, that is to say, living bodies, are still machines in their smallest parts *ad infinitum*.

These organic singularities permeate the world. All things are "the source of their internal activities"; "there is . . . nothing uncultivated, or sterile, or dead in the universe." Even the most stubborn stone "is full of animated bodies," nor is anything "so small that there are not in it animated bodies, or at least such as are endowed . . . with the *vital principle*." Like seeds, monads are not helpless victims of causality but rather themselves the sources of change. "Its present is big with its future," as Leibniz lyrically put it.[80]

Order and organization, then, are immanent in the materials, yet wisely disposed. Nature and morality have met once more. Yet, again typically, we can never *experience* the world as if that were actually so. Metaphysically speaking, we know that God made the monads so that they would move perfectly in sync with causal law. But they are spontaneous *to us*, insofar as we cannot—unlike God—see those infinities of causes and effects at work in any one motion. "Experience is unable to make me recognize," Leibniz wrote to his great correspondent, the Port-Royal theologian Antoine Arnauld, "a great number of insensible things in the body" which the "general consideration of the nature of bodies . . . might convince me." Our encounter with the world is "almost like the confused murmuring which is heard by those who approach the shore of the sea. It comes from the continual beating of innumerable waves."[81] We know there is an order to the waves, but the ocean gives us not information but noise. Theory can confirm that the wisdom of God organized the world we inhabit. But our experience of this world can never do the same. Retrogression stalls in the divine and infinite series that makes up human existence.

No surprise, then, that few people would actually believe in Leibniz's metaphysics, since their moral force (that is, their promise of freedom) depends on the impossibility of experiencing the true nature of things. There is a grain of truth in Voltaire's satire of Leibniz as Dr. Pangloss, insofar as it captured the tension between theoretical knowledge (everything is for the best) and experience (despite all evidence to the contrary). It is hardly a comfort to know that the aggregate good in the galaxy makes up for the real evil that we experience right now.

But the satire also misses the real substance of Leibniz's high-wire arguments. In his fantastic architecture of monads, seeds, and harmonic systems, Leibniz proposed to reconcile the paradoxes of Christianity, Cartesianism, and Epicureanism and create a new synthesis of order and freedom. In this synthesis, man's freedom is preserved, yet causality remains intact; God's direction is affirmed, but our choices still matter. As Leibniz put it in his *Theodicy*:

> [God] leaves [man] to himself, in a sense, in his small department. . . . He enters there only in a secret way, for he supplies being, force, life, reason, without

> showing himself. It is there that free will plays its game: and God makes game (so to speak) of these little Gods that he has thought good to produce. . . . Thus man is there like a little god in his own world, or *Microcosm*, which he governs after his own fashion. . . . He also commits great errors . . . but God, by a wonderful art, turns all the errors of these little worlds to the greater adornment of his great world. . . . The apparent deformities of our little world combine to become beauties in the great world.[82]

Here we can see an epitome of that effort to reforge the tools of Christian providence into positive instruments of philosophical and moral knowledge. Through providence, nature and morality are put back into conversation. This conversation was no longer the venerable Christian one, however, but a new one suitable to a new age.

## CONCLUSIONS: LOOKING TOWARD COMPLEXITY

To summarize, third-way thinkers in the late seventeenth century engaged in a massive project of reconciliation, trying to chart new ways out of old dilemmas using materials and tools that they borrowed liberally from the Christian archive. Providence in particular offered enormous resources for this reconciliation, forming a crucial substrate for the new natural philosophies of the period. From mechanism and simple models of natural processes, these natural philosophies turned ever more toward complexity, variety, and diversity as the frontiers of exploration. Providence helped them along this road, giving a shape and language for thinking about and describing nature's dynamic processes. Providence was also, in turn, a shelter under which the ideas of self-organization could grow. It made possible a huge amount of speculation on all of the themes that would figure centrally in the story of eighteenth-century invisible hands: the enigma of causality, the limitations of human understanding of it, the complexity of the world's orders, the impossibility of a purely geometric explanation of them, the interest in biological systems and their self-generating capacities, the speculations on the very tiny and the infinitely large, the power and limits of chance, and the difficulty of understanding human agency and choice in a world of complex order, not to mention many others. Providence as imagined by our third-way thinkers offered not just a pious way to approach these problems. It also offered an operating system for making them comprehensible at all.

At the same time, this was not Calvin's providence, the providence whose dispensations testified to the special place in history of the Christian church and whose mystery demanded the response of humility and faith. Rather, this was providence deracinated, plucked from the soil of tribal Christianity and woven into the fabric of a world-organizing epistemology, moral theory, and natural philosophy. Nor did these new uses of providence leave the concept unscathed. On the contrary, providence's regulatory role could be, at times, vague, heuristic, and more notional than concrete, a generalized principle of ordering whose exact mechanisms were ambiguous.

It is little surprise, then, that as that the language of self-organization grew ever more robust in the eighteenth century, it would include pious and impious variants alike. Most impiously, for example, the tendency toward order rather than disorder could be forced out of God's hands entirely. The notorious English deist John Toland did this, looking for inspiration in Spinoza's monism, from the ambiguity of the *deus sive natura* that is the "immanent . . . cause of all things." Toland wrote in his 1704 *Letters to Serena* of "the internal Energy, Autokinesy, or essential Action of all Matter." Even matter at rest is in "perpetual Motion," an idea that imagined a world where bodies are like springs, constantly pressing themselves outward into more complicated forms. Nature itself, for Toland, has become an immanent and dynamic principle of order. Less dramatically, Bayle too wondered what exactly this new providence accomplished. After all, he wrote circa 1705, "a pear that God made by a miracle and that . . . resembled in all of its qualities a pear produced naturally, would . . . have only the same qualities as an ordinary pear."[83]

Bayle had a point: at the end of the day, a pear is a pear. If adding God to the explanation brings no additional information, then why add him at all? For third-way thinkers, God and his design provided a deep structure but not an abundance of content. Over the next century, this compromise would play out in various ways, in fights about the nature of self-organization, as a system potentially supportive of, and inimical to, the workings of God in the world.

By the dawn of the eighteenth century, in short, European intellectual culture was buzzing with the effort to reinvent cause and effect and to redescribe the move from nothing to something. Content with neither older Christian models nor the new mechanism, theologians, scientists, and others conceived novel, even fantastic, visions of the world. Along with these came new fascinations with the immanent powers of materials and new appreciations for the dynamics of emergent order. The conversations and debates discussed in this chapter were conducted at the loftiest level of European culture. They also mirrored the experiences of many people, who at precisely the same time found themselves entangled in newly complicated social and economic orders, and thus faced an ever-growing need to come to terms with the complexity of the world.

CHAPTER 2

# Living with Complexity circa 1700

## WHEELS WITHIN WHEELS

In 1709 Daniel Defoe—self-proclaimed prophet-cum-champion-cum-chronicler of Britain's breathless embrace of modernity at the turn of the eighteenth century—published a history of "the strange Variety of Circumstances, Changes of Prospects, the Turns of Management, the Accidents and Niceties" that led to the 1707 union between England and Scotland. The union according to Defoe was hardly the outcome of mutual collaboration or deliberate planning. Rather, different interest groups on both sides of the border pulled each its own way with "vehemence," some opposing the union, others trying to imbalance it so as to serve their own ends with no regard for the whole; until, counterintuitively, "the Contraries that concurr'd in this Act, tho' from Clashing Interests, by the strange Circulation of Causes, work'd all together into another Extreme, which none of them design'd, and that was the Union." "By what strange Mystery, concurring Providence, like the Wheel within all their Wheels, center'd them all, in Uniting the Nations"—that mystery, Defoe declared, "is a secret History few understood."[1]

Wheels within wheels: in the previous chapter we have already come across this image, borrowed from the mysterious vision that introduces the prophet Ezekiel. It was an image with a special appeal to contemporaries in an age that was enthralled in equal measure with biblical and mechanical forces. The *OED* records in 1679 the first usage of "wheels within wheels" to denote "a complexity of forces or influences; a complication of motives, designs, or plots." In such usages, which peaked in the late seventeenth and early eighteenth centuries, the power behind the complexity of forces or designs was not necessarily divine. There were in fact two main uses of "wheel(s) within wheels" during this period. One denoted—as in the familiar-sounding words of a Kentish preacher in 1706—the "secret Providence, *a Wheel within a Wheel*, which doth Manage and Over-rule all those Worldly Events, that is not accountable to Humane Reason." The other described a secret cabal or conspiracy, as in the complaint of a Parliament member during the 1680 Exclusion Crisis that "there is still some such thing, as a Wheel within a Wheel; whether *Jesuits*, (for 'tis like them) or who, I cannot tell, nor how the Government is influenced." In essence both usages referred to the same thing, namely, an invisible power—be it divine or human, but still "a deep mystery"—that pulls hidden strings and fully determines the course of events.[2]

In Defoe's account of the Anglo-Scottish union, however, we find a different emphasis, or at least a different inflection, given to the image of the wheels within wheels. Once the "contraries" and "clashing interests" were led into a mysterious concurrence "by the strange Circulation of Causes," Defoe reiterated, "the Union grew up between all the Extremes as a Consequence; and it was meerly Formed by the Nature of Things, rather than by the Designs of the Parties."[3] While he used the language of cause and consequence, it was that of a strange circulation of causes, not linear in progression or effect, which led without design to a beneficial outcome. This unintended consequence, significantly, was formed *merely by the nature of things*. Providence, to be sure, still lorded over the whole process. And yet Defoe led the reader to imagine the providential oversight as quite distant, as the first cause of the nature of things as well as the determining cause of the final outcome. At the same time, his language allowed greater leeway for whatever happened in between, those multiple layers of disorder and uncertainty—accidents, contraries, and clashing interests—that were transformed through a mysterious causality-defying alchemy into the providentially decreed harmonious outcome. Defoe's wheels within wheels, it turned out, were something looser and more expansive than simple cogs in God's clock.

Defoe's complexity was unusual—most providential invocations of Ezekiel's wheels during those decades, as we suggested, were more traditional, more strictly mechanical, and less adventurous—but it was not unique. A similar vision, for example, might have informed a 1706 sermon that celebrated the surprisingly rapid spread of charity schools as an undesigned act of providence. The charity schools' opponents, it noted, "were made Instrumental in them almost without their own Knowledge," while "the prime Agents were most concealed"; from which the preacher (the future bishop of

Peterborough) concluded that this "unaccountabl[e]" outcome was "the Finger of God . . . a Wheel within a Wheel."[4]

More to the point, Defoe's musings were in line with the speculations about providence rife among those theologians, divines, and natural philosophers during the previous half century who sought what we termed in the previous chapter a third way. A providential third way was a compromise between determinism and fortuitousness, an intermediate level between first cause and final cause; be it Cudworth's mysterious notion of plastic nature, or Malebranche's severing of general providence from a now-retired notion of particular providence, or the pious Lucretians, followers of Gassendi, who combined an omnipotent providence with the self-propelling swerve of atoms and humans in the natural and moral worlds. The well-known nonconforming minister John Owen had wheeled in Ezekiel's chariot precisely for this effect as early as the 1660s. "In *Ezekiel*'s Vision of the glorious providence of God in ruling the whole Creation," Owen wrote, God was placed "as *governing, ruling, influencing* all second Causes." And those "*wheels within wheels* . . . which are all said to be *rolling* . . . in an unspeakable Order, without the least Confusion," could do so "because there was a *living powerful spirit,* passing through all, both *living creatures* and *wheels,* that moved them speedily, regularly and effectually, as he pleased; that is, the *Energetical* Power of divine Providence, *animating*, guiding and disposing the whole."[5] Every creature and every wheel has its own charge of "Energetical Power," and from all those self-motivating movements together, providential order ensues.

Given Defoe's habitual preoccupation with the paradoxes of providence, it is likely that he was well attuned to those recent speculations about a providential third way. But it is likely that he was also drawing on a very different set of experiences, ones on his mind at the same time. Defoe left us a clue to this connection in another piece of that same year, using the same image and indeed much the same logic. The topic of this journalistic essay was "this extensive Thing call'd CREDIT"; or, as he repeated, "this Machine call'd *Credit*—This is the Wheel within the Wheel of all our Commerce." Quite a different kind of wheel within wheel, then. Defoe's double reference within a single page to this novel thing "call'd Credit" reminds us that these were the heady early years of the so-called financial revolution, when the operations of credit were remarkably new and puzzling. Defoe was fascinated—spellbound—with this new presence in economic life. He therefore crammed his short essay with one mystified characterization of credit upon another. Credit has "a distinct Essence (*if nothing can be said to exist*) from all the *Phaenomena* in Nature," an essence that defies even "the greatest Alchymists"; "it is all Consequence, and yet not the Effect of a Cause"; it is "the lightest and most volatile Body in the World . . . a perfect free Agent acting by Wheels and Springs absolutely undiscover'd"; "its Motions are natural to it self"; "it comes to a Man, or a Body of Men, or a Nation insensibly. . . . It comes with Surprize, it goes without Notice"; and so forth until the resounding conclusion—"by this Invisible, *Je ne scay Quoi* . . . all our War and all our Trade is supported."[6]

Now consider Defoe's accounts of the operation of credit and of the emergence of the Anglo-Scottish union next to each other. Arising in contexts that could hardly be further apart, Defoe's two expositions converged unexpectedly. In both Defoe conjured the image of wheels within wheels for a complexity that defied straightforward relationships of cause and effect. In both he described situations in which unintended consequences, beneficial for the greater whole, emerged mysteriously from the unplanned, contradictory, self-propelled motions that were in the nature of the constituent parts. If in the former, therefore, Defoe's groping toward new understandings of complexity, causality, and agency appeared indebted to the reconfigurations of providence discussed in the previous chapter, the latter suggests that Defoe's thinking was equally informed by his experiences in the rapidly changing world of the stock exchange and the market.

In the present chapter we are interested in such experiences of complexity. In the late seventeenth and early eighteenth centuries, we suggest, more and more Europeans found themselves exposed to experiences that challenged any simple understanding of the world they inhabited. Chapter 1 focused on the late-seventeenth-century reconfigurations of a single problem central to the Western engagement with this world and the next, namely, providence. This chapter, by contrast, offers a broad investigation of the phenomenology of complexity circa 1700. It moves through various key developments in European social, cultural, and intellectual life, each of which brought complexity home for some. The development of markets and credit networks, innovations like lotteries and insurance, new sciences of probability and statistics, the rise of the professional gambler as social type, the intensification of consumer society, and the experience of the individual consumer: together, we want to propose, developments like these created a cultural environment conducive to the appreciation of invisible complex operations, of nonlinear relations of causes and effects, and of aggregates behaving differently than their parts. An environment, that is, susceptible to self-organizing stories. As Defoe directs us to do, we begin the tour of this environment of complexity with the stock exchange and the market.

## EXPERIENCING THE MARKET

In 1688, when a younger Defoe was still involved in speculating in the market rather than on paper, another man with a similar blend of business and literary aspirations published the first detailed description of a stock exchange. Joseph Penso de la Vega was a Spanish Jew living in Amsterdam, which had become during the seventeenth century the financial capital of Europe, complete with a dependable central bank, a leading international capital market, and an active stock exchange, decades ahead of the more-often-discussed "financial revolution" in Britain. Although stocks and speculations had been part of the Dutch scene throughout the century—including the notorious tulip mania of the 1630s—it was in the 1680s, during a period of considerable expansion and volatility in stock trading, that a public discussion of these activities began in earnest. Among its participants was de la Vega.

In a series of dialogues between a shareholder, a merchant, and a philosopher, de la Vega laid down a detailed picture of the Amsterdam stock exchange, interspersed with biblical, classical, and historical allusions. His title, *Confusion de confusiones*, foretold the main theme. "This business of mine is a mysterious affair," begins the shareholder; "even as it [i]s the most fair and noble in all of Europe, so it [i]s also the falsest and most infamous business in the world. The truth of this paradox becomes comprehensible, when one appreciates that this business has necessarily been converted into a game." The compulsive participants in this game—a form of high-stakes gambling, with loopholes for the knave and snares for the novice—are erratic and unpredictable. Even when presented with full information about the choices they need to make, "the commitments of the speculators change, and their decisions become uncertain," so much so that "there are many occasions in which every speculator seems to have two bodies so that astonished observers see a human being fighting with himself." The result is a chaotic, charged environment in which "rabbits become elephants, brawls in a tavern become rebellions, faint shadows appear to them as signs of chaos." This magnification of insignificant causes into momentous effects, the shareholder continues, is key to understanding "these absurdities, this confusion, this madness, these doubts and uncertainties of profit" that rule the stock exchange. In a place where individuals are found wrestling with themselves, where their actions are erratic, and where molehills routinely turn into mountains, relationships between causes and effects could hardly be expected to remain predictable and linear.[7]

For de la Vega's philosopher, indeed, this becomes the main theoretical objection to the new financial world. It is "incompatible with philosophy," the philosopher postulates ex cathedra, "that the bears [the pessimistic shareholders inclined to sell] should sell after the reason for their sales has ceased to exist, since the philosophers teach that when the cause ceases, the effect ceases also. But if the bears obstinately go on selling, there is an effect even after the cause had disappeared." And again: "Moreover, while philosophy teaches that different effects are ascribable to different causes, . . . at the stock exchange some buy and some sell on the basis of a given piece of news, so that here one cause has different effects." ("La philosophia enseña que los effetos contrarios es fuerça que procedan de contrarias causas: *Contrariorum contrariae sunt causae* y en esta variedad veo que siendo única la causa, son contrarios los effectos."[8]) A cause leading to multiple contrary effects, an effect that has no cause, a major effect with a minor cause: the goings-on of the stock market were incommensurable with philosophical precepts of causality.

Many passages in Joseph de la Vega's *Confusion de confusiones* are reminiscent, in cadence as well as content, of Defoe's later account of credit. Defoe's writings on the stock market, in turn, also often sound like de la Vega's, as when he reported on stock prices that "rise without Reason beyond the Nature of the Thing, and fall without reason below . . . leaving the Intrinsick Value . . . as a Useless and Insignificant Thing."[9] This is not very surprising, since both writers were able and willing participants in the same late-seventeenth- and early-eighteenth-century historical process that is only partly encompassed by the term "the financial revolution."

The financial revolution is discussed most commonly with regard to Britain after the Glorious Revolution, a period that saw the migration of Europe's financial center from Amsterdam to London. It refers first and foremost to the emergence of institutions of long-term public credit. Often modeled after Dutch precedents, these new financial practices—involving an active secondary market in both debt and equity securities and crowned by the foundation of the Bank of England (1694)—were essential for funding King William's unprecedentedly expensive wars. (The French, saddled with equally debilitating military expenses, were soon to follow suit.) These financial developments, in turn, fused with others equally important: the economic buoyancy since the Restoration, the political consequences of the 1688 settlement, the effects of the subsequent wars on the growth of manufacturing and trade in new markets, the influx of Huguenot and Dutch capital and skill. The combined result was a startling economic success. The expansion of English trade in the 1690s was greater than during the whole previous century. The blooming of finance capitalism was evident in the London stock market, which by 1695 saw trading in around 150 joint-stock companies—a major leap from the 15 companies of 1689, and one that lured in a much larger number of investors (some ten thousand by the early 1700s). Entrepreneurs tripped over each other in their eagerness to float new financial projects and monetary experiments in what Defoe in 1697 famously dubbed "*The Projecting Age*."[10] England, overtaking Holland, had become the economic powerhouse of Europe. This economic activity, moreover, was strikingly international, even global: be it the capital market, in which foreigners invested on a large scale in the debts of other nations, or trade, in which the steepest growth occurred in English reexports and extra-European trade.

Our point in these sixty seconds on late-seventeenth-century economic and financial trends is that a large and growing number of people were touched by matters of credit, trade, or finance, and that these experiences, like de la Vega's in the stock market and Defoe's in the new world of credit, could draw attention, more or less self-consciously, to questions of complexity, causality, agency, randomness, and chance.

Other aspects of the financial and economic world at that juncture added to this effect. This, after all, was a period in which even the correspondence between the metallic value of a coin and its market value was unstable—when (because of the acute coinage crisis) a guinea could cost as much as thirty shillings, thus turning a straightforward commercial transaction into a speculation. It was also a period in which the snowballing effects of compound interest became more widely recognized. The "seeming paradox" of "how such a trifle [in this case, Friendly Society membership contribution] can raise such vast advantages" was a manifestation of radical disproportionality that elicited, as in this example of 1715, both marvel and disbelief.[11]

Perhaps most prominent among new financial practices were the lotteries, a fad that began as a state-sponsored funding scheme in the 1690s and then spread "like a Plague" (thus a disapproving writer in 1695) throughout much of Europe. The lotteries, capitalizing openly on the operations of chance, will preoccupy us shortly. At this point we can suffice with the testimony of the Italian Gregorio Leti who, like de la Vega, discovered in Amsterdam the apex of confusion in the financial world. But whereas for de la Vega in 1688 it had been the Amsterdam bourse, for Leti in 1697 it was the lottery craze. "Since last year one hears nothing spoken about other than the lotteries, among all kinds of people," Leti complained—people who disregarded the fact that these lotteries were in truth little more than "a kind of children's game, a pastime for madmen . . . a trade without rhyme or reason (*un negoce sans raison*), a lure for the capricious."[12] The truth of the matter was that chance permeated institutions of the financial revolution in multiple additional ways: in tontine schemes, experimented with in both France and England, which involved gambling on the life expectancy of the investor; in mortuary tontines, which were in fact a form of life insurance; and of course in insurance more broadly, perhaps the most significant contemporary attempt to harness chance to the logic of finance. (Some sixty life insurance projects were floated in England between 1696 and 1721.)[13]

It was therefore a befitting sign of the times that when following the death of Louis XIV the French set out to reform their national finances and debts, inspired in part by the English example, the person chosen to lead the effort was John Law, a Scottish financial genius who had first made his reputation and his fortune through a legendary mathematical ability to beat the odds at the gambling table. The far-reaching consequences of Law's gambling spirit for France's finances will preoccupy us again in the next chapter. For now we wish to summon Law's testimony for a particular aspect of this new financial universe, central to its experience as a world of complexity and nonlinear causality: *credit*. In 1712 Law, still combing Europe for a potentate willing to heed his financial advice, announced the matter—which he deemed novel enough to be newsworthy—to the Duke of Savoy, whom he was urging to establish a bank in Turin after the model of the Bank of England. Through the extension of credit, Law explained, "the Bank increases the money because the credit has the same effect in commerce as if all the specie were in the coffers and the sums that it employs had the same benefit for the state, and for commerce, as if the quantity of money were increased by the same amount." The limitation of money had always been that "the same piece of money cannot serve in different places at the same time"; but this, as Law went on to prove with figures taken from England, was precisely the kind of alchemical magic that credit performed so well. It was thus equally a sign of the times that when Law did finally find a patron and disciple in the French regent, the regent promptly turned around and dismissed the court alchemists, whose services in these financial matters were no longer necessary.[14]

We have returned full circle to our first expert witness, Defoe, whom we have seen in 1709 enthralled by that mysterious credit that defied straightforward causal links and was effectively limitless. You might call it a "veritable illusion," a Dutch merchant puzzled in 1699, since one person can transfer it to another, "and this second transfers it to a third . . . and this can go on, so to speak, to infinity." (Recall the progress of military or investors' panic in instantaneous leaps and bounds in Montesquieu's *Persian Letters*.) This endless extendability made credit susceptible to dramatic swings. "When it is in flourishing Circumstances, a very little Money circulates immense Sums in Bills; but when a general Distrust spreads amongst the Dealers, it is quite the contrary." Small causes produce huge effects: positive ones in favorable conditions, disastrous ones in unfavorable conditions. On the latter, once more, no one waxed more eloquent than Defoe, who relentlessly warned against "the long Chain of Misfortunes that attend the Disaster of one Original Knave in Trade, who designedly engross[es] Credit" but then "runs away and leaves a Succession of Disasters from Seller to Buyer, and from Buyer to Seller; one breaks another; and he another *to the end of Trade*."[15] One knave or one accident could bring down the whole trading world ad infinitum. Could one imagine a more glaring demonstration of the lack of proportion between cause and effect?

Key to Defoe's mental image of this economic universe—what had made it a universe in the first place—was "the long Chain" of connections crisscrossing it from all sides. These largely invisible chains interlinking untold numbers of individuals accounted for the infinite generativeness or destructiveness believed to characterize credit. Indeed, as Thomas Haskell has memorably argued, the expansion of trade and financial markets during this period brought about a complexification of such chains of interdependence across expanding distances, together with a growing awareness of their pertinence to people's lives. A battle in Turkey, Defoe once wrote, could reverse the fortunes of trade and credit throughout Europe without the people affected having any knowledge of the event that brought this about. The same, indeed, could be said about an adversity down the street of which one had no more knowledge than if it had taken place in Turkey. "In large and Populous Cities" like early-eighteenth-century London (the words are Bernard Mandeville's in 1714), "where obscure Men may hourly meet with fifty Strangers to one Acquaintance," the chains of economic and financial interdependence were equally opaque.[16]

To be sure, contemporaries expected this complexity to be assuaged by new networks of information. "By the help of strange and Universal Intelligence," Defoe wrote in 1697, "a Merchant sitting at home in his Counting-house, at once converses with all Parts of the known World." Yet this information was inevitably partial and insufficient; and this even without interested parties deliberately spreading "designed uncertaintys"—an evocative oxymoron coined by one Richard Gorges in 1700. It increasingly dawned on contemporaries that they were bound in chains of causes and effects so complex that the actual causes of events—often events that affected themselves—remained invisible.

In the financial world, to quote Defoe once more, it was now the case that through "the strange and unheard of Engines of *Interests, Discounts, Transfers, Tallies, Debentures, Shares, Projects,*" and other such new instruments, people's lives could be affected "like Poison that works at a distance," and so they "hardly kno[w] who hurt them."[17]

"Poison" was a telling word choice. It is the ultimate undetectable agent, one that creates an abyss between cause and effect. It is also an agent that appears to work by itself. Was there therefore much difference between Defoe's unknowable poison working at a distance and his view of credit as that "perfect free Agent acting by Wheels and Springs absolutely undiscover'd" whose "Motions are natural to it self"? Or were these really two sides of the same coin—twin efforts to characterize effects without visible links to causes in the modern economy? When Sir Dudley North wrote in 1691 of "this ebbing and flowing of Money, [that] supplies and accommodates itself, without any aid of Politicians"—a half-comment hailed by historians of economic thought as perhaps the first intimation of a basic nostrum of modern liberal economics—was he thinking of money as truly self-moving or as subject to infinite chains of invisible causes that merely make it *appear* to be self-moving? In truth this does not really matter: the paradoxes of causality remain the same.[18]

Defoe himself was hardly slowed down by such logical niceties. His modus operandi, rather, was to mobilize one rhetorical device after another in order to signal the dawn of a mysterious age. "Let no Man wonder at these Paradoxes, since such strange things are practised every Day among us," he observed. "Trade is a Mystery," another trademark passage of Defoe's began in 1706, "which will never be completely discover'd or understood":

> It has its Critical Junctures and Seasons, when acted by no visible Causes, it suffers Convulsion Fitts, hysterical Disorder, and most unaccountable Emotions—Sometimes it is acted by the Evil Spirit of general Vogue, and like a meer Possession 'tis hurry'd out of all manner of common Measures; today it obeys the Course of things, and submits to Causes and Consequences; tomorrow it suffers Violence from the Storms and Vapours of Human Fancy, operated by exotick Projects, and then all runs counter, the Motions are excentrick, unnatural and unaccountable—a Sort of Lunacy in Trade attends all its Circumstances, and no Man can give a rational Account of it.[19]

Defoe's vocabulary for the operations of trade included "lunacy," "possession," "hysteria," "convulsion," and "disorder." Sometimes trade seemed to Defoe as the victim of eccentric and erratic human actions; other times it appeared eccentric and erratic in and of itself, positively Lucretian. Sometimes, conversely, it actually did obey the laws of cause and effect: even unpredictability was unpredictable.

To return to our main point. Surely one must make the necessary allowances for rhetorical flourish in such passages, as well as for the possibility that some insiders of the new financial world may have had a vested interest in keeping outsiders awed by its mysteries. Yet it is clear that quite a number of people in late-seventeenth- and early-eighteenth-century Europe were struck and at times overwhelmed by the complexity, the nontransparent causality, and the random dynamics of their new economic environment. For some these realizations helped prepare the ground on which alternative configurations of order as well as of relations of causes and effects could be imagined, including self-organizing ones.

The next chapter highlights the singular circumstances of the first major crisis of this new financial world that were soon to press a significant array of people to imagine just such alternative configurations of order. But if the opportunity and motive to think self-organizationally were just around the corner, the means were already all here. We therefore end this section with one economic theorist who already in the 1690s put all the pieces together. Like Defoe, his sometimes adversary, Charles Davenant was struck by the novelty of recent developments. Indeed, their assessments of the nature of credit and trade are often surprisingly similar. "Nothing is more fantastical and nice than Credit," Davenant announced in 1698 in his *Discourses on the Publick Revenues, and on the Trade of England.* Credit "comes many times unsought for, and often goes away without Reason." While " 'tis never to be forc'd," "when this Engine, which now seems so difficult to be stirr'd, is once put in Motion, it will move afterwards of its own accord." Defoe's vision was all here: credit as phantasm, its unpredictability, the difficulty in assigning causes to its movements, its self-propelling swerve. And the same was true for trade, the workings of which Davenant found to be equally "unaccountable." " 'Tis hard to trace all the Circuits of Trade, To find its hidden Recesses, To discover its Original Springs, and Motions, and to show what mutual Dependance all Trafficks have one upon the other"—those "Links and Chains by which one Business hangs upon another," often "by secret Fibres."[20]

But then Davenant took a couple of steps further. The passage that scholars reproduce most often from Davenant is the following:

> Trade is in its Nature Free, finds its own Channel, and best directeth its own Course: and all Laws to give it Rules, and Directions, and to Limit, and

> Circumscribe it, may serve the particular Ends of Private Men, but are seldom Advantagious to the Publick.

Historians of economic thought often cite Davenant's hydrological image for the self-propulsion of trade, destined to become a cliché by the late eighteenth century, in order to establish his precocious place in the genealogy of liberal economics.

But scholars rarely reproduce Davenant's next sentence, no less revealing of his full—and indeed precocious—vision of how the economy actually worked:

> Governments, in Relation to it [trade], are to take a Providential Care of the Whole, but generally to let Second Causes work their own way; And considering all the Links, and Chains, by which they hang together, *peradventure* it may be affirm'd, That, in the Main, all Trafficks whatsoever are beneficial to a Country.[21]

This, then, was Davenant's formula for a productive economic order. Take that mysterious self-moving trade (or credit), step beyond those inscrutable chains of secondary causes and effects, and allow them to operate of their own accord, while maintaining a more distant providential oversight. It is then that "peradventure"—that is to say, through the seemingly coincidental accumulation of random effects—an overall outcome would emerge, beneficial to the country "in the Main," though perhaps not for each and every individual.

Davenant's self-organizing economic vision, it is also important to note, was a radical departure from earlier wisdom. As Joyce Appleby has demonstrated, seventeenth-century economic theory had been robust and innovative, especially in the recognition of uniformities and laws in the workings of the economy. But this particular understanding amounted to "[a] concept of order in a market economy [that] necessarily implied a uniformity of human response . . . [and] encouraged an emphasis upon regularities and consistency." Seventeenth-century economic writers therefore "chose to ignore what was fortuitous, capricious, or socially conditioned about commercial transactions and to fix instead upon the regularities," assuming that only "predictable cause-and-effect relationships operated in the world of commerce."[22]

Thus, for example, consider this passage from the early-seventeenth-century pen of Thomas Mun, in Appleby's estimation the most astute economic analyst of the age:

> Let the meer Exchanger do his worst; Let Princes oppress, Lawyers extort, Usurers bite, Prodigals wast. . . . Yet all these actions can work no other effects in the course of trade than is declared in this discourse. For so much Treasure only will be brought in or carried out of a Commonwealth, as the Forraign Trade doth over or under ballance in value. And this must come to pass by a Necessity beyond all resistance.[23]

For Mun, the multifarious actions of individuals did not constitute order on an aggregate level, as they would for Davenant. Indeed, they constituted absolutely nothing. They were irrelevant, unable to disrupt an order that preceded and overruled them, an order derived from inevitable relationships of causes and effects impervious to all such disruptions. The difference is subtle but crucial. It was the same difference that distinguished the mainstream mechanical and clocklike wheels of God's providence from Defoe's wheels within wheels, with their new degrees of contingent, undetermined freedom. It was the same difference that separated the traditional understanding of divine providence from what we have called the providential third way, adding an intermediate space of uncertainty between accidental events and providentially determined outcomes. And it was the same difference that at the end of the century led Davenant, who newly envisioned an economic world where chance was endemic, where effects could not always be related to causes, where Lucretian swerves were permissible, and where benefits were gauged for complex aggregates rather than for each individual, to self-organization.

## PROBABILITY AND THE SECRET ORDER OF THINGS

Charles Davenant had one more announcement to make. With a hint of self-congratulation he introduced an innovation that allowed him "to argue upon the Revenues and Trade of *England*, in a way not commonly practis'd." The innovation was "the Art of Reasoning, by Figures, upon Things relating to Government," or "what is now call'd Political Arithmetic." The term "political arithmetic" had been coined by Sir William Petty, and it was to Petty's work in the 1660s that Davenant traced the origins of this revolutionary development. Davenant further provided a basic instruction to the budding political arithmetician: do not "argue from single Instances, but from a thorough view of many Particulars." This logic of aggregation was justified, "Mankind in the Mass being much alike every where," but to guarantee its accuracy its data "must be Compos'd of a great variety of Members." Aggregate reasoning eschews the individual case for the large number, ignores the particular for the general, and is best applied to a wide and varied range of singularities taken together. In passing, moreover, Davenant admitted that employing such aggregate reasoning in matters of state—as he was about to do himself—betrayed a certain loosening of the role assigned to providential oversight. After all, he recalled, "the Sin *David* committed in Numbring *Israel*, might be probably this, That it look'd like a second Proof of rejecting Theocracy, to be govern'd by mortal Aids and humane Wisdom." The objections of scripture to political arithmetic notwithstanding, "without doubt, it must very much help any Ruler."[24]

Davenant was not the first to revel in the secret patterns hidden inside masses of large numbers. Petty too was fascinated by the magical ability of mathematics to conjure order from the chaos of particulars that comprise human experience. With this magic, he made even the unpredictable growth and contraction of London—beset by plagues, itinerant beggars, and unhealthy humors—conform to some regularity, predicting that the city would reach its maximum population in the year 1800. Indeed, the whole world would fill to its limit within two thousand years, he discovered, when population density would provoke, "according to the *Predictions* of *Scripture*, . . . *Wars* and *great Slaughter*." Petty owed his professorship at Gresham College (founded by the builder of the Royal Stock Exchange!) to the intervention of another number-lover, the draper and freelance statistician John Graunt. Possibly with Petty's assistance, Graunt uncovered remarkable facts buried in London's mortality records. He published his curious results in the 1662 *Natural and Political Observations . . . made upon the Bills of Mortality*, offering a set of mortality tables and a set of interpretations of their significance. His dedicatory epistle listed some of the unforeseen discoveries revealed by aggregation:

> That the irreligious *Proposals* of some, to multiply People by *Polygamy*, is withall irrational and fruitless; That the troublesome seclusions in the *Plague-time* is not a remedy to be purchased at vast inconveniencies; That the greatest *Plagues* of the City are equally, and quickly repaired from the Country; That the wasting of *Males* by Wars, and Colonies do not prejudice the due proportion between them and *Females*; . . . That *London*, the *Metropolis* of *England*, is perhaps a Head too big for the Body, and possibly too Strong.

Aggregation uncovered the hidden patterns concealed under the bushel of information, patterns that often conflicted with common sense. All appearances to the contrary, London's plague of beggars were not actually starving, Graunt proudly announced. In fact, they "seem to be most of them healthy and strong," a fact that suggested (to Graunt anyway) that the state might be better served feeding and housing them outright, rather than penalizing them for their poverty.[25]

The astonishing powers of aggregation were discovered above all in lists of the dead. In part, of course, these were one of the few large data sets available to early modern Europeans. By 1660, parish records in France and England had been recording births and deaths for over a century, while in London, mortality bills were compiled beginning in the late sixteenth century to help keep track of the plague. In the Netherlands, the Dutch mathematicians and politicians Jan de Witte and Jan Hudde used the registers of annuity holders—which showed payouts to purchasers until their deaths—to compile their statistical work on life expectancies. Edmund Halley's famous tables, published in the *Philosophical Transactions* in 1693, used a smaller but more representative source, five years' worth of mortality bills from Breslau, a city in Silesia (now Wroclaw, Poland) whose fixed number of inhabitants provided, Halley thought, a stable picture of mortality in the aggregate. Using these, he tried to come up with proper annuity pricing that would take into account the likelihood of death.[26] Aggregate thinking demanded large quantities of information, and mortality was one of the few things quantified in the seventeenth century.

But mortality also symbolically condensed that unpredictable "single Instance," the counterweight to aggregate thought. Ecclesiastes already imagined death as the very essence of the unknown: man "knoweth not his time: . . . as the birds that are caught in the snare; so are the sons of men snared in an evil time, when it falleth suddenly upon them" (9:12). Death is God's territory, or the territory of fortune. It comes to everyone, but key to its meaning is the very unpredictability of its timing. It is under the shadow of sudden death that humans experience the world. Pastors prompt us to repent *now*, not after sowing our oats, for who knows when we may be carried off? The specter of sudden death was also integral to all thinking about contingency and accident. Great events depend on the "slightest things," we already saw Pierre Bayle write, since "one blow of a horse's hoof . . . would have saved the lives of millions of men who perished because of Alexander and would have spared the world an infinite number of miseries."[27] Just as death marks the limits of human existence, its timing marks the limit of human knowledge.

The patterns extracted from death thus savored of the miraculous taste of the secret order of things. That within two years London "repairs its loss of inhabitants" from even such a dread occurrence as the plague was only one of "many abstruse, and unexpected inferences out of these poor despised Bills of *Mortality*." Graunt's amazement at his inferences derived from their raw material: the sudden strokes of fortune and fate, the disease that would kill a neighbor, spare a son, take a wife, leave one confused and baffled by the exigencies of things. Precisely these sudden blows paralyze the mind, Antoine Arnauld and Pierre Nicole concluded in their 1662 *Art of Thinking*, the famous Port-Royal logic. Some people are "in a Panic dread" when they hear thunder and see lightning, fearing instantaneous death from the sky. "No sort of violent Death happens so rarely" as the lightning strike, they pointed out, and yet few ways of dying capture the imagination so vividly. Sudden death was not the only instance of the extraordinary power the accidental held over the imagination. For Nicole and Arnauld, the power of accident showed itself not just in death but in more benign arenas as well, not least in the lottery. "Twenty Thousand Crowns for one Crown, is not that a very great advantage?" the people cry, and "every one believes himself shall be that happy Person, upon whom this great Fortune shall show[e]r it self."[28] The sudden blow of the reaper's scythe and the sudden fortune of the lottery ticket were analogous: unexpected, unpredictable, and utterly compelling to the human imagination.

Practices of aggregation set the random and orderly into startling proximity. In doing so, they both described and created new forms of human experience ever more attuned to the uncertain. Phenomenally popular in the late century, the lottery was a perfect example, a new experience growing out of the power of large numbers. The Swiss theologian and critic Jean Le Clerc dated the fad to 1694, when a lottery proposed by one Thomas Neale—the groom porter to the king—brought a windfall to the English crown. The so-called Million Adventure and similar schemes sold hundreds of thousands of tickets to eager buyers, skimming money to float the state's war debt and repaying winners with long-term annuities. There was even a secondary market in tickets, where groups of investors banded together to purchase more expensive ones with better odds. Samuel Pepys reported the fervor in a letter to Newton, remarking

> that the late project (of which you cannot but have heard) of Mr Neale, the groom-porter his lottery, [*sic*] has almost extinguished for some time at all places of public conversation in this town, especially among men of numbers, every other talk but what relates to the doctrine of determining between the true proportions of the hazards incident to this or that given chance or lot.

By 1696, when Le Clerc was writing, Holland too was so "warm in these Projections" that everybody "hastens to bring in his Mony to those next to be drawn." This mania for lotteries was exactly the flip side of the fear of lightning: "though the Odds give more ground for Fear than Hope, yet every Man's Hopes are infinitely above his Fears," as Le Clerc put it.[29]

Here was (and is) the puzzle. Everyone *knows* that the lottery cannot possibly be a good investment. If it were, it would fail in its basic purpose, the generation of capital for ambitious state projects. And yet people simply do not seem to care, buying tickets they know are a bad deal. Curious about this, Le Clerc took the opportunity to reflect more generally on the nature and psychology of luck. "What is this *je ne scay quoy*, which denominates Men Fortunate or Unfortunate?" Le Clerc asked, adamant that fortune and misfortune alike were "dark Notion[s] of an odd unintelligible Quality." A person who wins a lottery is called lucky, but in reality, only a "Man that hath drawn the Great Lots . . . in several Lotteries successively" really deserves the title of fortunate.[30] Luck means *beating* the odds, not fulfilling them.

Yet surely even the one-time winner of the lottery feels pretty lucky. This feeling, for Le Clerc, was in error. The "single instance" tempts us to impute a profound significance to our own unique experience. The result is a constitutional superstition, where we ascribe events to a force greater than ourselves, whether destiny, fortune, or even God himself. But none of these can tell us anything about that singular event. Destiny has nothing to say about lotteries, for example, since there is no way "to give the reason of that Motion and Order" that produced the winning number. "Fortune" is merely a word posing as an explanation, one that "signifies nothing at all," saying only that what happened did, in fact, happen. But most pernicious of all is to imagine that God has something to do with luck, bad or good. Games are simply not important enough "that the Finger of God should commonly be thought visible in them." Otherwise, dice players, lottery speculators, and card sharps would all "engage God to declare for them by perpetual Wonder; and the Groom-porters, and Gaming-houses would have infinitely more Miracles wrought in them, than ever the Temple itself."[31]

Le Clerc's concerns about superstition and gambling echoed those earlier felt by Puritans, divided as to whether the casting of lots might accurately reveal God's intentions in the world. Some said yes, but already in 1623, Thomas Gataker had taken the negative position, insisting that if God really does "dispose every Lot according to the nature of the thing that is by it questioned," then one would expect that the same lot would "upon the second or third casting" fall out the same way. Anyone experienced in dice, he laughed, would see the absurdity. More significant, however, was the superstition that the "single instance" produced. The real target of Gataker's critique was his coreligionist's belief that "it is onely and immediately of God, that the Dice be so cast," such that each *single* cast happens because of "God's speciall direction."[32] The blasphemy was clear—gaming became a form of divination, each trick taken a confirmation of God's will in the world.

What was a moral issue for the Puritan became a much wider-ranging social and epistemological one in the age of Le Clerc, when the contrast ever sharpened between the experience of the single instance and the *meaning* of this experience in a world of aggregated orders. Happily for Le Clerc, however, he had access to tools that Gataker did not, tools that he imagined could narrow the distance between experience and aggregated order. These tools made a quick appearance at the end of Le Clerc's treatise, when he abruptly shifted from lotteries to the ingenious arguments of the Jansenist and mathematician Blaise Pascal. "Skeptics and Unbelievers" play life the way that they play lotteries, expecting the sudden windfall and cursing fortune when it fails to appear.[33] Pascal offered a sharp corrective, thought Le Clerc, by making clear the nature of the odds and allowing us to conform unruly experience to the reality of the world's orders.

Pascal pioneered probability theory, the development of which transformed European intellectual life in the period. It was Pascal who imagined the wager on God's existence, "a game . . . played at the extremity of [an] infinite distance where heads or tails turns up." It is a game we all *must* play, all must "of necessity choose." What will you wager? Pascal famously built this probabilistic answer:

> Let us estimate these two chances. If you gain, you gain all; if you lose, you lose nothing. Wager, then, without hesitation that He is. . . . There is here an infinity of an infinitely happy life to gain, a chance of gain against a finite number of chances of loss, and what you stake is finite. It is all divided; wherever the infinite is and there is not an infinity of chances of loss against that of gain, there is no time to hesitate, you must give all.

Pascal's wager put the doubting sinner in, as Ian Hacking describes it, "the same *epistemological* position as someone who is gambling about a coin whose aleatory properties are unknown." If the payoff is infinite (salvation), then even if the chance that God exists is vanishingly small, the rational decision—according to Pascal—would be to bet on God. The key, for us and for Le Clerc, lay in Pascal's reorientation away from the single believer to a kind of aggregated prudence. "Reason can decide nothing here," Pascal wrote, meaning that there is neither a deductively secure way to prove God's reality nor a way to confirm this reality by any given experience.[34] Instead, experience and reason must be suspended, and the problem reframed. The new frame is one of aggregation, and of decision making under uncertainty.

Pascal's mathematical apologetics grew out of a much more mundane (and profane) set of questions that had confronted gamblers and gamers for centuries. At issue were, among other things, the calculation of gaming odds, the fair pricing of tickets, the determination of the likelihood of individual dice rolls or card draws, and so on. Only in the 1660s and even more quickly in the 1690s, as Hacking has shown, were these investigations codified into a new set of analytical tools to deal with the uncertain. Pascal played a foundational role here, using the arithmetical triangle to determine the coefficients of a binomial expansion, for example, and thus the comparative likelihoods of different gambling outcomes. His 1654 work on the mathematics of points was generated in dialogue with Pierre de Fermat, who collaborated on the solution. Three years later, the Dutch mathematician Christian Huygens published the first full-length effort to treat the mathematics of probability, a project that grew out of discussions with both Pascal and Fermat, among others. It focused on the idea of "expectation," namely, the price one should pay to enter a lottery or a game. This idea of expectation, as Hacking and Lorraine Daston both argue, had a rich afterlife.[35] Most importantly, it grounded the early mathematics of probability in the model of the fair game, the game in which every player has equal expectations, skill does not matter, and an equality of outcomes is guaranteed only in the long run.

The quintessential example of this fair game was the lottery. In a fair lottery, every choice is symmetrical. You are as likely to win with ticket A as with ticket B. The expectation—or fair price of a ticket—can thus be calculated precisely. Knowing the size of the prize and the number of tickets, it is a simple matter to compare the values of lottery tickets. And this is exactly what lottery entrepreneurs did, listing their respective odds as a selling point in their advertisements. This was enough for one rather hostile writer (sometimes reputed to have been Defoe himself) to launch a short-lived "Weekly Paper, familiarly explaining . . . a Calculation of the Chance or Hopes any one has of getting a Price, or Prizes, in the Lotteries, as he is possest of any Number of Tickets." Given enough information, wary buyers could decide whether the risk of the ticket was worth the price of entry. Here, as in many areas, probability mathematics tried "to render us more rational in our hopes and fears," as Nicole and Arnauld put it, to allow people to regulate their experience of risk with the *more geometrico*.[36]

From the outset, then, probabilistic reasoning was about far more than games. It was instead, as Pascal showed, designed to teach us to live under the shadow of enduring risk, to live in a world of such aggregated complexity that the common sense offered by experience trembles and fails. When John Arbuthnot first translated Huygens into English—two years before the lottery mania really took off—he made the pedagogy of this mathematics clear. "The Force of Numbers . . . can be successfully applied, even to those things, which one would imagin are subject to no Rules," he wrote in his preface. In fact, "the Calculation of the Quantity of Probability might be . . . applied to a great many Events which are accidental, besides those of Games."[37] Later he applied these calculations to a very practical case, hoping that the magic of numbers might convert the terrors of ordinary experience into prudential acceptance of risk.

The case was smallpox. Smallpox was, recall, rife in Europe in the seventeenth and eighteenth centuries. Deaths from this terrible and disfiguring disease often topped 10 percent of *all* deaths in London in this period. In 1717, when Lady Mary Wortley Montagu described the Turkish practice of "engrafting" the virus, and its success in retarding the course of the disease, reactions were understandably cautious. Engrafting involved, after all, deliberately infecting the healthy with one of the most horrifying diseases around. Moreover, although it immunized most against future illness, it also inevitably killed or maimed some of those it touched. Arbuthnot was a believer in this practice. "A Practice which brings the Mortality of the *Small Pox* from one in ten to one in a hundred" and which "if it obtain'd universally would save to the City of *London* at least 1500 People Yearly," he wrote in 1722, left no doubt that "the same Odds wou'd be a sufficient prudential Motive to any private Person to proceed upon." Getting from one end of the sentence to the other, however, revealed the problem: Why should "any private Person" be convinced to take a risk whose guaranteed benefit applies only to the whole? Healthy patients who found themselves disfigured and dying by engrafting would hardly take much comfort from their statistical exceptionality.[38]

Early probability theorists tried to elude this more or less intractable conflict—between singular experience and aggregate order—through an ever more sophisticated mathematics. If you could really show how aggregation worked, not just in games but in the complex arena of the real world, they seem to have believed, you could make the individual problem go away.

In fact, there were real mathematical and metaphysical problems to solve in the move from game to world. In a game, for example, you know all the rules of its construction: how many sides the dice have, how many players there are, and what constitutes a win. But in real life, we know *none* of the rules. The Swiss mathematician Jacob Bernoulli confronted this problem in his *Ars conjectandi*, written after 1685, but only published in 1713. Bernoulli described the issue in a letter to Leibniz in October 1703:

> We do not know . . . how much more likely it is that a young person of 20 years will survive an old person of 60 than the reverse. This is so because . . . we do not know the number of cases in which a young person meets death before an old one or vice versa. For whence I began to think whether or not what was hidden to us a priori might not at least be known to us a posteriori from the outcome in similar examples many times observed.

What Mary Poovey calls "the problem of induction"—for this is what Bernoulli describes—had never really attracted much attention before the 1660s. In fact, many things intuitively obvious to us were at the time uncertain. It was unclear to Bernoulli, for example, "whether as the number of observations increases," the likelihood "also continually increases" that the observed statistical frequency is actually correct. Or, alternately, whether "I will eventually come to some degree of probability beyond which it cannot be made more probable to me that I have discovered the true ratio." If the second is true—if certainty is asymptotic—then "it is all over" with experiment.[39]

Modern readers "know" that more observations give a better picture of the real. Bernoulli wanted to believe it but was not sure how to show it. Leibniz's response could not have been consoling:

> Contingent things or things that depend on infinitely many circumstances cannot be determined by finitely many results, for nature has its habits, following from the return of causes, but only for the most part. Who is to say the following result will not diverge somewhat from the law of all the preceding ones—because of the mutability of things? . . . Given any number of points, infinitely many curves may be found that traverse them.[40]

Leibniz's point was an important one, not least because it revealed the metaphysical labor needed to build a science of aggregates. Death is an infinitely contingent thing, mutable over time. When we look at the past—and all probabilistic knowledge is retroactive—we only get a snapshot of information, bounded by our own limitations or those of our archive. But what if the archive is constantly changing? What if the order of things is amorphous or mutable?

Put another way, how do you know when you are seeing a representative sample, and when an anomaly? Are the mortality tables from 1348 to 1350 anomalous, for example? Or par for the course? Leibniz himself grew up in the shadow of the Thirty Years' War, when parts of Germany were entirely denuded of population by violence, disease, and famine. Breslau—the source of Halley's information—itself lost around 50 percent of its population. Indeed, that entire region along the Oder River saw populations fall by up to two-thirds and more in the period. Can the tables from this area offer representative information for humans generally? Or not? How long should the statistical time frame be to deal with the anomalous? Or can it deal with it at all? These kinds of questions fascinated Leibniz, with his own delight in the various and the contingent. But they had to be addressed for Bernoulli's mathematics to make any sense at all.

To answer them, Bernoulli came up with a thought experiment, one that illustrates both how every effort to rein in complexity and chance only exposed their presence in the world, and how strange and difficult it is to inhabit an aggregate order. Imagine, he said, that the world is like an urn. This urn is stuffed with black and white tokens, but we have no idea what their ratio is. Since we cannot simply dump the urn's contents out and count them, we have to reach inside and pull them out one at a time. Keep doing this, Bernoulli promised Leibniz, and "I will scientifically determine the number of observations necessary for it to be ten or a hundred or a thousand times more probable to you that the ratio" between the tokens falls inside a given limit than outside it.[41]

In fact, Bernoulli would never solve this problem of so-called inverse probability definitively (see chapter 5). But the image was a compelling one. It suggested, first, the world is a kind of black box, in which any *single* experience has near-trivial evidential value. And yet, second, information can be extracted from this world by dint of repetition, time, and calculation. The "mutability of things," in Leibniz's terms, is overcome by running the experiment essentially forever— "if the observations of all events were continued for the whole of eternity . . . then everything in the world would be observed to happen in fixed ratios"—so that changes in the world (should they happen) can be captured inside processes of iterated calculation.[42] Better mathematics did not solve, it seems, the problem of the single instance. Aggregation and large numbers offer insight into a determinate order that underlies all things, in short, but it is an order invisible to ordinary experience.

This conjunction—determinate order but invisible to us—was as much metaphysics as empiricism. "All things under the sun, which are, were, or will be, always have the highest certainty," Bernoulli proclaimed in near harmony with a Hobbes or a Spinoza: "even in the most accidental and fortuitous" events, we must "acknowledge a certain quasi-necessity and, so to speak, fatality."[43] For Bernoulli, though, this was a comforting token of God's organization of things. Statistical reason conjures hidden order out of massy experience and gives what cannot be sensed in the ordinary course of things a determinate form. In the midst of the everyday, Bernoulli felt sure, operated an order invisible yet divine.

Those who tarried with chance tended to share these commitments. In 1713, the gamester Pierre Rémond de Montmort was emphatic that "to speak precisely, nothing depends on chance," since the "Author [of nature] works in a general and uniform way" suitable to "the character of an infinite wisdom and foresight." The Huygens translator John Arbuthnot did not stop there, and detected in the statistical register clear signs of God's providential order. At issue was the apparent anomaly earlier discovered by John Graunt. In his mortality tables, Graunt had noticed a curious thing, that while we imagine births to be equally divided between men and women, in fact, "there be more *Males*, then [*sic*] *Females* . . . by about a thirteenth part." Arbuthnot found the persistence of this anomaly astounding. Chance events like births, he reasoned in 1710, should distribute evenly—thus the ratio should always converge on 1:1. But it did not. For eighty-two years, the record showed a constant deviation. The extreme unlikelihood of this—Arbuthnot calculated it at $\frac{1}{2}^{82}$—made the anomaly into one of the "innumerable Footsteps of Divine Providence to be found in the Works of Nature," proving "that it is Art, not Chance, that governs."[44] Statistical analysis revealed the secret necessity of things opaque to the eyes of ordinary experience.

Probability was, however, very carefully balanced on the same edge between accident and fate that we found in the self-ordering dynamics of providential materialism. Historians of probability—Ian Hacking, Lorraine Daston, and Thomas Kavanagh, among others—tend to focus on its fatalist moments. Hacking, for example, has announced that " 'mechanical' determinism" was the "accompaniment" to the seventeenth-century investigation of chance; Daston that "the philosophical climate of opinion during the seventeenth and eighteenth century grew ever more resolutely deterministic"; Kavanagh that the universe of probabilistic science was "rigorously determined." To a certain extent, they are right. Yet the focus on what these theorists *believed* misses what their science actually *does*. For a "staunch determinist," as Daston herself acknowledges, Bernoulli's urn was ambiguous, suggesting "only chance connections between the inaccessible causes and the observed effects." Moreover, Bernoulli never lost sight of the curious suspension of cause and effect implied by probabilistic analysis. He surely believed in a strict causal law, but his analytic method suspended its operations, postponing full causal explanation—the achievement of "a perfected science [that] would no longer need probability theory"— into an eternal future.[45] Probabilists had their cake and ate it too. They believed in a well-designed, causally determinate world at the same time as they designed tools that *described* a world in thrall to the powers of chance.

This state of affairs meant that the nature of chance (the very object of their study) was oddly hard to grasp. Thinkers circulated restlessly, as Hacking has noticed, between an embrace of real aleatory processes and an insistence that chance merely represents the limits of human knowledge. Either could be emphasized depending on need. In the first edition of his *Doctrine of Chances* (1718), for example, the Huguenot refugee and mathematician Abraham de Moivre stressed that chance was a real thing, since "chance alone by its Nature constitutes the Inequalities of Play, and there is no need to have recourse to Luck to explain them." In the third edition (1756), however, de Moivre decried the "atheistical" implications of elemental chance and insisted that it was "a sound utterly insignificant." Chance "imports no determination to any *mode of Existence*; nor indeed to *Existence* itself, more than to *non-existence*; it can neither be defined nor understood."[46] This ambiguity was constitutive. Probabilistic analysis could be used to distinguish the purely random from the regular, but because it said nothing about the causes of this regularity, it could make no definite claim as to the real constitution of the world's orders.

Figures like the urn were thus productive as thinking devices exactly *because* they said so little about the real organization of things. You don't have to know exactly how and why things work the way they do to develop and test hypotheses about them. This meant, however, that while God *might* be the designer of things, no proof of this could be found in an urn. Nicholas Bernoulli—Jacob's nephew and one of the stars of the Bernoulli mathematical family—saw this clearly. John Arbuthnot, remember, had "proved" God's oversight from the slight overabundance of male births. Bernoulli saw the problem here. In a letter to Montmort, Nicholas pointed out that Arbuthnot's "proof" supposes that we are dealing with *two-sided dice*, that birthrates *should* distribute evenly. But why suppose that, Nicholas asked? We have, after all, no idea what kind of dice lie inside the urn. "Imagine 14,000 dice each with 35 sides, where 18 are white and 17 black," he asked, and do the calculations. After doing the math, he concluded that the probability that "among 14,000 infants the number of males would not be larger than 7363 and not less than 7037" was about 43.5 times more likely in any given year than a scenario where the "number of males falls outside of these limits."[47] In other words, if the dice are already stacked for males, irregularity would lie with an *equal* birth rate. The problem, of course, is that we do not have any information at all as to the number of sides, except what we observe from the cases. The urn is mute about what is anomaly and what regularity.

If the world is an urn, then, its orders cannot help but seem to emerge magically. Perhaps God is there, organizing things. Perhaps instead order is mere happenstance. Or perhaps order is an immanent process, the operations of self-organizing matter and spontaneous complexities. Experience and experiment, the point is, can make *no distinction* between these causes. As the science of an uncertain age, probability theory was not exactly the science of self-organization—often its developers rejected exactly this—but it was a science that loosened the ability to feel secure about order's origin and purpose. It made stable deductions impossible and let orders emerge in unpredictable ways from the aggregate behaviors of chaotic elements.

## CHANCE AS A WAY OF LIFE

In an age beset by complexity, some saw problems, others opportunities. Even as our mathematicians invented tools to manage chance, people used its unpredictable dance to their own advantage. Gambling was hardly a new activity in the late seventeenth century, of course, but it may have witnessed a renaissance in the period. The earliest gaming manuals and the oldest surviving gaming tables both date back to these decades, and a chorus of self-appointed moral guardians in England, France, and elsewhere sounded the alarm about the evils of this habit.[48] What was doubtless new, however, was the appearance of a different kind of gambler, the professional gambler who employed the mathematics of probability for the systematic calculation of odds and risks. Defoe described this "*Novelty*" as "Gaming by Rule."[49] John Law was one such gambler. Another, in France, was the marquis de Dangeau. When the marquis died, in 1720, he was eulogized by the legendary secretary of the Académie des Sciences, Bernard de Fontenelle, for his "supreme feel for gambling":

> He had penetrated the entire algebra, that infinity of relations between numbers which reigns within its various games, as well as all the delicate and imperceptible combinations of which they were composed, and which are often intermeshed with such complexity that they resist even the most subtle analyses. . . . He applied theories only he could understand and solved problems only he could pose.

Imperceptible combinations and an infinity of relations: the complexity that Dangeau reputedly conquered resonates with other complexities already evoked above. Decades later, Edward Gibbon expressed even more clearly his appreciation for this achievement of Dangeau, the man who "saw a system, regularity and connection, where others only perceived the wanton caprices of chance." Gibbon was partly wrong. As we have seen, others too had begun to perceive system, regularity, and connection. Most of the earliest probability treatises were how-to gambling manuals, recall, from Huygens's *Laws of Chance* to Montmort's *Essay d'analyse sur les jeux de hazard*, manuals that put at least a veneer of system on the workings of dice, cards, and lots.[50]

From a moral point of view, gamblers were not an easy bunch to embrace, and the vice that attended the gaming parlor was a favorite topic of social critics. Even critics, however, felt the pull of the gambling life in this new age of complexity. For all of his finger-wagging at immoral gamers, for example, the nonjuror clergyman Jeremy Collier seemed sympathetic to their basic point of view. His 1713 dialogue *The Essay upon Gaming* gave the gambler a chance to speak his own mind, for example, and what he said was hard to refute. How different, the gambler wonders, is gambling from real life?

> If you recollect your self you'll find Wealth and Condition depend mostly on Chance. . . . And thus by accidental Events, Poverty and Riches are transplanted, and shift their Seat; and a Blast of Wind, than which nothing is more uncertain, drives good Fortune from one Hand to another. And since Casualties dispose of Things at this arbitrary Rate, since the World is but a kind of Lottery, why should we Gamesters be grudged the drawing a Prize?

Life is a lottery, its course shaped by sequences as unpredictable as any dice throw. The gambler invoked not just the traditional wheel of fortune—the loss and gain of fortunes or the sinking of ships—but also a powerful image of randomness pervading all, even existence itself:

> For to go to the Bottom, even Peoples coming luckily into the World seems a great Contingency; it depends on the Marriage of their Parents, or rather on the Marriages of all their Ancestors; and what is it which brings about these Engagements for Life? Oftentimes nothing but a casual Visit, some random Conversation, and forty other things, which never came under Foresight or Design.[51]

Anticipating *Tristram Shandy*—and published, through a fine Shandean coincidence, the very year that Laurence Sterne was born—the *Essay upon*

*Gaming* suggested that everything, even marriages supposedly based on love and choice, might well be a game of chance.

In this light, it is remarkable that, although the upright interlocutor in Collier's dialogue frowns at gaming, he does not bother to refute the gambler's assertion of the universal presence of chance. In this Collier joins the other voices we have heard who appeared resigned to accepting the role of chance in everyday life. Even if mathematics can aid the "human spirit," Montmort conceded, "this virtue called prudence will never have more than uncertain rules," since "an infinity of obscurities" attends every moment of human existence. Even disciplines traditionally dedicated to the eradication of chance—judicial astrology, for example—found themselves under its sway. Hence William Browne's 1714 translation of Christian Huygens was dedicated to the physician Richard Mead, who, the translator exclaimed, showed that "the State of our Bodies on Earth, are Subject to surprizing Alterations and Changes from the various Positions of those two heavenly ones, the SUN and MOON." Mead's medical astrology combined exactly the two desires of the probabilistic age: on the one hand, the urge to show that "natural Effects of the same kind are owing to the same Causes," and on the other, the concession that there are a "Thousand . . . Varieties" that can hinder the "Regularity of Appearances." Human health depends on the simple motions of gravity and tides, he claimed—except when it doesn't, except when it is affected by, among other things, vapors forced from the bowels of the earth, passing comets, erupting volcanoes, and bolts of lightning.[52] Gambling is not just part of life. It *is* life. Rules apply in the aggregate, but perhaps not to you.

The coincidence of rule and chance helps us understand the odd fact—pointed out by Daston and others—that while Europeans had already calculated mortality rates in the late seventeenth century, their calculations were never quite translated into the financial instruments sold to manage risk. Edmund Halley's mortality tables aimed to set a rational purchase price on annuities, for example, a project that interested de Witt and de Moivre as well. Indeed, that the value of annuities should be pegged to life expectancy had been known since Roman times! Yet despite the surge in annuity schemes circa 1700, sellers of these financial instruments paid scant attention to statistical information. Partly because they were often loans disguised to avoid still existent prohibitions on usury, annuities took no account of the age or condition of their purchasers well into the eighteenth century. Even life insurance sellers relied on rules of thumb and fixed rates, selling shares with little regard for determined actuarial principles. These sellers were not ignorant of probability, however. The Dublin office of a mortuary tontine—a kind of early life insurance—was explicit: "NO Man can expect to Undertake any thing in this World with Certainty, Probability being the only grounds we proceed upon in the ordinary Affairs of Life."[53] The problem, however, was that a probabilistic life was fundamentally ambiguous. It was simply unclear—as probabilistic thought expanded into various domains—whether to put the stress on the orderly if secret logics of the aggregate or on the tumultuous and uncertain experiences of individual people. The more robust the mathematics for controlling chance became, the more pervasive the sense of chance became.

These tensions between chance and order even extended their tendrils into that most orderly of institutions, the law. Legal historians tell us that English common law witnessed the gradual emergence from the last quarter of the seventeenth century of a new kind of liability, a "strict" liability for accidents that was broader than previously. Who, if anyone, is liable in the event of an accident that was unforeseen and not intentional? For centuries such cases had been brought as "trespass actions," which took into consideration intent and fault. Around 1700, however, lawyers began to elaborate a new general principle: that a man was "answerable for all mischief proceeding from his neglect or his actions, unless they were of unavoidable necessity."[54] Suppose an ordinarily calm horse suddenly kicks a passerby, yet its owner was free of intent to harm and even perhaps tried to prevent the injury. (This scenario was in fact largely that of the turning-point case in 1676, the significance of which gradually dawned on contemporary juries thereafter.) Previous legal practice inclined to deny the owner's liability, on the grounds of lack of intent or prior knowledge. But the new development of liability in negligence was independent of trespass. In a real sense, people's responsibility was now extended to events and encounters that as far as they were concerned were truly random: say, a gnat flying into the horse's ear. As liability was expanded and thus divorced from questions of intent, motive, and foresight, common law was reconfigured to accept that chance events not only happen but are a regular aspect of life, and thus in need of a uniform legal principle of compensation. Uncertainty rules, but rules still apply.

The literary critic Sandra Macpherson has suggestively demonstrated the centrality of this new logic—what she calls the "tragic" logic of responsibility, holding people inescapably accountable for harms they did not intend and could not predict—to the fiction of the by-now-familiar Daniel Defoe. Responsibility for Defoe, she argues, whether in *Journal of the Plague Year* (1722), *Roxana* (1724), or elsewhere, is "profoundly accidental" rather than "agential and intentional."[55] But Defoe did not only describe this accidental life; he also offered cures. Early on in his writing career, in the *Essay upon Projects* of 1697, Defoe already understood how to tame chance through aggregation, how large numbers might impose order upon randomness. His goal was to devise a method whereby "all things which have Casualty in them, might be Secur'd." The key to regulating random events ("casualties") was to *spread the risk*: that is, to have "a Number of People entring into a Mutual Compact to Help one another, in case any Disaster or Distress falls upon them." The prototype was that of friendly societies, or insurance schemes, which Defoe wanted to ground in sound mathematical principles. "One thing is Particularly requir'd in this way of Assurances," he explained: that "none can be admitted, but such whose Circumstances are, at least in some degree, alike, and so Mankind must be *sorted* into Classes; and as the Contingencies differ, every different Sort may be a Society upon even Terms." For aggregation to succeed in neutralizing chance events, the individuals comprising the aggregate should all be equivalent—we might say statistically equivalent—with regard to the possible hazards they face. Sailors, shopkeepers, soldiers each have their own risks, and thus should form their own societies, "Seamen with Seamen, Soldiers with Soldiers, and the like," based on the calculations of their own distinctive odds in life's lottery.[56]

In point of fact (and not surprisingly at this point), it would take another century before friendly societies moved their risk-hedging practices onto what we would regard as solid actuarial footing. Until then, the moral obligation incurred by the individual case would trump the ethics of the aggregate. Defoe had already seen this tension in 1697, framing it in terms providential. Although "all the Contingencies of Life might be fenc'd against" by the aggregation of risk, Defoe wondered if such mechanisms might not thwart God's design for each of us:

> I don't pretend to determine the Controverted Point of Predestination, the Foreknowledge and Decrees of Providence; perhaps, if a man be Decreed to be Kill'd in the Trenches, the same Foreknowledge Order'd him to List himself a Soldier that it might come to pass; and the like of a Seaman; but this I am sure, speaking of Second Causes, a Seaman or a Soldier is subject to more contingent hazards than other Men, and therefore are not upon equal Terms to form such a Society.

Here the irreconcilability between individual and aggregate destinies was made explicit. If divine providence had once lent a moral dimension to cosmic order, as we saw in the previous chapter, techniques of risk management pried these two things apart. Calculating the odds of disaster, and limiting their consequences for survivors, improved the conditions of a collective mankind.[57] But this improvement was independent of any particular moral qualities of the actors, each now as deserving of his fate as another. The immoral sailor and soldier would benefit as much as the virtuous in Defoe's scheme, life and death now cut loose from the wise ways of God.

Finally, then, consider Defoe's *Essay upon Projects* together with Davenant's *Discourses on the Publick Revenues, and on the Trade of England* discussed a few pages ago. These two interventions, on rather different topics, were published within a year of each other. Both insisted on the invaluable contribution of aggregate reasoning to social improvement. Both insisted that the logic of the aggregate was separable and different from the logic ruling individual instances. Both dwelled on the accurate procedures for aggregation (though one emphasized the need for a diverse range of particulars, the other the necessary statistical equivalence of all particulars). Both were uncomfortable about the possible affront that such reasoning might pose to God's will. Both tried to circumvent this difficulty through a providential third way: that is, by positing a degree of freedom for secondary causes, an intermediate level in which contingency inheres and where it is not God but man that has a duty to intervene. Both were thoroughly convinced that they were onto a new and valuable insight. Both were writing in 1690s London, capital of the financial revolution, bearing the marks of this innovation-happy "Projecting Age," as we have tried to lay them out. And both made remarkably astute prognostications of what was to come.

## THE ASS IN THE GARDEN OF FORKING PATHS

The individual and the aggregate, then, are in such scenarios analytically irreconcilable. Probabilistic thinking sheds little light on the individual component within the aggregate—the sailor who might still drown, the fortunate win in the lottery, the singular death by lightning. These individual components of a system are of course themselves subject to rules, different rules that govern their own actions and fortunes. Having demonstrated at some length the emergence from the late seventeenth century of new ways of thinking about systems as aggregates, we now wish to shift attention to the individuals. When complex and probabilistic thinking addressed *human* systems, the idea of self-organization required a concomitant investment in imagining individuals as if they were free agents, self-moving and unpredictable in their actions. To complete the picture of the emergent intellectual and cultural environment conducive to self-organizing thinking, the final section of this chapter shows how in the late seventeenth and early eighteenth centuries individuals became more plausibly imaginable as self-moving agents with a greater degree of freedom. To be clear: we are not suggesting that suddenly all constraints were removed as Europeans submitted themselves to the inexorable march of freedom. Rather, the new forms of agency and apparent degrees of freedom came with their own new constraints. But more and more people did experience new possibilities of agency and choice, and these, in turn, spurred their conceptualization of themselves and others as elements in systems that could indeed self-organize.

One indication of this expansion of individual agency was a new linguistic environment. We can start with the word "agency" itself. As is often pointed out, "agency" encompasses two different and even opposite meanings: agent as an emissary or instrument of an external power, and agent as a self-moving initiator of action, emissary of one's self. In English, the former meaning appears from the late sixteenth century. By the middle of the seventeenth century, agency is found to denote an autonomous moral subject, though there was still some way to go before agency was detached from a subservience to a distinct external authority. In political philosophy, as C. B. Macpherson noted long ago, an important turning point was Hobbes's "reduction of human beings to self-moving and self-directing systems of matter" while "dispens[ing] with any postulate of moral purpose imposed from outside."[58] Edward Andrew has insightfully identified a key aspect of this transition in the late-seventeenth-century separation of the term "consciousness" from "conscience," which made it possible for human agency to coexist in tandem with "the God within" and yet remain distinct and independent from it. This distinction was key to Locke's *Essay Concerning Human Understanding*—another milestone in agency thinking—and was carried over through its translations into other European languages: Christian Wolff coined *Bewusstsein* (consciousness) in his German Locke to distinguish it from *Gewissen* (conscience), while Pierre Coste's French rendition stuck with an inelegant differentiation between *conscience* and *con-science*.[59]

Locke, in turn, drew on what appears to have been the first effort to delineate this agency-reinforcing separation of consciousness and conscience, undertaken by Ralph Cudworth in his *True Intellectual System of the Universe* of 1678. But recall that it was in the same book that Cudworth also developed his notion of "plastic nature" as an ingenious if enigmatic form of providential third way, endowing matter with an "inward principle" that countered excessive materialism and excessive determinism. The degree of freedom that Cudworth's purposive consciousness or agency provided human individuals vis-à-vis the divine will and its decrees paralleled that which he allowed to matter through his creative "plastic nature." Across the Atlantic, Harvard College's acting president Urian Oakes had just made pretty much the same point regarding the "Agency of second Causes." "Created Agents," Oakes stated in a 1677 sermon titled *The Soveraign Efficacy of Divine Providence*, were endowed by God with "active power," since "a *cause* cannot be a *cause* without an *active power*, or *sufficiency* to give *being* to this or that Effect."[60] Here, then, was a new, more proactive understanding of agency, one that complemented the intensified preoccupation with looser configurations of providence.

These shifts in the uses of "agency" and "consciousness/conscience" were part of a broader cluster of English linguistic innovations in the second half of the seventeenth century, enriching a possible self-organizing vocabulary and in particular expanding the semantic fields of individual agency. Two new word usages link back to the loosening of causal determinism in the aggregate. "Probability," in its mathematical meaning as the relative frequency of the occurrence of events, is first recorded in the *OED* in John Arbuthnot's 1692 preface to his translation of Huygens's work on chance. This was only two years after the first recorded appearance of "coincidence" in the sense of "notable concurrence of events or circumstances having no apparent causal connection": a sense central to a world ruled by probability or third-way providence.[61] Next to these two usages referring to relations in the aggregate, three other innovations appeared at the level of the singular: "spontaneous," "caprice," and "whim." "Caprice" and "whim" are first recorded as nouns in 1667 and 1697, respectively, and as generalized dispositions of the mind in 1709 and 1721. Spontaneous had originally meant "of free will," and only during the period discussed here did it come to mean "without effort or premeditation."[62] (Hobbes provides the first example in the *OED*, dated to 1656.) All three words signified forms of unpredictable and unpredetermined behavior that characterize individuals endowed with "agency," individuals now ready to be cast in the role of self-moving components of self-organizing systems. The self-motivated individual was the bearer of another important neologism of this period, one that will preoccupy us later: "self-interest."

Such a cluster of linguistic changes stands at the interface of the two levels between which our argument oscillates: the high, articulate conversation that is fodder for the intellectual historian, and the broader sociocultural practices of interest to the cultural historian. Keeping our eyes on the latter for another moment, we can add one more late-seventeenth-century lexical shift, which at first sight seems unrelated to questions of agency. The word "convenience" derived etymologically from the Latin *convenientia*, connoting meeting together, agreement, harmony, conformity, or suitableness. In the fifteenth and sixteenth centuries, accordingly, "convenience" or "conveniency" conveyed harmony and conformity to an external given order. During the seventeenth century "convenience" gradually shifted to a more open-ended suitability to any purpose. This drift evolved by around the 1670s into a new meaning, connoting "material arrangements or appliances conducive to personal comfort" (*OED*), which became the common one by the early eighteenth century.[63] Convenience thus became an intermediate step between necessity and luxury; and if the pursuit of luxury was condemned, the purpose of "conveniences" was left morally neutral and open ended.

And the connection to our argument? The shift in the meaning of "convenience" directs our attention to a fundamental sociocultural transformation with important implications for the emergence of a new configuration of agency at this particular juncture. (It will also explain the Anglo-centric focus of these paragraphs.) In order to see this, we need to examine more closely the possibilities that inhere in free agency, as distinct from free will.

Recall once more the story of Buridan's hapless ass, dying of hunger at the fork in the road between two equally enticing bags of oats while seventeenth-century philosophers were debating its options. The ass in fact raises not one but two questions. The first concerns its ability to self-propel: is the ass truly—spontaneously—self-moving, so that it can choose one bag over another in the absence of an external cause? The second question concerns not the ass but the bags of oats: can they be truly equivalent, so that the ass is confronted with a fully symmetrical choice? Spinoza and Bayle, as we have seen, disagreed on the former question (Spinoza preferring to see the ass expire than to grant it the possibility of free will, Bayle allowing the ass to resort to lots or chance to determine its course of action), but accepted the symmetry of the ass's options. Leibniz, by contrast, focused precisely on this second assumption, simultaneously denying that one can ever encounter "an indifference of perfect equipoise" while also granting the ass, epistemologically, "an indifference in freedom."

This "indifference," the seeming equivalence of options in free choice, is as important to "capricious" or "whimsical" behavior as is spontaneous self-motion. Judeo-Christian free will had always meant the freedom to choose between good and evil, right and wrong; that is to say, the freedom to make moral choices between weighted options that a prior determination had already placed in a strict hierarchy to each other. But in order to imagine free agents exercising unencumbered free choice, one needs to envision individuals roaming around a Borgesian garden of forking paths, encountering symmetrical choices and marking their progress one bag of oats at a time. The analogy to the ass is somewhat overdrawn here, since from the point of view of the individual making a choice, even a trivial choice (say, which computer should I purchase for writing these words), the options are rarely experienced as truly equivalent. But from the point of view of the external observer there is, to repeat Leibniz's formulation, an epistemological "indifference in freedom." The eventual choice of the free agent appears unpredictable and thus, from the perspective of the system, *arbitrary*. And once it became commonplace to think about the actions of individuals as unpredictable (which is not the same as inexplicable), or arbitrary from the perspective of their surroundings, these actions could be taken as equivalent to coin tosses, and their aggregates as fodder for probabilistic analysis, or, indeed, for description as self-organizing.

The late seventeenth and early eighteenth centuries, we want to suggest, were a period of rapid expansion in the experiences of what we may call "indifferent agency," namely, the agency of choices between symmetrical options that can be described as arbitrary. Why? Let us take cues from two contemporary commentaries, one light-hearted, the other heavy-handed. The first is a 1711 *Spectator* essay of Joseph Addison's, which took its own cue from Buridan: "Some ludicrous Schoolmen have put the Case, that if an Ass were placed between two Bundles of Hay, which affected his Senses equally on each side, and tempted him in the very same Degree, whether it would be possible for him to eat of either." Professing ignorance about "the Ass's Behaviour in such nice circumstances," Addison arrived at his real goal, which was to "take Notice of the conduct of our own Species in the same Perplexity." So where did Addison find people facing this perplexity?

> When a Man has a Mind to venture his Money in a Lottery, every Figure of it appears equally alluring, and as likely to succeed as any of its Fellows. They all of them have the same Pretensions to good Luck, stand upon the same Foot of Competition, and no manner of Reason can be given why a Man should prefer one to the other before the Lottery is drawn. In this Case therefore Caprice very often acts in the Place of Reason, and forms to it self some groundless imaginary Motive, where real and substantial ones are wanting.[64]

Unsurprisingly, perhaps, it was in the new financial world, with its array of speculations and gambles undeterminable by reason, and more specifically in the lottery, that Addison observed such equal-weight options that encouraged free agents to make arbitrary choices, choices for which he mobilized that newly popular word "caprice."

Our second cue, pointing to developments beyond the financial revolution, comes from what John Lambe, chaplain to the king and future dean of Ely, had to say about agency in 1684. Lambe's sermon, *The Liberty of Human Nature*, opened with a rejection of beliefs in both fate and determinism. The real liberty of human nature comes into play, he explained, precisely where God's laws are at a distance. "Most of the Laws of God are *general*, only the great Lines and Heads of Duties, but the application of Particulars"—namely, how to apply God's general rules to specific circumstances—"are left to the decision of Human Reason." Consequently "we feel a *Vital* Principle within our selves, a perfect Liberty of Choice." We exercise this perfect liberty, this "absolute Authority over our own Actions," in our "indifferent" choices in this "world of various Contingencies"—that is, in a world of unpredetermined forks in the road.[65]

And yet Lambe, ex officio, was in a moralizing, not philosophizing, mode: "your Liberty"—we can see him wagging his finger—has become "your excuse for Luxury." The cat was out of the bag. What bothered this late-seventeenth-century divine was *luxury*: the loaded contemporary code word for the insatiable pursuit of goods in an expanding marketplace. The "various Contingencies" in which people found themselves, Lambe further explained, those moments when they faced forks in the road, occurred "in the midst of different Objects." Luxurious objects, presumably. But luxury to man is not really like oats to the ass: "Do ye not know," thundered Lambe's pleading conclusion, "that the Liberty of Man is such an Authority over his Will and Actions, that he may *always be indifferent* to things indifferent"? Here, then, was an original twist to Buridan's tale: the ass, now confronting luxuries not necessities, stays at the crossroad not because it has no free agency but because, having it, it realizes that the best choice is no choice at all.

For us, the significant lesson from Lambe's creative response to the moral challenge of symmetrical choices is the framework in which he placed these choices, revealed when he hurled at his audience the charge of "luxury." That context was the fundamental transformation, accelerating rapidly in Lambe's lifetime, that brought into being a new consumer society. Luxury was what you called the new temptations offered by this expanding market society when you made a stand against them. But as the Scottish Lord Advocate Sir George Mackenzie bemoaned in 1691, looking back toward the end of his life, "Luxury and Avarice" were "in this Age" masked with an "insinuating common name," when "we now by a kind and gentle word, call [them] *Convenience*."[66]

We are finally in a position to understand the significance to the development of agency of that new word usage "convenience," an innovation that

Mackenzie noted in real time. "Conveniencies" indicated consumer society's new regime of individual choice, those goods that were neither predetermined as necessaries nor prejudged as luxuries: commercialism's third way, one might say. The mark of these new possibilities for indifferent agency was also evident in another lexical shift, in the uses of the term "comfort." In the sixteenth and seventeenth centuries, as John Crowley has shown, "comfort" had still referred primarily to psychological and spiritual, not physical, circumstances, often in reference to providential blessings. Only from the late seventeenth century did its meaning expand to include material goods and physical surroundings. This shift diminished its moral underpinning and the imperative to correspond to an external order, such as social hierarchy, leaving the choice of its particular accessories, once again, morally neutral and open-ended, a commercial third way.

"Convenience" and "comfort," then, signaled new possibilities for individuals to practice seemingly arbitrary choices in an evolving consumer society. Students of this key development in the history of Western Europe have helpfully illuminated why changes in consumption patterns during this period entailed an expansion in what we have called indifferent agency. Jan de Vries follows many eighteenth-century commentators in characterizing this shift, which he sees first in the Dutch Republic and then more strongly in England, as one from "Old Luxury" to "New Luxury." Instead of conservative hierarchy-reinforcing high-cultural norms, this "modern, or proto-modern" pattern of consumption was now based on "the innumerable choices of an enlarged population newly endowed with discretionary income." "The new commodities were not simply replacements of goods formerly made at home," concurs Maxine Berg:

> Their key attribute was their availability in a range of patterns, styles, qualities, and prices. . . . They unpicked the uniformity and clear social hierarchies previously imposed by sumptuary legislation, and made individuality and variety an option to much broader parts of society.

Schematically put, in contrast with earlier generations of consumers that had sought goods that bespoke their anchoring in generations of skills and traditions, eighteenth-century consumers were surrounded by goods deliberately designed and marketed to showcase their novelty and variety.[67]

Defoe, the never-failing barometer of changes in socioeconomic practice, was well aware of how unusual this pattern of consumption was. In those parts of Europe that remained in the grip of premodern consumer patterns, he wrote in 1706, including Scotland, Spain, and Portugal, "Habits are National, known, constant, and without or with but small Variation;" consequently, "every one knows what to wear, Ascertains his Expence, and wears Garments to their due Extents." Consumer behavior in England at the beginning of the eighteenth century could hardly be more different: "The Clothes thrown by in *England* not for their being worn out, *but merely for their being out of Fashion*, is incredible, and perhaps are Equivalent to the general Cloathing Expense of some Nations."[68]

New social institutions emerged as catalysts and products of this new regime of consumer choice. Central among them were the retail shops, which emerged in new designs with street fronts, newly introduced glass shop windows, and elaborate displays of quantity, novelty, and variety. Their purpose was to appeal to browsing patrons, a form of shopping that in itself was new, precisely by impressing upon them the many symmetrical choices on offer. ("It is a modern custom, and wholly unknown to our ancestors," Defoe asserted, "to have tradesmen lay out two thirds of their fortune in fitting up their shops.")[69] Symmetrical choices and consumers with indifferent agency were, moreover, both the cause and the effect of the evolution of advertising, a significant phenomenon in both France and Britain from the second half of the seventeenth century. ("Advertisements . . . inform the world where they may be furnished with almost every thing that is necessary for life," Addison mused in 1710. "If a man has pains in his head, cholics in his bowels, or spots in his clothes . . . if he wants new sermons, electuaries, asses milk, or anything else . . . this is the place to look for them in.") Appropriately, among the earlier instances of printed advertising was the marketing of lotteries, those public games of chance that left contemporaries, in Addison's memorable estimation, in the same perplexity as that of Buridan's ass.[70] (See also figure 2.)

No consumer sector embodied this new regime of choice, materially as well as metonymically, better than clothing, or "fashion." The first instance of what has been described as consumer frenzy in fashion occurred during these same closing decades of the seventeenth century, in the so-called calico craze. Certain kinds of fabrics imported from the East Indies, offered in an unprecedented variety of colors, patterns, textures, and prices, became phenomenally popular in France and in Britain seemingly overnight. The duchesse d'Orleans, for one, found the enthusiasm for this range of choices unsettling. "I don't know why people have so many different styles of dress," she complained in 1695; "I wear Court dress (*le grand habit*) and a riding habit; no other; I have never worn a *robe de chambre* nor a mantua; and have only one *robe de nuit* for getting up in the morning and going to bed at night." But the

FIGURE 2. A South Sea Bubble playing card (King of Hearts). This depiction makes explicit the equivalence between shopping (in this case, for china ware) and participating in the financial world of stocks. The equivalence is emphasized further by placing both kinds of choices in the hands of *women* (women who are also described as part of an indifferent variety). Courtesy of the Baker Library, Harvard Business School, Harvard University.

gripe of the middle-aged duchess could barely be heard over the din of commercial entrepreneurs and eager consumers who scrambled to exploit the potential of "Consumption, *of which, there can be no end*, if there be the means" (these words are from 1691). "Note this for a constant and General Rule," the East India Company instructed its factors in 1681:

> That in all flowered Silks you change the fashion and flower every yeare, as much as you can—for English Ladies, and they say the french and other Europeans—will give twice as much for a new thing not seen in Europe before, though worse, then they will give for a better silke of the same fashion worn the former yeare.[71]

Commerce, explained a self-proclaimed expert in 1713, supplies "the Necessaries or Conveniencies of Human Life," which, from a small number of different "Generals" (food, physic, wearing apparel, furniture, ornament, art) "the Variety may afterwards be extended to a kind of Infinity."[72]

*A kind of infinity*: this was indeed a new consumer experience. In a garden of forking consumer paths, strollers were encouraged to make repeated choices between equivalent bags of oats, choices that were morally neutral and open ended. It was the unprecedented experience of a "modern" fashion regime that became such an eighteenth-century cliché, with its glorification of novelty and variety, its ephemerality, and its rapid mutability from one moment to the next. "Fashion is Fancy," warned one observer of the calico craze, and anyone who tells you different—that is, that such choices do matter—is trying to sell you something.[73]

Unsurprisingly, while liberating for some, this new regime of multiple choice was seen by others as a threat: a threat to order, to hierarchy, to morality, to religion. The language of "fashion" thus became a hackneyed cause célèbre of eighteenth-century social criticism, interlaced with connotations of random, whimsical, unpredictable behavior. "Fashion is . . . always mending, but never improving," went one 1714 essay that can stand here for many more; "to change for the sake of Changing, is to submit to the Government of Caprice." This criticism directed its fury at, among others, a new social type, or stereotype, that in the last third of the seventeenth century emerged from virtual obscurity to become a familiar embodiment of this unwholesome tendency of the age: the *coquette*. The coquette, Theresa Braunschneider has shown, signified first and foremost a woman who experienced the recently expanding range of options and resisted any constraint upon her choices, characterized instead by constant motion and boundless, capricious desire. The language of "luxury" too was enlisted to resist the proliferating possibilities of equivalent choices. After all—recall John Lambe's sermonizing—what were the nervous attacks on luxury if not a gigantic effort to reload the dice, that is to say, to reintroduce hierarchies of judgment into those "indifferent" choices of consumer society? "A Bed of Down," went one antiluxury rant rejecting the assumption of symmetrical choice, "may be unhealthy, and consequently a piece of *Luxury*, when one of harder Make, tho' altogether as costly, may *not* be so."[74]

Indeed, even in art, a cultural domain that had previously been subject to a powerful external hierarchization of value, the strict standards of academic taste now had to jostle with efforts to cater for personal taste, or agency of choice, in a public marketplace. A wonderful embodiment of this change in the world of art is Antoine Watteau's last painting, *L'enseigne de Gersaint* (1720–21).

The very purpose of this painting was a gesture to consumer choice: Watteau designed it for Gersaint's storefront in order to attract passersby—that is, as an advertisement. (Within weeks, however, the painting was removed from Gersaint's door and sold off, an object rather than vehicle of commercialization.) Even more telling was the scene it depicted (figure 3). Watteau packed the shop with floor-to-ceiling displays of dozens of artworks for sale, interspersed with other luxury goods such as mirrors and a clock. He populated it with conspicuously fashionable patrons inspecting selected art pieces as they would pieces of cloth. He manned it with attentive shopkeepers showing their wares, while other shop employees handle the artwork like merchandise. And, most famously, he loaded it with the portrait of the deceased Louis XIV, formerly commissioned by the king himself, now carefully packed sideways, perhaps to satisfy the commission of a new anonymous buyer, no differently than a landscape, a classical scene, or the nude nymphs inspected by the clients on the right. With all these, Watteau's representation of commercialized high art was as evocative an indication as any of the new indifferent agency of the eighteenth-century consumer, facing equal-weight choices that bypassed all other hierarchies.[75]

The new free agent, then, found in the expanding consumer market a fertile ground for the exercise of indifferent choices; choices that from the point of view of an external observer were unpredictable, capricious, and thus arbitrary. In 1724 a perspicacious clergyman noted just that. Unlike choices for which "a real Standard," namely, "the intrinsick Difference of Virtue and Vice," provides "a *real* Dissimilitude or Inequality," the clergyman explained, "Modes [are] purely arbitrary," and thus "in [their] Case, the Difference may be only accidental and imaginary." He continued:

FIGURE 3. Jean-Antoine Watteau, *L'enseigne de Gersaint* (1720–21). Photograph: Schloss Charlottenburg, Berlin/Giraudon/The Bridgeman Art Library.

> In Things of their own Nature *arbitrary*, or indifferent, such as those, which relate to Modes or Fashions, the Reasons of our Approbation or Dislike, of our using or discontinuing the Use of them, are mutable, according to the different Time, Place, or Disposition, wherein we may be.[76]

How arbitrary were consumer choices? Imagine, for instance, being required to choose from the following catalogue of textile goods, which was included in a 1706 merchants' manual from Leipzig: mouris, salomboris, bettelles gulconda, tansjeps, parcallen, goa concherulaes, gerberys, aloes socratina, cuttannees, indico lahore (garbled), turmerick, cardamons, succutums, ophium, izzarees.[77] Though this was a handbook innocent of satirical or ironic intent, the effect was comic. Such a list was a common rhetorical device that employed excess to convey limitless variety and abundance. But at the same time it also displayed the futility of order and the impossibility of hierarchizing through reason: choices from such a list, for virtually all consumers, were inevitably indifferent.

Signs of the arrival of this indifferent agent are easy to find during this period, in both commentary and practice. The Parisian essay *Sur les jeux de hazard* of 1708 recommended extending the logic of games of chance to "civil life" for precisely this reason. Unlike physical objects, whose movements are predetermined by force and direction, it explained, humans often do not recognize what is best for them, and even when they do they often do not act accordingly. Rather, "caprice determines their actions much more than reason," and thus "the consequences of their freedom [of action] cannot be judged but only guessed." The Oxford-based minister Digby Cotes observed similarly in 1713 that man is "the only irregular part of the whole Creation," since he "almost always act[s] at random." Man is unique among all creatures, a midcentury essayist reiterated, because of "his Oddities and Caprice," as a consequence of which "Man's Life is spent in liking and disliking, in chusing and refusing, the same Things."[78]

In terms of practice, the consequences of a new kind of agency can be seen, for instance, in Lori Branch's suggestive argument about a revaluation of spontaneity in religious practice during this period (related, of course, to the changes in the meanings attached to the word "spontaneous"). The putative disavowal of religious ritual in the Enlightenment, Branch writes, needs to be considered against a counterpart transformation "in which spontaneous emotions and behaviors"—"rituals of spontaneity" from free prayer through effusions of joy, weeping, or poetic inspiration that came to be associated with the culture of sensibility—"became increasingly, almost irresistibly, compelling."

Self-propelling, unpredictable actions of individuals, Branch argues, became central to collective religiosity.[79]

Or, rather differently, to collective anxiety. Thomas Laqueur has documented the "modern" wave of panic about masturbation, the beginning of which he pinpoints to in or about 1712.[80] Surely we can see in the sudden explosion of angst about masturbation a revealing response to the slippery slope of individuals' unfettered, uncontrolled, spontaneous, self-moving, whimsical, excessive, imagination-driven agency?

Furthermore, if masturbation was free agency metastasizing to pathological excess, its broader semantic field was perhaps that of addiction. So here is one final linguistic shift that seems relevant to our argument within the same decades, in the uses of the word "addiction." Before the seventeenth century, "addiction" referred to the state of being bound or indebted to another: a lord, or the devil. During the seventeenth century the word first took on transitive and intransitive uses for actions performed by others or by oneself on oneself. But its modern meaning only emerged toward the end of the century, when it became in all its derivatives a reflexive term, referring to self-inflicted actions that diminished one's free agency.[81] One could now become addicted to alcohol, or tobacco, or books. Like masturbation, these were instances of individual self-propelling choices—typically exercised within the expanding possibilities of consumer society—taken to such a pathological excess that they undermined the very agency that had enabled them in the first place.

There is another, perhaps surprising, manifestation of the arrival of indifferent agency during this period, one that takes us back to a higher intellectual conversation: the late-seventeenth-century movement for religious toleration, which was intertwined throughout with assumptions about free agency and free choice. Given the troubles brought about by religious differences, William Penn wrote in 1685, surely it would be excellent if all mankind were of one religious persuasion. "But the pleasure of that Harmony is a thing to be wisht, rather than yet expected. 'Tis Fact [that] we differ, and upon a point wherein *Unity* is out of our Power." From this, Penn, like other contemporary champions of religious toleration (including Locke, whom Penn closely echoed), concluded that religious choice must be accepted. "We ought in Charity to presume, that all men think they chuse the best way to *Heaven*, especially where the choice is against the Stream, and draws Loss or Disgrace after it."[82] Not only should we tolerate other people's choices, but we should assume that they all have equal claims to being good choices.

This argument for toleration evolved from a Reformation debate on the doctrine of *adiaphora*, or things indifferent; that is to say, actions, beliefs, ceremonies, or objects unnecessary for salvation and consequently left to the discretion of the individual or, more usually, the magistrate. The doctrine in and of itself was thus indifferent to the question of individual free agency. John Locke, however, developed in his several tracts on toleration a more radical position, one that grew out of his confidence in individual self-direction. No authority can tell which things are truly indifferent, he argued, since to the believer *nothing* in religion is truly indifferent. And since human fallibility prevents us from being certain that our choices are correct, the conclusion must be to allow unfettered freedom of choice in matters of religion. (Tolerationists differed on whether this latitude extended as far as Catholics or atheists.) By denying the doctrine of indifference as it applied exclusively to trifles, Locke now paradoxically extended it to all religious matters.[83]

For David Wootton, one of Locke's modern editors, "underlying [Locke's argument for toleration] one can surely see the values of a market society. Locke assumes that consumers should have the right to make choices." Thus Locke could wonder, say, whether "it be reasonable that he that cannot compell me to buy a house should force me his way to venture the purchas of heaven."[84] The same analogy, juxtaposing religious and consumer choice, was even more explicit and forceful in Pierre Bayle, that other towering late-seventeenth-century advocate of toleration. Writing in 1686 from his religious exile in Holland in reaction to the revocation of the Edict of Nantes, Bayle began much like Penn a year before: "'T were to be wish'd that all Men were of one Religion; but since this is never likely to happen, the next best thing they can do is tolerating each other." In order to persuade readers of the need to accept the equivalence of religious choices, Bayle proposed the following analogy.

> Wou'd People take this Course, the Diversity of Persuasions, of Churches, and Worship, wou'd breed no more Disorder in Citys or Societys, than the Diversitys of Shops in a Fair, where every honest Dealer puts off his Wares, without prejudicing his Neighbor's Market (*sans traverser la vente d'un autre*).[85]

Here was the full-blown connection to consumer society. As long as they are presented honestly, the choices of religion are like those between goods in the market place: choices between equivalent "wares," none of which have a special a priori claim on the free agency of the consumer.

Moreover, not only did this individual consumer—here even in the marketplace of religion!—resemble once more the self-moving operator presupposed by an ideal-type self-organizing system; Bayle also gestured at the aggregate order of the system as a whole. Given the statistical improbability of all these random agents lining up in unison, diversity in the aggregate was inevitable. Crucially, however, it was also harmless. Even better, far from breeding disorder, religious pluralism actually had the opposite effect. Religious diversity, Bayle explained, produced "an honest Emulation" or "Strife" between different sects as to which would be the most virtuous in the eyes of God and "which [would] out'do the other in promoting the Interest of their Country." Hence Bayle's triumphant conclusion: "Now it's manifest, such an Emulation as this must be the Source of infinite publick Blessings." While each sect is preoccupied with its own promotion, the aggregate outcome of religious multiplicity and competition is beneficial, "producing a harmonious Consort of different Voices, and Instruments of different Tones, as agreeable at least as that of a single Voice."[86]

Bayle was not alone in his leap from the necessity of toleration for every free agent to the benefits of free religious choices of individuals for society as a whole. Enforcing religious uniformity is "impracticable," the staunch tolerationist Sir Charles Wolseley claimed in 1668. Furthermore, it is also bad policy: "So many divided Interests and Parties in Religion, are much less dangerous than any, and may be prudently managed to ballance each other, and to become generally more safe, and useful to a State, than any united party or interest whatever." Penn, likewise, insisted that the benefit of toleration was "a *Ballance at home*." There is no historical evidence, he pointed out, "*That there is danger in building upon the Union of divers Interests*." To the contrary, "they are neither few, nor of the weakest sort of men, that have thought the *Concord* of *Discords* the firmest Basis for Government to build upon. The Business is to *Tune* them well, and that must be by the skill of the *Musician*."[87] Wolseley's vision of a "prudently managed," divide-and-rule balance of religious "Interests and Parties," and Penn's of a skilled musician fine-tuning the "Ballance" of "divers Interests," were certainly not self-organizing. But the recognition of the value of variety and multiplicity of diverging individual choices was a step in that direction.

Defoe, himself a dissenter with a stake in toleration, closely echoed Bayle in recommending open religious competition. "Let the Strife be who lives best, and the Contention of the Clergy who shall preach Best, and, by this make as many parties and Factions as they please"; the result would be "a Communion of Charity and Civility between the parties," "a Union of Interest." In fact, as J. A. W. Gunn pointed out long ago, the post-Restoration English debate on religious toleration encouraged some to recognize—in Penn's words on another occasion—that "the word INTEREST has a good and bad Acceptation."[88] It was precisely during this late-seventeenth-century debate that the language of "interests" was introduced with a beneficial connotation into political analyses of competing groupings in the aggregate, a move of which we will see much more below.

Finally, here is a variation on this theme from Locke. "The mindes of men are soe various in matters of religion," Locke scribbled in a manuscript addition to his 1667 *Essay concerning Toleration*,

> that where men are indifferently tolerated . . . they are apt to devide and subdivide into soe many litle bodys & always with the greatest enmity to those they last parted from or stand nearest to, that they are a guard one upon an other & the publique can have noe apprehensions of them.[89]

For Locke, the harmonious balance resulting from such localized negative tensions among multiplying subgroupings—Penn's "interests," engaging in Bayle's and Defoe's "strife"—was a natural outcome, not a managed one. In this vision we see, first, the individual random operators necessary for conceptualizing a self-organizing system, and then the further step of combining them into the narrative of the self-organizing aggregate order itself—the kind of narrative that will preoccupy us in the remainder of this book.

Thus we conclude this quick discussion of religious toleration, the broader subject of indifferent agency, and this chapter as a whole. Its goal, you recall, has been to lay down an array of key developments in the late seventeenth and early eighteenth centuries that changed the texture of life for many and expanded the horizon of the imagination for some. New experiences introduced contemporaries again and again to greater complexity, opening the door also to a greater appreciation of its potential benefits. One could begin to imagine individuals with new possibilities of free choice, systems with new possibilities of aggregate order, invisible and nonlinear causality with new possibilities of relating causes to effects, a providential world order with new possibilities of leeway commensurate with divine oversight, and overall new possibilities of order emerging from disorder. The essential components of the vocabulary of self-organization are all here, and some contemporaries have already put them together into self-organizing narratives. It now remains for us to return to the particular historical moment flagged in the prologue, a moment in which unusual and extraordinary circumstances rendered such narratives especially useful, and consequently pushed the language of self-organization into the open as a repeated, insistent, broadly resonant public phenomenon.

CHAPTER 3

# Man-Made Apocalypse: The Public Emergence of Self-Organization

We are ready now to go back to the four figures of our prologue—Montesquieu, Vico, Mandeville, Pope—all of whom, you will recall, articulated self-organizing thinking in one way or another in the 1720s and early 1730s. In some of their writings, at least, we find clues as to why: that is to say, to the specific circumstances of this particular decade that may have driven their self-organizing experiments.

Take for example Montesquieu, who in a way is the odd person out in our opening gambit, since his possible engagement with self-organization in the *Persian Letters* was not about social or natural order, like the others', but about the carefully contrived yet seemingly spontaneous emergence of literary order. Flagging this particular text was not really necessary. We could have readily put forth other passages written by Montesquieu in the 1720s that would have mingled more easily with our other guests—for instance, his reflections on the decline of the Roman Empire, published in France and in immediate translation in Britain just as Pope was completing the *Essay on Man*. Counter to what some assume, Montesquieu asserted (under the influence, perhaps, of a long recent sojourn in England), the divisions inside Rome were not the cause of its decline. On the contrary, what in fact spells the end of the spirit of liberty is when everyone in a republic is completely tranquil. "Union, in a Body Politick," Montesquieu insisted, "is a very equivocal Term: True Union is such a Harmony as makes all the particular Parts, as opposite as they may seem to us, concur to the general Welfare of the Society, in the same Manner as Discords in Musick contribute to the general Melody of Sound (*concourent à l'accord total*)."[1] How similar to Pope's virtually simultaneous couplet penned at the same moment: "Till jarring int'rests of themselves create / Th' according music of a well-mix'd State." But the reason to begin with Montesquieu's earlier work is that the *Persian Letters*, in both timing and content, points an unmistakable finger at the particular historical context that propelled self-organization past a quantitative and qualitative threshold to its first public appearance as a meaningful European cultural phenomenon: namely, the crisis of hyperactivity in financial speculations that took over much of Western Europe around the year 1720.

The most familiar memory of the financial events around 1720—to this day a favorite of many an economic pundit—is the "never-to-be-forgot or forgiven" South Sea Bubble, denoting the first-ever stock market crash, which took place in London in late summer 1720.[2] In fact, however, these financial goings-on were more protracted in time and more international in scope. In France, the maverick Scotsman John Law—"a kind of wizard whom one day we will perhaps believe as imaginary as Merlin"—had already begun putting into place his many-tentacled financial "System" in 1716. Starting with a private bank, Law expanded his efforts through the Mississippi Company (established to exploit French holdings in North America, and floated on the stock market through aggressive propaganda) to a monopoly over coining and issuing money combined with complicated maneuvers to liquidate the royal debts and expand public credit. By the time the overreach of the System led to its collapse in early 1720, it had reportedly drawn to Paris tens of thousands of people from all over France and the rest of Europe who wanted to try their hands at this newfound way of amassing wealth.[3] In Britain, the "South Sea Scheme"—a proposed privatization of the public debt by commuting it into shares of the chartered South Sea Company, which drew investors in 1719 and 1720 into a rapidly spiraling share market, including many who had previously placed their confidence in Law—was but the most conspicuous of many financial schemes floated during those years. Thomas Gordon, echoing Defoe, described these years as the "Projecting Age." In the aftermath of the collapse of the South Sea stock, it became a pastime among critics to compile long lists of real and imaginary financial projects recently directed at an apparently infinitely gullible public. And simultaneously with these financial activities in Paris and London, though often overlooked, was a rash of speculative activity throughout Western Europe—including in Amsterdam, Vienna, northern Germany, and Portugal.[4]

These events left deep impressions in the *Persian Letters*, reverberating throughout with Montesquieu's thinly veiled exasperation at John Law's innovations. One letter describes Law as a dealer in balloons full of wind, which he peddled during "a wandering life in the company of the blind god of chance." Another revealed what Law sold to the French regent as a mathematical system as in truth but a chaos of randomness and chance, analogous to "the fortuitous conjunctions of the stars" and akin to gambling (a famous forte of the Scotsman).[5] Something was rotten in the kingdom of France, which emerges from the *Persian Letters* as beset by acute disorder.

It was a disorder, moreover, that bred more and more chaos in its wake. Montesquieu is especially suggestive on this point, as he recounts how the frenzy of investors led to the rapid making and unmaking of fortunes, and finally to an investors' panic and thus the collapse of the stock market. "Fortunes have been made so unexpectedly that even those who made them can't believe it; God does not create men from nothing with greater speed."[6] The effects of Law's scheme amounted to the instantaneous creation of something out of nothing: a profoundly disruptive, nonlinear sort of event that challenged the creative powers of God himself.

The dynamics of investors' panic are revealingly if elliptically elaborated in a letter from Rica the Persian to a Jewish doctor, pooh-poohing his cabalistic belief in the power of "arranging certain letters in a certain order." (Is there here also a punning gesture at the contrived (dis)order of Montesquieu's own series of letters?) Suppose, says Rica, there is a battle in which one side emerges victorious. The Jew "refus[es] to admit even that it has a cause" and, while remaining "blind" to natural, rational causes fully "sufficient to produce this effect," seeks instead an explanation in "magic spells," "the appearance of an invisible power," or other "supernatural" occurrences. Thus far, the letter reaffirms an apparent confidence in straightforward causality. But Rica then whimsically subverts this confidence, while gesturing back to the financial crisis that followed Law's schemes. Rica asks the Jew to imagine that one army was defeated because of "panic (*ces terreurs paniques*) among the soldiers, a form of terror which you have such difficulty in explaining." Rica offers an alternative explanation for the panic:

> Do you believe that in an army of a hundred thousand men it is impossible for there to be a single coward? Do you think that his faint-heartedness could not cause another man to lose heart? That the second, deserting a third, could not soon cause him to abandon a fourth? That is all that is necessary for a whole army at once to give up hope of victory, and the more numerous it is the more easily will it give up.[7]

What started as a defense of causality ends up highlighting its *limits*. In a mass panic—the evident analogy is the contagious frenzy that overtook investors in Paris at that very moment—a nonlinear progression of causes takes over. It leaps from the most minor trigger—the single cowardly soldier, the single investor who loses confidence—to a disproportionate major consequence: the immediate collapse of a full army, or of the stock market. And the greater the multitude, the faster the effect. In truth, then, how distant is this account from the Jew's "invisible power"?

For Montesquieu, in sum, the disorder at the time of the *Persian Letters* was manifest not only in social and political disruption but also in the disruption of the mechanisms of order: through randomness, through erratic actions of the multitude, through the reliance on chance in public affairs. As a consequence, whether in the creation of something from nothing, or in the unleashing of large-scale effects of minor causes, one significant casualty of John Law's financial schemes was causality itself.

To be sure, this suggestion is speculative. And yet if Montesquieu *was* prodded by the unprecedented financial havoc of 1720 to confront the question of the emergence of order from disorder, he was not alone. He was not alone, first, in the realization brought home by these events—amplifying and perhaps confirming the doubts that had already niggled at Joseph de la Vega, Daniel Defoe, or Gregorio Leti—that prevalent models of causality and mechanisms of order may be insufficient to describe a complex world, not least in the realm of finance. Indeed, as we shall see in a moment, contemporaries repeatedly described their experience of these events as one of *chaos*: an inexplicable chaos driven by the capricious actions of multitudes of individuals. Many were led by these experiences to question prevalent assumptions about causality and order. And here is the second way in which Montesquieu was not alone. For some European spectators these doubts prompted alternative speculations about the origins and nature of order, leading to explorations in self-organizing thinking.

Which brings us back to Pope, who wore the connection between these financial speculations and his own speculations about the origins of order conspicuously on his poetic sleeve. Here again is Pope's sharpest formulation of the logic of self-organizing order, emerging spontaneously from the chaotic pulls of different individuals:

> Hear then the truth: " 'Tis Heav'n each Passion sends,
> "And diff'rent Men directs to diff'rent Ends.
> "Extremes in Nature equal Good produce,
> "Extremes in Man concur to general Use." (*To Bathurst*, lines 161–64)

As Vincent Carretta has shown, Pope published the *Epistle to Bathurst* as a political intervention carefully planned to chime with current events in early 1733, but the immediate context within which it must be understood was the South Sea Bubble. The memories of the events of 1720 were thrust again to the forefront of public debate during the 1732–33 investigation of "the Charitable Corporation," yet another project to relieve public debts that the opposition identified with the by-then-notorious South Sea scheme. The leader of the campaign against this new scheme was none other than Lord Bathurst, otherwise an unlikely addressee for Pope's verses. The *Epistle to Bathurst*, timed to hit the streets precisely on the day before the opening of the parliamentary session, is suffused with references to people and events involved in the 1720 fiasco.[8]

Furthermore, not only was the 1720 financial crisis the general background for the poem as a whole, it was also the specific context of its particular verses devoted to self-organization. What precedes these verses is a long passage about the persistence of political corruption, which begins with the words "Much injur'd Blunt! Why bears he Britain's hate?" (135). John Blunt had been a director of the South Sea Company, a key player during the events of 1720 who was often compared to John Law, and subsequently one of the main figures prosecuted and punished for it. Pope did not doubt Blunt's culpability but protested his being singled out when the nefarious practices underlying the South Sea Bubble, with its cheating and gambling, were still alive and well. And why do people participate in such practices, even reasonable people like Pope, who had himself invested considerable sums and efforts in that financial roller coaster ride of 1720? "All this is madness, cries a sober Sage, / But who, my Friend, has Reason in his Rage?" (153–54). Pope devotes several lines to this madness, shaped as it was by individuals' ruling passions that conquered their reason, before delivering his final judgment: "hear then the truth." The truth, significant enough for Pope to cross-reference unusually several times, was the revelation that in fact the frenzied disorder of the erratic actions of so many people, driven by their passions, ultimately comes together into an overall harmony for "general use." Pope's observations on mass behavior during the South Sea Bubble led him directly to the discovery of self-organization.

For Pope, then, the line connecting the events of 1720 to his thoughts on the emergence of beneficial order from the disorder of the multitudes is unmistakable.[9] Montesquieu also had these disorders freshly on his mind, but we can only speculate as to whether they did indeed influence his chaotic and perhaps self-organizing plan for the *Persian Letters*. For Vico and Mandeville, we have offered no evidence that their own ruminations on the same general problem were triggered or catalyzed by the South Sea Bubble or the Mississippi Scheme, other than the suggestive timing.

It should be said, however, that in the case of Mandeville the argument from timing is quite revealing. This is because Mandeville had already published an earlier version of *The Fable of the Bees* in 1714, and the short poem that had spawned it, *The Grumbling Hive; or, Knaves Turn'd Honest*, as early as 1705. A careful comparison of the *Fable* of 1714 with the expanded (and now standard) version of 1729/1732, and those in between, shows how much Mandeville's thinking about social complexity had developed in the meantime. In 1714 Mandeville had already unfolded both his anthropology of man and his basic insight summarized in "private vices, public benefits." But it was left for the later editions to spell out the dynamic through which this beneficial outcome emerges, that which we would describe as a dynamic of self-organization. Thus, the key passage we quoted in the prologue, on the need to keep in mind the greater vision of multiple concatenated events to see how good springs from evil, was added by Mandeville in 1723. Even more revealing are Mandeville's changing uses of the word "multitude." In the 1714 *Fable* "multitude," when employed to connote more than simply a synonym for "many," is always used negatively: a multitude is a mob. In the 1729 edition, however, we can also find Mandeville using the word—now, more typically, in the plural form "multitudes"—with a new connotation of a necessary and beneficial precondition to aggregate complexity.[10] Our attention is thus drawn once more to the decade and a half after 1714 as the period in which Mandeville appears to have come to combine his famous moral vision with a model of society as a self-organizing system.[11]

Our speculations regarding one or another of these particular individuals may be off the mark. Be that as it may, the following pages establish, we trust, a much broader case for the relationship between the European financial events in the years leading to 1720 and a concentration of self-organizing thinking in their wake. This chapter begins by mapping the challenges these events raised for European notions of causality and order. It then charts various ways in which contemporaries responded to these challenges. One of those responses—not the most prevalent one, but a radical new departure with a significant presence—involved the conjuring up of self-organizing stories, stories that seemed to contain the difficulties of the moment especially well. The previous two chapters have shown the development of the *means* that had made such conjuring acts possible: namely, recently emergent ways of thinking about God and providence, about complexity and causality, about aggregation and large numbers. The unusual constellation of events around 1720, which we discuss now, provided an *opportunity* and generated a pressing *motive* for disparate individuals in disparate places to weave together self-organizing narratives, drawing on those conceptual tools and cultural resources already in place. It is because of these specific circumstances, in short, that the 1720s turned out to be the decade in which self-organization crossed the threshold as a full-fledged cultural-cum-intellectual phenomenon in Western Europe.

### IN THE BEGINNING THERE WAS CHAOS

Many in England, France, and beyond in the years around 1720 were enthralled by the long spells of financial hyperactivity, during which bouts of steep rises in the financial markets were followed by abrupt collapses. "If you Resort to any publick Office, or place of Business," an English pamphleteer scoffed, "the whole Enquiry is, *How are the Stocks*? if you are at a *Coffee-House*, the only Conversation turns on the *Stocks*, even the Scandal of the *Tea-Table* is forgotten."[12] The drama was considerable: in both booms and busts contemporaries insisted that what they had just witnessed was unprecedented and extraordinary.

At first glance this does not appear surprising, if one imagines the scene like those poignant black-and-white documentaries of the Great Depression, replete with pictures of widespread misery and impoverishment. One is helped in filling in the details of this mental image by a barrage of contemporary lamentations invoking that unique, universal, unmitigated distress. "None can be insensible,"—one example following the bursting of the South Sea Bubble can stand for many—"that there is a sudden run of general distress, affecting innumerable persons and families, beyond any instance within the reach of memory, and possibly within the reach of history."[13]

As it happens, however, this is not an accurate picture: 1720, it turns out, was not at all like 1929. The economic historian Julian Hoppit has reevaluated the British case—the most acute, and that which has produced the greatest number of distress narratives—and concludes that the economic and social repercussions of the South Sea Bubble were far less severe than these emotional contemporary reactions have led us to believe. People's losses were not always as dramatic as reported. In fact, if one held on to the South Sea stocks from January 1720 to December, through the crash of the market, one would have *gained* a healthy rate of return of 50 percent. Investors were largely the wealthier members of society, who risked only a fraction of their wealth and could afford to take a hit, not the common men or women whose losses might have turned them into the street. The disruption of the wider economy was limited, as evident from the telling fact that there was no overwhelming surge in the number of bankruptcies but merely a gentle blip. There is no substantive evidence of a major disruption of trade, industry, or agriculture; though international credit flow, affected by the Mississippi Scheme and other international developments as well as by those in London, was indeed significantly disrupted. Nor was the bursting of the Bubble as unexpected as contemporaries and historians have asserted (it arrived, Pope wrote to a friend, "*like a Thief in the night*"). Many observers had in fact assessed quite accurately and publicly the dangers of this speculative bubble and predicted its collapse, whereas the decisions of others to take the risk were often well informed and well considered.[14]

So while the Bubble, together with the other financial tremors of that year across Europe, was a significant crisis, and some people surely did suffer significant losses, its overall direct impact seems insufficient to explain why it unleashed such pervasive puzzlement and angst, often bordering on the hysterical.[15] We would like to propose another possible reason, suggested by a wide trawl through the many hundreds of publications generated by these events (for which purpose we focus primarily on the British case[16]): the persistent concern with the problem of *causality*. The goings-on that the bevy of hack writers and self-appointed analysts observed in the financial marketplace appeared curiously resistant to available frameworks of rational description and causal explanation. This resistance was rather unsettling, not least because few had doubts as to why individuals were doing whatever they were doing in the stock market. It was news to no one that investors, in the words of one writer, "know no other Principle than to get Money, and the Rule and Byass of all their Actions is Interest."[17] And yet acknowledging the individual pursuit of naked self-interest, it turned out, did not go very far in making sense of what was going on.

This inability to link causes and effects was more unsettling still since this was, after all, the era of Isaac Newton, widely believed in his time to be a once-in-a-millennium genius of causal explanation; one "who stands up as the Miracle of the present Age"—thus Addison's *Spectator*—and "can look through a whole Planetary System." Through the whole planetary system perhaps, but not through the chaos of the stock market. "*I can calculate the motions of erratic stars*," Newton's frustration in response to the South Sea Bubble is famously quoted, "*but not the madness of the multitude*." (Newton himself lost some money in the Bubble, but probably not as much as is often claimed.)[18] Curiously, there is no actual source for this widely reproduced Newton quote.

Perhaps it was put in his mouth by a generation who felt this is what Newton *must* have said, since for them he stood for causal certainty and explicability. This remained true, as we shall see further in a moment, even if in fact Newton himself was not quite as certain about causes as in his public image.

Rather, the basic experience of people observing these events was one of profound and unsettling disorder. In France, already in 1718, the *parlementaires* of Paris had not minced words. John Law's still-evolving System, they warned then, "produce[d] disorderliness in everything" and created "a chaos so great and so obscure that nothing about it can be known." And if the *parlementaires* might have had in mind mystification from above more than self-generating chaos from below, this was certainly not true of the English divine who opened his finger-wagging sermon in November 1720 by lamenting "the great disorder and confusion which an eager desire of wealth has of late occasion'd amongst us." The problem, "the unhappy sources of the [present] calamities," was that the stock market made possible the "irregular pursuit" of wealth, by facilitating an "irregular traffick" and opportunities for "irregular and indirect courses to acquire it." It was not the desire for wealth in and of itself but rather all this rampant irregularity that kept our sermonizer up at night. To all intents and purposes, the collective outcome of what the writer Charles Gildon called "the giddy Fluctuation of the People," of the unpredictable "Humour and Whimsey, that made the Stocks run up to that prodigious Degree" and then "according to its changeable nature" sent the market back down "as fast and as unaccountably as it run up," was simply random.[19]

Jonathan Swift, poet, satirist, and sometime South Sea investor, penned these verses (as well as many others) to try to capture this particular aspect of the Bubble:

> The Bold Encroachers of the Deep,
> Gain by Degrees huge Tracts of Land,
> Till *Neptune* with one gen'ral Sweep
> Turns all again to barren Strand.
>
> The Multitudes capricious Pranks
> Are said to represent the Seas;
> *Breaking* the *Bankers* and the *Banks*,
> Resume *their own* whene'er they please.[20]

On one side in Swift's image stand the bankers and the "bold encroachers": the rational operators of the financial scene, whose actions are ordered and measured—they who "gain by degrees." However, the real power to shape the outcome is not in their hands but rather in those of the multitudes, whose destructive tidal wave is unleashed like "capricious Pranks," randomly and unpredictably, "whene'er they please." The main force here is not greed, with which the bankers and encroaching speculators are amply endowed; it is the unfettered agency of the multitudes, their unpredictable capriciousness, irregular whimsy, and giddy fluctuations, resulting in randomness and chaos.

If the purpose was to convey chaos, the pun was inevitable. The Bubble was biblical Babel: the ur-example of disorder brought about by multitudes of self-propelled individuals. "Did the *Tower* of *Babel* make more Confusion of *Tongues*," went one example, "than the *Babel* of the *Stocks* has done to *Fortunes* in all Nations?" Talk about "our *South-Sea Babel*" also brought to mind another important aspect of the biblical story: it was man-made chaos in defiance of God.[21] The question of God's involvement in these events, or the lack thereof, will preoccupy us in a moment. At this point suffice it to note that the contrast with divine harmonious order could be mobilized further to underscore the unfettered human agency behind the chaos. Thomas Gordon, for instance, addressed a moral tale "to All South-Sea Directors" that began with a lawgiver in antiquity who had founded morality on a contrast between heaven above and earth below. It would be unreasonable "if, while we observ'd so just and beautiful a Regularity above, we could suffer our selves to be so wild and *eccentric*, as I may say, in our Motions below." This man-made chaos was more puzzling for being freely self-inflicted rather than the consequence of external pressures. "It will hardly be credited in future Ages," exclaimed another, that this great nation "was brought almost to the Brink of Ruin, in a few days, without the Calamities either of a Foreign Invasion, or an Intestine War." In the past, echoed a third, we had coped with "the malicious contrivances and unjust attempts of our enemies; but the present distresses of the nation *are directly self-made*."[22]

Wild, eccentric, capricious, irregular, giddy, changeable, unaccountable, self-made: small wonder that Newton reputedly summed it all up as the incalculable "madness of the multitude." His contemporaries, it seems, could not agree more. The multitudes were "like Men possest and frantic." "Have you really all gone mad in Paris?" asked Voltaire in response to the investors' frenzy already generated by Law's actions in 1716; "it is a chaos I cannot disentangle." The son of Robert Harley, the earl of Oxford who had invented the South Sea Company a decade earlier, reported from London in May 1720: "The madness of stock-jobbing is inconceivable. This wildness was beyond my thought." And a month later: "Nothing is so like Bedlam as the present humour which has seized all parties, Whigs, Tories, Jacobites, Papists, and all sects." A Dutch observer described these events "as if all the Lunatics had escaped out of the madhouse at once." This last tableau is especially rich: it conveys the kind of chaos generated by a multitude of individuals operating each on his or her own steam in random directions. This was "the *many-minded Multitude*," Sir John Meres wrote—another evocative image, which he then resorted to Latin to explain: "the *incertum Vulgus*, of whom there is scarce any Thing certain, but this, *Quod scindet in contraria* [that which tears up in opposite directions]." Two and a half centuries later P. G. M. Dickson, the usually temperate doyen of Bubble historians, described the behavior of those early-eighteenth-century investors, "both in the boom and after its collapse," as "apparently hysterical and ungoverned."[23] For observers then and now, they acted erratically and randomly, as if possessed, not directed by any distinguishable rationale or logic.

Was it possible, then, *pace* Newton's reputed exasperation, to disentangle this seeming chaos according to what contemporaries believed was Newtonian causal logic? (That is, the logic they derived from Newton's celestial mechanics more than from his curious theologies.) Was it possible to satisfy the hope of John Theophilus Desaguliers, the foremost popularizer of Newton's work, that *The Newtonian System of the World*—the title of his didactic poem—would ultimately be able to theorize also disorders like the South Sea Bubble, in which "The Coin, to Day, shall in its Value rise, / To morrow, Money sinks and Credit dies"? For many the answer was a resounding no. This, for instance, was the implication of the following two stanzas from what is probably the best known literary reaction to this "*Madness of the Crowd*," Swift's endlessly reprinted poem "The Bubble":

> There is a Gulph where thousands fell,
> Here all the bold Advent'rers came,
> A narrow Sound, though deep as Hell,
> CHANGE-ALLY is the dreadfull Name;
> Nine times a day it ebbs and flows,
> Yet He that on the Surface lyes
> Without a Pilot seldom knows
> The Time it falls, or when 'twill rise.[24]

Newton was present in these lines through the reference to the regularity of the tides. It was his gravitational theory, after all, that famously succeeded in explaining the tides better than anyone before. And yet the tides in Exchange

Alley do not follow Newtonian logic. They therefore leave the hapless adventurer at the mercy of their unpredictable ebbing and flowing.

An essayist who went by the name of Cleverkin developed the same contrast in more detail. "Sir *Isaac Newton* had reduced that long amazing Phænomenon the Tide to Order and Rules of Geometry," Cleverkin began his "whimsical" device, "but I have not yet heard of any venturous *Cantab.* that has aimed at accounting for, and reducing to, any settled Principle, the Motions and Phænomena of the *South-Sea*" with its "seeming Irregularities." (The association of this financial venture with an oceanic body of water of course invited such quips.) Perhaps, Cleverkin surmised, in analogy with Newton's chain of causes traceable to the moon, the behavior of his countrymen, "whom, for the future, I shall call the Men of the Moon," was likewise based on their "observ[ing] the different Position of the Stars, before they buy and sell Stock." The ironic elision of science with astrology, reminiscent of Rica's move in Montesquieu's *Persian Letters*, underscored the extent to which the financial developments in France and England were in fact *not* subject to the laws of mathematics and science. Reflecting on the laws associated in his mind with Newton, Cleverkin admitted: "But as every Thing grows older, something new is observed; and though a Rule in general be very true, yet there is a Difficulty in bending Particulars to it." What Newtonian laws could not handle was exceptions, exceptions like the newly observed French System and the English Bubble.[25]

No matter how partial contemporaries' actual understanding of Newton's achievements was, they were confident in the lessons these achievements taught about scientific models of causality. The *Director*, a short-lived post-Bubble periodical, began one of its many riffs on the South Sea Scheme chaos with "a nice Question" for its "learned Readers": "In what Part of the Agency of Nature lye the Secret Wheels of Retrograde Motion. And by what unknown Powers do Things move, when the World stands with the Bottom upwards?" On the one hand, it stated, we know a lot about "the Laws of nature," even the more elusive ones ("retrograde motion" was a well-known example of an elusive astronomical phenomenon). "We find them steady and masterly; that they are govern'd by a most exact Sovereignty, and a regular Hand"; most importantly, we find "that Consequences are obedient to their Causes and that the same Causes are, *generally speaking*, directly tending to the same Consequences." In the laws of nature, consequences can be depended upon to follow causes in a linear and stable fashion. Recent events, however, revealed the limits of these laws. "But, since we see in the World new Phaenomena rise up every Day"—this writer too was struck by how novel the developments of 1720 had been—"and that the Springs and Motions seem to be acted by Principles differing from these which we call Natural," we ultimately remain ignorant as to "the Secret *primum Mobile* of these Unaccountables." Unyielding to familiar models of causality, recent events remained "secret," "unaccountable," and "unknown."[26]

Instead, in this financial world regularity crumbled into chance and accident. The "blind fortuitous manner" of the stock market, observed Pope's friend the philosopher George Berkeley in 1721, rendered it a "public Gaming Table." After all, the hero and alleged wizard of this speculative world was John Law, the ultimate professional gambler, whose "incomparable readiness in Numbers made him a perfect judge of the hazards and advantages of all Plays."[27] Most people, however, were overwhelmed by these hazards, seeing their fate subject to accidental vagaries, the "capricious Pranks" of Swift's multitudes, inflicted "whene'er they please." They frequently underscored the randomness of these events with the language of *accident*. Could the directors of the South Sea Company, wondered one pundit, reasonably "expect that the least ill News from abroad or Accident at home, would not Tumble [the South Sea stock] down headlong?" John Trenchard and Thomas Gordon, in their renowned *Cato's Letters*, wrote as if they were responding directly to his question. "Common Sense could have told them that Credit is the most uncertain and most fluctuating Thing in the World, especially when it is applied to Stock-Jobbing." Consequently, and inevitably, credit

> always tottered, and was always tumbling down at every little Accident and Rumour. A Story of a *Spanish* Frigate or of a few Thieves in the dark Dens of the *Highlands*, or the Sickness of a foreign Prince, or the Saying of a Broker in a Coffee-House; all, or any of these contemptible Causes were able to reduce that same Credit into a very slender Figure.

Another commentator used the revealing yet seemingly gratuitous phrase "unprovided Accidents," bringing to mind the new legal doctrine of strict liability. We have heard others who earlier had already begun to recognize this inherent, unpredictable instability of credit. Credit—thus an elderly director of the South Sea Company, republishing words he had written as early as 1697—is "liable to be overturn'd by the least puff of Wind or panick Fear amongst us."[28] The least ill news, a rumor, an accident, any such "contemptible Cause" could bring down the whole financial market, and linear causality right there with it.

Like Montesquieu's Rica in the *Persian Letters*, who compared the investor frenzy in Paris to the single cowardly soldier spreading panic by nonlinear leaps and bounds through an entire army, many pinpointed their anxiety about current financial events on the puzzling disproportion between causes and effects. An investors' bubble bursts, an English periodical contributor explained, when,

> bringing a prodigious Number of Sellers to the Market, one Man selling alarms another, and makes him sell, and thus the Stock has run down insensibly, till all the People are put in a Fright; and such has been the panick Fear, that it has brought great Confusion along with it.

It was as if Montesquieu were ventriloquizing this English pundit. The South Sea Scheme, wrote another, was concocted as a means to raise public credit, "but it was only a Monster like the Shadow magnify'd on the Wall by the Light of a Candle, which was no ways proportionable to its Object."[29] The dynamics of disproportionality could hardly be more vivid.

No less a figure than the chancellor of the exchequer, John Aislabie, mobilized the deceptiveness of the relationships between cause and effect for his own public defense. "My Lords," he speechified in Parliament, "there was something very extraordinary in the Consequences of this Affair." The extraordinariness lay in the fact "that the more the *South-Sea* Company were to pay to the Publick, the higher did their Stock rise upon it." This positive feedback loop then spiraled out of control, generating an outcome out of any proportion to its cause (economists call this dynamic a "vicious circle").

> From that Time it became difficult to govern it; and let those Gentlemen that *open'd the Floodgates* wonder at the *Deluge* that ensu'd as much as they please, it was not in one Man's Power, or in the Power of the whole Administration to stop it, considering how the World was borne away by the Torrent.

The drops, by themselves, turned into an ever-expanding deluge: "The *S. Sea Scheme* was become *ungovernable*."[30]

It was Pope who a couple of years later penned the most memorable description of a financial feedback loop from this period: "Money upon money increases, copulates, and multiplies, and guineas beget guineas in sæcula sæculorum." The phrase "money begets money" and its variants, it turns out, had gone through a reversal that is quite telling. In the early modern period, as Deborah Valenze has noted, it referred to those aspects of usury that had traditionally carried negative connotations: the critics of usury, wrote Sir Francis Bacon, say "That it is against Nature, for *Money* to beget *Money*." Later in the eighteenth century, however, this phrase assumed the more modern economic overtones of a positive feedback loop. It falls to us to add that a survey of ECCO, EEBO, and the *OED* suggests that the early examples of this usage for the self-propelling quality of money were concentrated in the years following the South Sea Bubble, including not only Pope's, or the common wisdom example in a proverb book of 1732, but also a programmatic example from Daniel Defoe, that master crafter of a new linguistic toolkit to suit an emerging economic environment: "Money begets Money, Trade circulates, and the tide of Money flows in with it."[31]

Few were haunted by the mysteries of causality surrounding the events of 1720 as much as Defoe. Early on, wondering about the sudden meteoric rise of the French stock market, Defoe came up with a Franco-centric explanation for the conspicuous incongruity of causes and effects:

> Th[e] warmth of the *French* Temper, which prompts them to push things up to the Extremity, was certainly the Reason, I mean the Original Reason of the sudden Advance of these things, for as yet there was no weight in the things themselves, that could bear any proportion to the New Credit they assum'd.

Once the British stock market followed suit, Defoe could no longer rely on the idiosyncratic extremities of the French. Furthermore, now the very parallels between events in Paris and in London themselves defied causal thinking. They were obviously similar, and happened only a few months apart, yet the causal link between them remained a mystery. In a later pamphlet Defoe reflected again on the French and British schemes, between which "the Parallel goes on still, both were overturn'd by their own Bulk, the unperforming Machines blew themselves up by the Force of their own Motion." And what caused this self-propelled combustion? "The Overthrow of Mr. *Law* and his Schemes" arrived "when but a minute Accident gave them a Check" and brought about their collapse; which likewise happened "to Sir *John Blunt* and his Schemes, when they received a Shock from a like Accident as the other." Defoe returned to the parallel a few pages later: "Come we then to the *Circumstances* which overthrew our *South-Sea* Fabrick here, which are in Proportion the same: That which gave them the first Blow was likewise a Trifle, and from which even the longest head among them did not expect the Consequence that happen'd."[32]

Pondering the meaning of 1720 in both France and Britain, then, the most indefatigable commentator on the financial world of the early eighteenth century ended up singling out nonlinear relationships of causes and effects, set in motion by unpredictable trifles. We are back to accidents. And given the centrality of accidents to Defoe's interpretation of both events, is it a coincidence that Defoe wrote the two novels in which the literary critic Sandra Macpherson finds his new logic of "profoundly accidental" responsibility, *Journal of the Plague Year* and *Roxana*, in the early 1720s?

Pushed *ad extremis*, nonlinear thinking could end up with the most challenging scenario of disproportionality, familiar already from our first chapter: begetting something out of nothing. When an account of John Law's schemes talked of "People, never known in the World, and sprung from nothing" into high places, when another defined "the Word BUBBLE" as a way "to get a Million or two out of Nothing," or when Berkeley lamented how "Men shall from nothing in an instant acquire vast Estates," the nothing-to-something image can be seen as a figure of speech. This may be harder to do in the following verses, comparing John Blunt to Moses:

> So MOSES smote the barren Rock,
> (An Emblem of the SOUTH-SEA Stock!)
> Which touch'd with his cælestial Rod,
> Pour'd a full Stream, the Work of God.

By now, the South Sea Bubble was envisaged as a challenge not only to Newton but also to God himself. This was troubling. Britons in this period of unprecedented disorder were like madmen, exclaimed a pamphlet titled *The Lunatick*, since they posited that something can emerge out of nothing without the intervention of God, but "if there was, at some time, Nothing, how could Something ever be?" Edward Ward, typically, was one to make the point without mincing words:

> Our cunning *South-Sea*, like a God,
> Turns Nothing into All Things.

As Montesquieu's Rica noted too, nonlinearity was shading into blasphemy.[33]

In such reactions to the events of 1720 one readily recognizes echoes of older anxieties about the insubstantiality of this brave new financial world, as articulated from Joseph de la Vega onward, now magnified and reverberating more widely. The men of 1720 were described as "pursuing gilded Clouds, the Composition of Vapour and a little Sunshine; both fleeting Apparitions," or as dealing not "as formerly, with *Substances*, but with *Shadows*." Voltaire, in his response to Law's early schemes, had already wondered: "Is this reality? Is this a chimera?" When Defoe, picking *The Chimera* for his title, dramatized the crisis of Law's Mississippi Scheme as "an inconceivable Species of meer Air and Shadow, realizing Fancies and Imaginations, Visions and Apparitions . . . the Substance is answer'd by the Shadow," he was in fact recycling his earlier language for the day-to-day operation of credit. (Credit, he had written a decade earlier, "acts all Substance, yet, is it self Immaterial: it gives Motion, yet it self cannot be said to Exist: it creates *Forms*, yet, has it self *no Form*; . . . it is *the essential Shadow of something that is Not*.")[34]

And lastly, these multiple anxieties had a scapegoat: the stock-jobber. The annals of collective abuse may not readily offer another occupationally defined figure subjected to such a unanimous and vicious condemnation, and for much more than simple greed. Thomas Gordon was not atypical in the almost hysterical tones: "These Monsters," with their "irregular Method of acquiring Riches," "stand single in the Creation." Stock-jobbers—a term that was used from the late seventeenth century not only for stock market speculators but for those who engaged in any form of professional brokerage—were put forth as the epitome of everything that was wrong with these newfangled financial activities, and by extrapolation with these modern times. They bought and sold stocks with little relationship to their real value ("the *Vermine* call'd *Stock-jobbers* . . . put a counterfeit Value where there is no real one"). Their actions embodied randomness and insubstantiality ("They have invented the most idle chimerical Notions, their giddy Heads could devise"; "The *Jobbers* . . . none can wager on their Durance, / Some cry the *New*, some *Old Assurance*"). The "*Ductile Jobbers* in the *Alley*" introduced the hazards of games of chance into the stock market ("they are in reality only a sort of Wild *Gamesters*," "whimsical Fellows . . . [who] deal in this *imaginary* Commerce"). They were those who conjured something out of nothing, through "*hocus pocus* Management"; not least, their own standing (those "*Almighty Stock-jobbers* have risen to be so from small, and often sudden Beginnings"). They produced the dynamic that magnified effects out of all proportion to their causes ("What will Posterity think of us, when they hear of a whole Nation made the Game of a few insignificant stockjobbers?").[35]

The stock-jobber, in short, became the synecdoche for disorder. All the concerns that troubled people so were funneled onto the head of the hapless stock-jobber: man-made disorder, erratic behavior, random and jittery events, unreliable lack of substance, disproportionate and perhaps imponderable relations between causes and effects, the apparent collapse of linear causality. "Stocks, and *Stock-Jobbing* [are] a Mystery born in this Age," engaged in "invisible Traffick." "What we properly call *Stock-Jobbing*," went another quasi-lexical definition, a novelty only "in our Age introduc'd," was undertaken "without any Regard to Reason or Computation" or even to "any proportional Gain."[36] Stock-jobbers nullified reason, calculation, and proportionality. This, it seems, was enough to animate the extraordinary zeal with which the whole choir now ganged up on them.

## WHAT WENT WRONG

Disaster, even merely perceived disaster, focuses the mind. By now it is clear that the exasperation attributed to Newton in reaction to the stock market frenzy at the time of the Bubble was hardly the most original thought associated with his name. The previous section has shown that many across Western Europe shared in the puzzlement and anxiety of these years. "The Eyes of all the World seem to be turn'd this way," John Applebee's *Original Weekly Journal* reported from London on the first of October 1720: "The sudden Fall of our Stocks, without any visible Reason for it, is the Surprize of the World."[37] Dramatic events had transpired—chaotic, disordered, capricious—for which there was no visible reason, and thus no evident cure.

But what about *invisible* reasons? Those, unsurprisingly, were invoked aplenty. Faced with the challenge of understanding and rationalizing the events that came to a head in 1720, contemporaries asserted their conviction in invisible causes and invisible dynamics of various kinds. For us, the most significant set of reactions is of those observers who found themselves conjuring up narratives of order emerging unexpectedly on an aggregate level from the very chaos itself. With them we will close the circle back to the four writers of the prologue to part 1 of this book, demonstrating how and why the financial turmoil of these years propelled European self-organizing thinking beyond a qualitative and quantitative threshold. First, however, we would like to look quickly at the range of other responses to this crisis of causality, not only because they were more prevalent but because they demonstrate that there are always alternatives to self-organization. We draw here primarily on the English reactions to 1720, which display pretty much the whole gamut of possibilities. It is against this range of other reactions that the supple potential of the innovative self-organizing idiom is thrown into sharp relief.

The prolific English writer Charles Gildon, for one, was interested in this panoply of reactions. Written as he beat back the poverty and blindness of his declining years, Gildon's last publication was a lengthy pamphlet in the manner of Chaucer about a group of strangers en route to Canterbury recounting their tales, tales that revolve around the follies of the South Sea Bubble. The interlocutor who obviously represents Gildon's own view, a parson, quotes at length the same report from Applebee's journal on the lack of visible reason for what had just transpired, "the best Account, said the Parson, I have met with of this Affair." The parson seems to agree with the reporter that there was "not the least Reason" that could explain why people behaved the way they did "upon no better Security, than the giddy humour of a changeable People" but then turns this observation on its head. "For my part," the parson explains, "I view this wonderful Event, with other Eyes than most People; to me there appears in a very visible Manner, the finger of Providence itself." Insisting several more times on "*the visible Hand of Providence* in that Affair," Gildon took the very lack of a visible cause as in itself "very visible" proof that these events were brought about by direct and purposive divine intervention.[38]

At the same time, however, Gildon's parson admits that he is none too comfortable with this solution. "I am not for Coining of Providences upon every slight Occasion"; it was only the extraordinariness of this "wonderful" event, lacking any other visible cause, that drove the parson to invoke divine intervention. Another of Gildon's characters, asking whether anyone "could . . . fix the giddy Fluctuation of the People" in the Bubble (in which the speaker supposedly lost a lot of money), makes an even stronger point. "Nothing but a Divine Power can do that," he asserts, "nor, even that Divine Power itself, without a visible and amazing Miracle."[39] For such chaos to have an overall purpose required, even for God, an extraordinary (and again "visible") miracle. All this was beyond the pale of routine providence.

God's ministers did not all agree. The Presbyterian Edmund Calamy, for example, found "such a Shock to Public Credit," from a providential point of view, quite routine. "There were (it is well known) heretofore Disorders in the Natural, the Moral, the Civil World, as well as more lately." Ultimately, he calmed his flock, "the Times [at] present are ordered by the same GOD, and under the Conduct of the same over-ruling Providence, as former Times were." Appearances may have led them to believe that "the proper Order [is] oddly inverted" and that "things fall out or happen by chance," but in truth nothing they witnessed was a novelty or extraordinary. Rather it fell under the jurisdiction of the same divine providence that had always guaranteed order in human and worldly affairs. In a similar vein, Digby Cotes could barely restrain his impatience. "To imagine," he sneered, "that the same Infinite Wisdom and Power which hath *made a Law for the Rain, and a Decree for the Sea*"—hello Newton, God-circumventing theorist of the tide—"hath left Mankind to act at random, and without Order and Government"?! Curiously, however, it was only a few years earlier that Cotes himself had imagined precisely that. As we have already seen, in 1713 he cogently explained why, in contrast with "Brute Creatures, [who] by the meer Instinct of Nature, should observe so much Order in all their Actions," man "should almost always act at random, and be the only irregular part of the whole Creation." In 1721, by contrast, Cotes was busily denying such randomness in human behavior. He therefore titled the later sermon, appropriately, *God the Author of Peace and Order*."[40]

Or take one more late-1720 sermon, from the pen of the dissenting minister George Smyth, in truth a repetitive exhortation on the matters of causality and providence. When we encounter calamities, Smyth wrote—by which he meant the South Sea Bubble as well as the outbreak of the plague in France—we might think "that second Causes act with an irresistible and unbounded Force." In fact, nothing could be further from the truth. "Those Calamities which at any time befall a Nation, or a People, are to be consider'd by them as the Effect of God's mighty Hand." And again: "If we be smitten, it is not by blind Chance, or an undirected Concurrence of second Causes, but by that mighty Hand which holds or superintends all human Affairs." Recent events, like any other, were surely not driven by chance or by a concurrence of random, "undirected" causes. "The excluding the Agency of God," Smyth warned, and thus "overlooking the divine Providence, as though it were wholly unconcern'd and intermeddled not with humane Affairs," was not only "absurd"; it was also a dangerous way of thinking that must inevitably "destroy the Order and Harmony of the World, and bring all Things into the utmost Confusion." Reasserting the place of providence was thus everyone's immediate and important duty.[41]

So one common response to the chaos of recent events was the invocation of providence in the manner long familiar in early modern Christian Europe. And yet one cannot read these pulpit orators and their many colleagues in 1720–21 without getting the sense that they were protesting too much. Even as they asserted time and again the undisputed rule of providence, the events unfolding around them also forced them to admit, often unwittingly, that in truth the world did seem like "a mutable Scene, in which we have multiply'd Evidence [that] there is nothing certain." These words, perhaps unexpectedly, were also Edmund Calamy's. Similarly Smyth, even as he admonished against the "utmost Confusion" that must ensue if divine providence were taken out of human affairs, also described the world of 1720 as "a giddy World" full of "Madness and Folly": but didn't this mean, by Smyth's own logic, that it was a world already forsaken by providence? Or take Berkeley, who combined a firm providentialist belief in "the Finger of God" with an admission on the next page that in the stock exchange "Mony [*sic*] is shifted from Hand to Hand in such a blind fortuitous manner, that some Men shall from nothing in an instant acquire vast Estates without the least desert." Blind randomness and the something-from-nothing scenario ended up sitting side by side with God's providential hand. In the House of Lords, Thomas Greene—soon to become the Bishop of Ely—delivered a sermon titled *The End and Design of God's Judgments* in which he too combined belief in providential oversight, as confirmed by history, with an admission of puzzlement about the current events, "by which we have been all of a sudden strangely impoverished in the Midst of Plenty, our Riches having made themselves Wings, and flying away no body knows whither." Faith in providence reassured Greene of a purpose in these sudden goings-on, but the logic of their cause and effect remained "strangely" inexplicable. Even Gildon's parson, asserting as he did in multiple ways "the visible Hand of Providence" in the South Sea Bubble, also resigned himself at some point to attributing the chaos to "some providential Infatuation": providential it was, but at the same time it was also infatuated—that is to say, as random as everyone had said.[42]

Overall, the disorder and chaos presented the interpreters of God's ways with a considerable challenge. "Since our Text of H. Scripture has arrived in this Nation," declared one with a flair for hyperbole, "we *Divines* perhaps never had such a Harvest of experienc'd Indignation, and Outcry at the *Vicissitude* of this World" as now after the Bubble, "an *Æra* of the strangest, of the suddenest Changes."[43] (It probably did not help that the Church of England was itself seen as deeply implicated in stock market affairs, a fact so memorably captured in the satirical print that propelled William Hogarth's career, *The South Sea Bubble*.) The capriciousness of the multitude, the man-made disorder, the role of randomness and chance, the difficulties in linking causes with effects—these aspects of the recent experiences could not easily be sermonized away by the simple reassertion of old-style providence.

Perhaps, then, providence had little to do with what had transpired? This was certainly the opinion of the Presbyterian minister John Evans, whose sermon on Guy Fawkes Day in 1720 was intended to refute such attempts to rope in the divine. "The present distresses of the nation are directly self-made," we have already heard him declare. "Men have evidently involved themselves in the present calamity, by their own unparallel'd covetousness. . . . They have no reason at all to complain of God, nor indeed of any so much as of themselves." (These words gain special pathos from a fact that Evans kept well hidden, that he himself had lost most of his savings as well as his wife's fortune in unwise

South Sea stock speculations.) The invisible causes of recent events should be sought not with providence but with men.[44]

Such a move from divine to human responsibility proved increasingly irresistible for many people seeking explanations, though perhaps not quite in the manner that Evans intended. Rather than finding fault in the avarice of everyone, they opted for conspiracy. All that had transpired, in this popular narrative, was orchestrated by the invisible hands of a small group of men: a coterie of self-interested individuals who craftily pulled invisible strings to create chaos for their own selfish ends. The Bubble was "not altogether owing to natural Causes, but was rather the Result of very artificial Engines, and secret Springs, that were set to Work, both at Home and Abroad, by those who had no small Concern in the Success of this Affair." The financial collapse was in truth the result of "the fatal Conspiracy" of the directors of the South Sea Company, who "made use of unnumerable Stratagems to deceive the People." "It cannot be imagin'd"—thus another—that "all these Things have been brought to pass by a meer fortuitous Conjunction of Atoms, as *Epicurus* suggests the World happened to be made; that all has happen'd without the Agency of Rogues, and the Craft of Knaves." "The deluded Multitude," another wrote, pointing an accusing finger at Chancellor of the Exchequer Aislabie, "who were playing all along in the Dark," might have believed they dealt with one another "upon the Uncertainty of Chance." But the truth of the matter was that "you [were] the Projector and Architect of the Machine, the chief Engineer in moving it," the invisible puppeteer pulling the strings.[45]

Although some remained skeptical that the South Sea directors could have orchestrated so many disconnected random actions ("Could they fix the giddy Fluctuation of the People? . . . How absurdly ridiculous therefore is all this Noise and Clamour against the Directors"), most seemed more than happy, even relieved, to accept that what had the appearance of chaotic disorder was in fact the result of "premeditated precipitancy and confusion." By 1721 this was the most widely accepted narrative in Britain for the turbulence of the previous year. The directors were therefore prosecuted with extraordinary vigor, their assets were publicly inventoried and auctioned, and a committee of secrecy of the House of Commons circulated reports proving their guilt. Petitions from across the country were sent to Parliament to demand their punishment for their "wicked Arts and Delusions" and "continued Series of unparallell'd Contrivances and Practices," which were the true "Secret Springs" of the recent events. Here again is that repeated phrase for invisible causes. It was as if all the exasperation of incomprehension were now channeled into the vindictive persecution of the South Sea directors.[46]

With hindsight, some—and not necessarily only friends and supporters of the directors—admitted that this scapegoating had perhaps gone too far and distracted attention from a deeper understanding. This was Pope's opinion, as we have seen, when he wrote: "Much injur'd Blunt! Why bears he Britain's hate?" A 1726 retrospect presented the onslaught on the directors not as a reasoned response to the frenzied pattern of 1720 but as itself a continuation of the same inexplicable pattern: "As the Infatuation [in the financial market in 1720] was supernatural, so was the Anger against these unhappy Gentlemen, and their Punishment unpresidented [*sic*]." Likewise, when Thomas Gordon—"Cato"—looked back in 1723, he admitted the mistake of many earlier reactions, including his own, in which "the whole iniquity of the *South-Sea* Scheme was charged, as if contrived by the Men in Power." As Gordon had written earlier, this was a classic case of mistaking symptoms for causes. "If the flagrant Frauds of that Affair have taken rise from some *undiscover'd Springs*, and yet only appear'd in the Persons of the *Directors*; the bare sacrificing of those servile Wretches, will hardly content or stop the Resentments of an exasperated People."[47]

Gordon's goal, however, as he had laid it down already in his 1721 tract that went quickly into seven editions, was less to reveal the "*secret Springs* and *Machines*, by which so much Fraud has been set on foot and perpetrated," than to figure out how the disorder they engendered could be cured or safely contained. Gordon proclaimed his inspiration from Zaleucus of Locris, the first Greek lawgiver. Zaleucus had urged his people to take a lesson from the regularity of the heavens and realize the need to eliminate "so wild and *eccentric*" irregularity as characterizes human behavior, for which he presented them with a recipe: the famous Locrian Code. "Now the nearest Pretence that we have of imitating this Divine Regularity, must be by the Provision of good and wholesome Laws: Laws, establish'd by the *Wisdom*, and not the *Caprice* of the Legislators."[48] Not an invisible hand, then, but a very visible one: that of the benign legislator who combines both meanings of the word "order" to legislate chaos and caprice out of existence. As we shall see, this vision of an omnipotent and omniscient legislator, who guarantees harmony by positive intervention from above, returns often—more commonly in the eighteenth century through the image of Lycurgus the Spartan, not Zaleucus the Locrian—as a sometimes practical, sometimes wishful-thinking alternative to self-organization.

Moreover, if human laws could not contain the chaos, at least not at present, perhaps mathematical laws could do so better? As one surveys the gamut of reactions to the South Sea fiasco, one is struck by how many interventions interlaced their rhetoric with detailed numbers: with tables, sums, accounts, percentages, calculations. Never before, perhaps, was a public crisis so painstakingly mathematicized. Yet the actual events seemed every time to exceed the reach of numbers and the orderliness of tables. Thus, when the author of a garden-variety Bubble pamphlet that boasted the certitude of precise numbers and sums, enumerated over many pages to the last shilling, opened with a frank admission of having just witnessed "the Madness of the People" together with "the surprizing Alternation in the Value of *South Sea* Stock, and the unexpected Fall of *Publick Credit*," the dissonance was unmistakable.[49] In juxtaposing the confident arithmetic of long-term calculations with the random disarray of the singular events of the moment, the very layout of this pamphlet—boring, unimpressive, and indistinguishable from dozens of others—reproduced, rather than eliminated, the limitations of predictability and causality that were flagged throughout this financial crisis.

Pride of place among the numerological tribe, those who sought to rein in the disorder and tame it through numbers, must go to Archibald Hutcheson, MP, lawyer and autodidact political economist, who developed a veritable obsession with the South Sea affair and has recently been described as its best contemporary analyst. Hutcheson insisted that it was "absolutely necessary" that this affair "should be searched to the Bottom, and the Disease be perfectly understood, before any Remedies can be applied," and assiduously went about this task in more than a dozen publications in 1720–21, with titles such as *A Collection of Calculations*, *A Computation of the Value of South Sea Stock*, and so on. Hutcheson nailed to his mast an uncompromising commitment to linear causality—"The same Causes do naturally produce the same Effects"—and to its mathematical, one might say Newtonian, representation. And yet despite this conviction, buttressed as it was by a barrage of calculations, Hutcheson too could not keep out the occasional admission of the unexpected, the random, and the chaotic; as when he suddenly called on providence's benevolence for help in "all the cross Accidents which befall us" or when he summed up the South Sea stock as "this blazing and astonishing Meteor."[50] (We shall see later that "meteor" in this period connoted singularity and unpredictability to a greater extent than today.) Many a reader was not fooled by these efforts at containment through arithmetic. At this juncture, one warned, it was futile to insist on "knocking our Heads against the Stone Walls of our Decimals and Fractions, and cashing up, with innumerable Figures, the Merits of the Cause." The rise and fall of the South Sea stock, wrote another, "was a Mystery that Figures were incapable of demonstrating." "The Frenzy rose so high," observed a third, "that even Mathematical Demonstrations (of which this Subject was always capable) were slighted, and the Authors of them treated as if they had been Quacks or Ballad-Singers." Newton's exasperation at the unruliness of the chaos once more comes to mind.[51]

Nobody facing this exasperation, however, was more determined to beat back the chaos with the aid of methodical planning than Sir Humphrey Mackworth. Mackworth, an ailing Welsh entrepreneur and former member of Parliament with a particular interest in matters of finance, dedicated the final bout of his public activity to the energetic promotion of "*a new Columbus's Egg . . . reserv'd for a miraculous Deliverance of this Nation from Ruin.*" The powers of Mackworth's miraculous cure—which turned out to be a new form of specie, based on the security of the public revenue of the kingdom, with stable, non-oscillating value—lay in its promise to eradicate *all* random, unpredictable, and disorderly elements from the realm of finance. The problem with the current state of affairs, Mackworth declared, was that "*any [financial] Scheme that depends on the Opinion of the People*"—that is, on the varied and unpredictable desires of a great number of people—"*must be very precarious, and ready to vanish with every popular Breath.*" But the body politic is like "*a great Machine or Engine*, where *all the Wheels must go together, or all stand still.*" Forget, then, about the relative freedom of wheels within wheels: the necessary solution must entail the realignment of all parts of the machine and the restoration of the total unison of their actions. Only once this was attained—following, naturally, Mackworth's proposals—only then would ventures like the South Sea Company find themselves "*on a better Foundation than they have hitherto stood*; a Foundation *that shall never fail, nor be precarious on popular Breath.*" Mackworth's insistence, underscored by obsessive underlining, that if and only if randomness was eliminated, everything else would inevitably, mechanically, determinedly, fall into place (even the "South-Sea *Stock may rise again* without using any *Art* or *mysterious Calculations*")—such insistence was as clear an indication as any of the desperation an aging financier felt in the face of a financial world suddenly gone awry.[52] (It so happens that a decade earlier Mackworth was caught red-handed fixing the results of a lottery he had organized: another way, one supposes, to eliminate the randomness of chance.[53])

## ORDER FROM DISORDER SPRINGS: THE POTENTIAL OF SELF-ORGANIZING THINKING

We are now approaching our destination. Mackworth wished to beat back random, purposeless, disorderly chaos. Others wanted to contain it through mathematical tabulation or legislative decrees. Others still tried to explain—or explain away—the apparent disorder as the deliberate product of invisible yet familiar causes, be they human conspiracy or divine intervention. What all these reactions had in common, and what they all together reveal about the pressures of this particular historical moment, was the pressing impulse to grasp, deny, or rein in the most unsettling aspects of the 1720 financial crises: the unpredictable multitudes, the man-made disorder, the vulnerability of events to chance, the exposed limitations of linear causality.

Against this background we finally arrive at a group of voices who responded to the same challenge differently, and are therefore of the greatest interest to the narrative and argument of this book. This group extends the logic that we suggested was shared by the foursome of our prologue in the decade following 1720: Montesquieu, Vico, Mandeville, Pope. Disorder for them was not only a key problem but also a key part of the solution. Rather than explain them away, these voices embraced precisely the chaotic elements—the aggregate of erratic, independently operating, and self-interested individuals, the nonlinearity and invisibility of relations of causes to effects—as the premises for new understandings of higher-level order emergent from disorder itself. This radical breakthrough did not come in the form of an overwhelming torrent but rather more as a scattered trickle of people experimenting with what we have called the family of self-organizing conceptual moves. Typically their efforts were disparate and unrelated, drawing less on each other than on the generative resources that had developed during the preceding half century, as discussed in the previous chapters. Some of their efforts were tentative or incomplete. Some were hedging their bets. A few may appear to belong to this group only by coincidence. But taken together, these voices, crossing a threshold in terms of the articulateness of their vision and of the concentration of such visions put forth within a few years, add up to an unprecedented materialization of self-organizing thinking as a distinctive, meaningful, cumulative presence in European intellectual and cultural history.

We can begin from the end, from those who reflected on the dramas of 1720 from the distance of several years later. These had also the benefit of hindsight, and thus the knowledge of the perhaps surprising fact that order reestablished itself remarkably fast once the dust had settled, with much greater ease than the near-hysterics of 1720 could have led anyone to expect. Several retrospectives on both sides of the Channel were unambiguous in their resort to self-organization as a framework for retelling those events. Pope, writing as we have seen in the early 1730s, is a case in point. So is Defoe. Having been so troubled in 1720 by the mysteries of the nonlinear relationships of causes and effects, as we have seen, by 1728 Defoe returned to the topic with a calmer and more confident assessment. The context was a discussion of overproduction. If particular products, "whether prudently, or rashly," glut the market, he wrote now, the result, despite appearances, is not "a Decay of Trade." Rather, having made "Trade a Bubble . . . , it returns upon them like the late *South Sea*, and every thing goes back from its imaginary to its intrinsick Value." Although Defoe still noted that individuals paid a price during the ups and downs of the crisis of 1720, the South Sea Bubble had now become for him prime evidence for a more optimistic vision: an inevitable and beneficial aggregate process whereby—here is the key phrase—"Trade is returned to its natural Channel, after an imaginary and casual Start out of it by the Accidents of foreign Commerce" or other "rash" accidents, as the case may be. Production and trade are by nature self-correcting, or one may say self-organizing. By the late eighteenth century, as we shall see, this particular formulation, of trade overcoming accidental obstructions to resume its flow in its natural level, will become one of the most common—indeed clichéd—images for the type of self-organizing dynamic that produces a state of equilibrium.[54]

In France, the marquis d'Argenson, whose father had preceded John Law as minister of finance, mobilized the very same image for his own retrospective evaluation of Law's schemes.

> It is, however, important, to remark that the finances of France were soon re-established, notwithstanding the catastrophes of the bank and the *Visa*; so true it is that in matters of finance, public credit and circulation find their own level, like the water of the sea, after storms and tempests.

With hindsight, it turned out, 1720 was describable on both sides of the Channel as a disruption soon corrected by the natural, self-organizing flow of credit and trade.[55] Interestingly, shortly thereafter d'Argenson also found himself extending the same logic even further back, all the way to the Dutch republic of the seventeenth century. "There was an understanding of the Dutch between themselves," he wrote in admiration from his 1720s vantage point, "like an ant-hill or a beehive, where each insect acts in accordance with its own instinct." From the unfettered "multiplicity of interests" that was thus allowed to thrive there "ensued immense effects for commerce" and "a great accumulation for the needs of society; but this was never achieved by orders from above, nor even by magistrates who obliged every individual to follow the views of their leader."[56] D'Argenson moved here one step further. Self-organization was not only a mechanism for the restoration of order after an extraordinary disruption.

For the exemplary commercial republic, he now realized, it had been a highly beneficial aggregate social dynamic all along.

Nobody looking back at 1720 was naturally more invested in demonstrating how well things had turned out than John Law himself. It is thus telling that Law's *Histoire des finances pendant la Régence*, an apologia that he wrote and rewrote compulsively during his 1720s exile after the collapse of his System, is peppered with gestures at the dynamics of self-organization. Money became for Law, as Thomas Kavanagh has aptly put it, "a demiurgic force thriving on its own movement." Credit was even more powerful, an aggregate much greater than the sum of its parts: in Law's words, credit "multiplies by itself, and in so multiplying also multiplies the common good." The continuous self-propelled flow of credit and money counteracts disorder, "since I call disorder that which disrupts the harmony and affinity that all things must have between them, which connects them in a way that they mutually reinforce each other to produce the public good." Order and harmony are aggregate effects of the self-organizing powers of credit and money (powers, incidentally, that in Law's view were superior under an absolute ruler, who can best guarantee their smooth and equal flow). As in the case of Defoe, Law had already formed the germs of these ideas before the years of financial hyperactivity: recall Law's memo to the Duke of Savoy in 1712. But it was now, in the aftermath of 1720, that both seasoned economic theorists-cum-practitioners pulled together these former fragmentary insights toward a more fleshed-out vision of self-organizing order.[57]

Nor did one have to wait for the retrospectives. One can actually find multiple instances of self-organizing idiom, prefiguring—however hesitantly, incompletely, or fleetingly—many key themes of the rest of this book, in the midst of the 1720 maelstrom itself.

Even as the events of 1720 were unfolding, for example, the auditor of the Scottish exchequer, Sir David Dalrymple, stumbled on a perhaps surprising model of prices in a market that "regulates it self," based on people's freely formed opinions of quality and promise. As long as designing men do not meddle with price levels, this natural dynamic that determines them—Dalrymple called it "the Rule"—is capable of absorbing and correcting the temporary perturbations caused by "Accidents," be they the accidents of scarcity or novelty or whatever. Dalrymple's immediate response to the wild rise and fall of the South Sea stocks thus approached almost inadvertently a model of a self-organizing and self-correcting price system, set in motion by the free flow of multiple individually formed opinions.[58] Furthermore, Dalrymple also argued against the view of stocks as gambles, which fall therefore outside notions of predictable causality or liability, and thus cannot incur legal obligations in contractual exchange. Not so, said Dalrymple: "*Answer'd*, That a Chance may be bought too dear as well as other Things. *There is a Method of putting a Value upon any Chance that can happen*."[59] Invoking the emerging calculus of probability, Dalrymple insisted that chances could be calculated and evaluated even if individual events were unpredictable. In short, Dalrymple was intimidated neither by erratic and accident-sensitive market fluctuations nor by the element of chance. Seen as an aggregate system, the stock market could still be expected to behave in a calculable, self-correcting manner.

The status of the South Sea contracts also preoccupied an anonymous 1720 pamphleteer who was himself a stockholder. While disagreeing with Dalrymple in some details, when it came to the dynamics of the recent developments in the stock market he reached a similar conclusion. The current situation, this writer asserted, must be recognized as unique and unprecedented. Just as the Romans "for many Ages had no Law to punish *Parricide*; it never entring into their Minds," so the ancestors of the Bubble generation could not have imagined "the Case of the *South-Sea* Contracts, [in which] Hazard and Chance is part of the Consideration, and the Value the Parties set on that not being to be known."[60] Now, however, unlike these ancestors and much like Sir David Dalrymple, the author could indeed imagine transactions that rely on hazard, chance, and nonlinear leaps in value. His contractual solution, however, was different. Whereas Dalrymple relied on the new science of probability to calculate those elements of chance and thus bring them back into the fold of existing contract law, this pamphleteer remained skeptical about the claims of mathematicians to make sense of the South Sea frenzy, recommending instead to treat such chance-prone contracts as equivalent to gambling debts that are free of the constraints of proportionality.

This writer went on like so many others to pin the blame for the bursting of the Bubble on the South Sea directors. But in explaining *why* their actions led to this result, he said something strikingly new. The problem with the directors' manipulations was that they caused all the individual operators in the market to undertake the same actions in unison. The directors were therefore guilty not of creating disorder but rather of eliminating it: "It is a fundamental maxim in Stock-Jobbing, that *when great numbers have the same view, no Benefit can be made*."[61] Randomness and chaos are not a threat to the stock market. Rather, it is the very randomness and chaotic nature of the actions taken by "great numbers" of operators, each pursuing their own individual path, that guarantees its beneficial operation! What this writer presented as "a fundamental maxim in Stock-Jobbing" was in fact an abbreviated gesture at a radically new view of the market as a self-organizing system emergent from the independent and heterogeneous actions of many individuals. The success and stability of such a market depends on that very chaos that so many other contemporaries were busily trying to make disappear.

Both this anonymous writer and Sir David Dalrymple, then, were moved in 1720 to acknowledge the role of chance as an unprecedented aspect of the stock market that now became acceptable, even legitimate.[62] This realization, in turn, pushed them to take the more far-reaching step toward the insight of self-organization, when tentatively, almost in passing, they speculated about how this unpredictable randomness of the market was in itself the basis for a new kind of aggregate order. Both, then, fit well the broader pattern we are proposing. The next couple of paragraphs, by contrast, present a text from 1721 that deals only in passing with the financial troubles of the previous year but that spins out in more detail precisely those elements that come together in the telling of self-organizing stories. Tamworth Reresby was a military officer whose *Miscellany of Ingenious Thoughts and Reflections* was written under the acknowledged shadow of the South Sea Bubble. His sometimes confused musings that make him something of a poster boy for this chapter lay in a section titled "Providence, Fortune, Chance."

Reresby maps the various possibilities for conceptualizing providence, fortune, and chance. On the one hand he posits the stoical paradigm of fortune, which "maintains a Fatality and unchangeable Sequel of Events, which proceeds from a Necessity, emergent from, and inherent in the Things themselves." On the other hand is the Christian paradigm of providence, which Reresby finds to be equally deterministic, even as it attributes "the Necessity of all natural Contingencies to the free Decree of *God* himself, who executes unavoidably what he freely ordains." In both, ultimately, "there be no such Thing as *Fortune* or *Chance* in the World," only what our own ignorance misleads us to label as fortune or chance. But in between these two poles Reresby identifies a space for a third way, a less deterministic possibility that he traces back to the writings of Quintilian, which allows man to accept chance while forming and executing schemes "in Spite of all intervening Accidents." If we accept this understanding of fortune, then "we may demonstrate *Politicks* as well as *Mathematicks*, and build certain Rules upon the Contingency of human Actions, which will as little agree with the *Stoical* as the *Christian Doctrine.*" Here, then, was a man who in 1721 fused the recent South Sea Bubble experience together with the lessons of the providential third-way experiments increasingly prevalent since the late seventeenth century, as discussed in our first chapter. The result was an experimental theory that simultaneously accepted disorder at one level—that is to say, the undetermined and accident-prone contingency of human actions, "loose, irregular, and fickle"—together with order at another level, visible in the rules that govern the aggregate and that promise to make politics as predictable a science as mathematics. This maneuver brought back hope, as well as restored one's faith. Even "if the best methodiz'd Affairs are turn'd out of their Course, and baffled, the wisest Projects defeated, the most flattering Hopes confounded . . . and the humblest Condition comes to be exalted" (an unmistakable echo of the Bubble)—even then one must believe that God "has cast all Accidents into a certain Method" and hope that with his help one "shall be able to apprehend the Regularity of those Accidents, which once seem'd so confus'd."[63] Reresby's effort to reconcile accident and order was as good a prefiguration as any of so many subsequent moves toward self-organization.

By the end of his tract it is unclear whether Reresby is fully swayed by the Quintilianic position or retreats to a certain reliance on direct divine providence. But this does not really matter. It is not surprising that stabs at self-organization in the 1720s, undertaken in the immediate aftermath of a national crisis that had generated almost hysterical concern, were raw, tentative, and incomplete, and all the more so if their bottom line was a Panglossian belief that disorders like the recent ones all end up for the better. What is important to us is that such experimentation was a distinctive phenomenon repeated multiple times at this particular historical juncture. It bears reminding that we are not especially interested in identifying the *first* such conceptualization, or the genealogy of an idea. Our goal is to point to an emergent cultural phenomenon, the echoes of which will then reverberate with increasing intensity later in the century.

That such an emergent phenomenon did occur in the 1720s is attested by, among other things, contemporary efforts to dismiss it. Take one George Flint, writing in the shadow of the South Sea Bubble against "the true System of *Lunaticks*." "Suppose a Shipwreck'd Man thrown upon an Island," he mused with Defoe's *Robinson Crusoe* also freshly on his mind; an island which is "all an entire Wilderness, uninhabited by any Thing but Beasts." And suppose further that he encounters there "a beauteous Structure the most exquisite Hands ever built." According to the present-day lunatics, "he must suppose that this was built by Chance, by a mere accidental Jumbling of Things, call'd Atoms, together." Surely, Flint sneered, surely "*Homer* composed no Iliads, nor *Virgil* Eneids; Letters jump'd by Chance together, and form'd themselves into these Poems." Flint does not specify what particular context or author provoked his outburst against these Lucretian scenarios, but it quickly becomes a shrill tirade against those unspecified lovers of disorder: "*We love not Order*," they say, "*but will have Disorder; Huzzah! Boys Disorder for ever*." To counter their influence Flint appeals urgently to his audience "to own the God of Order." It does not take a huge leap to recognize in Flint's imaginary foes not only the echoes of the disorder of 1720 but also those of contemporary speculations about chance and randomness as a basis of order. It was the same environment, we recall, that brought George Smyth to preach the role of providence against those who might see in the Bubble the operations of "blind Chance, or an undirected Concurrence of second Causes."[64]

So was the self-organizing logic "in the air" in the 1720s? Some of the evidence is unmistakable, and explicitly linked to the moment. Many other signs of what we propose as this emergent cultural phenomenon are more circumstantial. Our proposal to link Montesquieu's preoccupation with Law's schemes to his mysterious "*chaîne secrète et, en quelque façon, inconnue*," as itself an experiment in self-organization, is obviously speculative. Whether Vico's experiments with self-organizing providence at the same time bore any relation to these events of 1720 cannot be ascertained. When Francis Hutcheson inserted into the 1726 edition of his *Inquiry concerning Beauty, Order, Harmony, Design* a new observation about the "mysterious effect of [musical] discords: they often give as great a pleasure as continued harmony," the apparent echoes of order mysteriously emerging from disorder may have been purely coincidental. Still, the fact that the same musical image was picked up by both Pope and Montesquieu does suggest that at this particular moment it carried a certain resonance.[65]

The following year, the clergyman John Maxwell published the first English translation of Richard Cumberland's *De legibus naturae* of 1672. In this response to Hobbes, Cumberland had dwelt at length on the relationship between individual actions and public welfare, which he presented as a manifestation of divine order established through the natural mechanics of linear causality. On this particular point, however, Maxwell, the translator of half a century later, remained unsatisfied. "I think our Author is abundantly *too general* in this *Chapter of the Nature of Things*," Maxwell added a short note. "He should . . . have shewn more *particularly*" how it could happen "that the very Actions which promote the private Interest of any particular Person, do in all, at least in all common Cases, necessarily tend to the Advantage of the Publick." Was it a coincidence, then, that in 1727 Maxwell singled out for additional clarification the dynamic that transformed the disorder of many self-interested actions into a beneficial aggregate? Of course, Maxwell's note may have simply registered the impact of Bernard Mandeville's recent provocative thoughts on the very same question. But then wasn't Mandeville's own thinking about the aggregate dynamics of self-organization probably shaped by the events circa 1720, as attested by the precise chronology of the additions to the successive editions of the *Fable of the Bees*?[66]

And here is one more suggestive coincidence. This book's opening device introduces four European figures who didn't walk into a bar. But at the risk of overplaying our rhetorical hand we could have had five, adding to the Frenchman, Italian, Dutchman, and Englishman also an American. His name is Benjamin Franklin. At age nineteen the precocious Franklin, recently arrived from America to England on what turned out to be a fool's errand, found himself working for a living at a London printer's shop. There he typeset the third edition of William Wollaston's *Religion of Nature Delineated*, a good example of third-way providentialism from 1722, which wished to square "the conjunctions and oppositions of [men's] interests and inclinations" with the existence of "an unseen governing power." In response, Franklin wrote *A Dissertation on Liberty and Necessity*, a heterodox fancy that appeared without indication of author, publisher, or bookseller. It made the young Franklin notorious with London libertines and the more mature Franklin apologetic for this "erratum" for the rest of his life.

Franklin's essay assesses the role of man as a free agent in the cosmic plan. "As Man is a Part of this great Machine, the Universe, his regular Acting is requisite to the regular moving of the whole." Man's liberty is absolute "and his Choice influenc'd by nothing." But in facing innumerable choices at every fork in the road man cannot know "all the intricate Consequences of every Action with respect to the general Order and Scheme of the Universe, both present and future." He must therefore be "perpetually blundering about in the Dark," having "but as one Chance to ten thousand, to hit on the right Action," and thus 9999 opportunities to get it wrong while "putting the [universal] Scheme in Disorder." And yet, miraculously, the world remains orderly and harmonious. How come?

> It is as if an ingenious Artificer, having fram'd a curious Machine or Clock, and put its many intricate Wheels and Powers in such a Dependance on one another, that the whole might move in the most exact Order and Regularity, had nevertheless plac'd in it several other Wheels endu'd with an independent *Self-Motion*, but ignorant of the general Interest of the Clock; and these would every now and then be moving wrong, disordering the true Movement, and making continual Work for the Mender.

Nevertheless, in our world the mender or clockmaker does not need to intervene actively. This Franklin took as conclusive proof of the existence of an omnipotent God who overrules man's free agency and thus his potential for virtue. We take it, rather, as another suggestive indication of the spread in the 1720s of self-organizing narratives, in which wheels within wheels (more or less) are imagined—even in denial—as possessing an independent swerve and being ignorant of the general interest of the system, thus creating localized disorder that does not disrupt the harmony of the whole. Small wonder that this tract got Franklin an audience with Bernard Mandeville.[67]

In sum, although each one of these instances, taken in isolation, may appear as a freestanding coincidence, when considered in the aggregate they are cumulating fast enough to demand consideration as a collective phenomenon. And here is one more, one last example from a completely different domain, before we close the 1720s with a return to economic-cum-financial thinking, circling back in the final section to the matters that set the first part of this book in motion and that provide a telos for its later parts.

This final noneconomic example deals with international relations, or, in the language of its anonymous 1720 author, with "the ballance of power . . . its necessity, origin and history." As in the case of Maxwell and Cumberland, this tract also introduces an earlier text as its point of departure, namely, the theologian and French royal tutor François Fénelon's "Essay on the Ballance of Power, [which] ranges [i.e., arranges] all the different Dispositions of Power that can happen." Following Fénelon, the 1720 author identifies four possible configurations of international affairs: a single state greater than all others combined, a state greater than all others individually but not more powerful than their combination, smaller states that combine to counterbalance a bigger one, and a near equality between several great states. Our author assures his readers that the latter possibility is not only the happiest configuration; in the long run it is also the inevitable end point of all others. Reason as well as the historical record proves that no matter what efforts different states have made to overpower others, these efforts every time "have ended in this Fourth Class," through dynamics of alliances or overgrowth-leading-to-internal-collapse that the tract explains at some length. Every state is driven only by self-interest, and without a view of the whole, and yet the whole naturally orders itself into an equilibrium. This equilibrium is not static. The unpredictable actions of particular states destabilize it, resulting in "the Caprice of the Ballance" or "the Inconstancy of the Ballance." Yet in the long run stability is restored, as they all succumb to "that natural Inclination, that centripetal Force as it were, to be of the Fourth Class."[68] International relations thus turn out to be a self-organizing system, reaching equilibrium in and through the caprice of princes.

Only this is not in fact what Fénelon had originally said. Fénelon, writing around 1700, while commending the balance of roughly equal powers of the "Fourth Class" as the author of 1720 reported, had had no intention of leaving this felicitous outcome in the invisible hands of free-floating or natural dynamics. To the contrary, Fénelon believed that if states were left to their own devices, "*the most powerful will certainly at length prevail, and overthrow the rest, unless they unite together to preserve the Ballance*." Fénelon's purpose was to offer not dispassionate reflections about a natural state of affairs but rather advice on policy making, namely, how to guarantee through proactive interventions that same balance that the 1720 author believed would emerge by itself. If there was any guaranteed outcome in Fénelon's view, it was the *overthrowing* of any balance, which could only be preserved through constant vigilance. In between the late-seventeenth-century master and his proclaimed follower of 1720, then, the theory of the balance of powers acquired the dynamics of self-organization.

To end with one more twist: the previous paragraph includes a quote from Fénelon's essay on the balance of Europe, taken from a contemporary English translation published at the time by a Scottish advocate named William Grant. In his added commentary to the translation Grant too had something to say about the dynamic that produces said balance. The balance of powers, Grant observed, the principles of which had been elaborated so well by Fénelon, "may be demonstrated to have arisen necessarily, and from the Reason and Nature of things, as a Cure of what happen'd amiss in human Affairs." But as we already know, Grant was in fact going beyond—and even against—Fénelon in thus asserting (in Fénelon's name!) that if happenstance sends human affairs off course, this random erring will be cured by the natural self-corrective dynamic of the balance of powers. It is therefore quite suggestive, we find, that Grant published his translation of Fénelon's essay, together with these reassuring comments about the inevitability of order emerging from erratic disorder, in that very same generative year of 1720.[69]

## APPARITIONS OF INVISIBLE HANDS

This chapter has established that the 1720s were indeed an early—we think the first—historical juncture in Western Europe in which self-organization emerged as a recognizable cultural phenomenon; that is to say, the scene of a significant cluster of speculative ruminations about the emergence of aggregate high-level self-organizing order from lower-level disorder. This wave of experimentation, we have argued, was triggered by the singular financial troubles that opened that decade in much of Western Europe. At the same time it drew on game-changing developments that had gradually evolved during the previous half century, threads from which Europeans could now begin spinning self-organizing stories. Although much of this speculative activity, understandably, took place in the context of reflections on financial and economic order, we also saw multiple instances of these experimental self-organizing narratives spilling into ruminations on order in other domains.

Keeping in mind the role of the Smithian invisible hand in our mise-en-scène, we would like to bring the first part of this book to a close with a look forward, through a final batch of 1720s texts that remind us how precocious these early experiments in self-organization were. This final section therefore focuses on three largely forgotten discussions of economic theory. From one perspective, they are among the very best examples of that surge of interest in the potential of self-organizing models in the 1720s. From another perspective, all three prefigured, sometimes uncannily, the political economy of half a century later.

The first appeared in the year 1720 itself, with the ambitious title *The System or Theory of the Trade of the World*. Its author was an Englishman of Huguenot descent, Isaac Gervaise, a longtime merchant and director of the Royal Lustring Company. It so happened that in 1720 this monopoly-based company lost its charter, and Gervaise laid down in his only known tract a remarkably modern-sounding plea for keeping trade "natural and free." The argument was based on a careful analysis of the international trade equilibrium and of the self-regulating mechanism for the global distribution of gold and silver, which Gervaise referred to as "the grand Denominator." Consider, for instance, Gervaise's detailed description of the latter (which can only be appreciated at its full length, for which we beg the reader's indulgence):

> When a Nation has attracted a greater Proportion of the grand Denominator of the World, than its proper share; and the Cause of that Attraction ceases, that Nation cannot retain the Overplus of its proper Proportion of the grand Denominator, because in that case, the Proportion of Poor and Rich of that Nation is broken; that is to say, the number of Rich is too great, in proportion to the Poor, . . . in which case all the Labour of the Poor will not ballance the Expense of the Rich. So that there enters in that Nation, more Labour than goes out of it, to ballance its want of Poor: And as the End of Trade is the attracting Gold and Silver, all that difference of Labour is paid in Gold and Silver, until

> the Denominator be lessen'd, in proportion to other Nations; which also, and at the same time, proportions the number of Poor to that of Rich.

As this and other like-minded passages make clear, Gervaise's theory was based on an unflinching model of the flow of currency, labor, and national riches as an interconnected self-organizing system.

At the end, having applied the self-organizing logic several times over to the social equilibrium between classes within nations and to the international balance of trade, Gervaise arrived at his final conclusion:

> if Trade was not curbed by Laws, or disturbed by those Accidents that happen in long Wars, etc. which break the natural Proportion, either of People, or of private Denominators; Time would bring all trading Nations of the World into that Equilibrium, which is proportioned, and belongs to the number of their Inhabitants.

Here, in 1720, we find in effect the sanguine vision of Smith's invisible hand, more than half a century *avant la lettre*.[70]

Gervaise, as one economic historian put it, was "a violet that blushed unseen for two centuries," having had no traceable echo until his rediscovery in the twentieth century. This is more or less true.[71] But it is equally important to note that he was not a lone blushing violet. Our second text, published anonymously several months later, appears to have remained completely unnoticed by historians of economic thought. This is a shame, since *A True State of Publick Credit; or, a Short View of the Condition of the Nation, with Respect to Our Present Calamities* also deserves a place on their honorary list of the precursors of classical political economy. Written out of concern for "the Fall of *South-Sea-Stock* [that] has ruined *Credit*," it was preoccupied with the restoration of healthy credit, a goal it claimed was beyond the reach of legislative intervention or administrative decree. "Credit, however it is sunk, can, from its Nature, only rise, with a general Care and Industry, *by its own Ducts and Modes*." Credit according to this writer was inherently a self-organizing system. Parliament, therefore, should engage in "no compulsive or unnatural Laws," but only set up favorable conditions, such as public registration of landed properties or a fund for public credit, and "the rest will be perfected by a Machine, *which will move it self*." And again: credit "must be left to itself," and if it is, within a generally beneficial environment, it "will disperse it self, and find out just as many Channels as will be useful." "If the Legislature will but lay the Foundation Stone, the Superstructure will not want any other Assistance in raising."[72] Curiously, these repetitive assertions of the self-organizing nature of credit were all crammed into the dedication of this tract (signed 25 April 1721) but were not repeated in the body of the text. Was it because they were a metareflection that did not affect the writer's practical suggestions, or because they were a belated realization arrived at between penning the text and its eventual publication? Be that as it may, here was a remarkably clear economic vision responding to the perceived financial meltdown with an insistence on the self-organizing emergence of aggregate order, tending without external intervention to the public good.

Finally, the best example is that which we have left for last. It is also an apt one with which to close the first part of this book that opened many pages ago with Montesquieu, since during this period Richard Cantillon was among Montesquieu's close friends. It is more apt still, since nobody we have encountered in this chapter, with the exception of John Law himself, was more closely entwined with and affected by the events of 1720 than Cantillon. Indeed, if scruples are not a precondition for heroism, Cantillon may well be considered the unsung hero of that year of financial bubbles. Cantillon was an Irish émigré who had become one of Paris's most successful private bankers, and who had subsequently turned down an invitation from John Law—a personal friend and sometimes rival—to become his deputy in running France's finances. Instead Cantillon positioned himself to make not one but several financial killings through canny short-term speculations—aided by insider information?—in 1719–20: first, twice over, in the stocks of the Mississippi Company (in which his brother happened to be a central figure); and then again, after cashing in at the peak and moving his winnings out of France, in the South Sea Company.

In the late 1720s, beginning in part as a defense in a court case related to the Mississippi scheme, Cantillon wrote his *Essai sur la nature du commerce en général*, which was simultaneously a critique of Law's System, though he never mentioned Law by name, and a comprehensive view of the nature of economic systems. This treatise remained unpublished until 1755, long after Cantillon's mysterious death—his London house burned down in 1734, but rumors persisted that he had actually staged his own death and fled to Surinam—although it is likely that the manuscript had circulated earlier. (Otherwise it is hard to explain how full sections of Cantillon's prose found their way, unacknowledged, into Malachy Postlethwayt's well-known *Universal Dictionary of Trade and Commerce*, published in part before Cantillon's *Essai*.) And it is in this text little noticed outside circumscribed circles, written by an extraordinary figure in the international financial world of 1720, that we find what economists from William Stanley Jevons, Cantillon's late-nineteenth-century "rediscoverer," to Friedrich Hayek have flagged as "the most original of all statements of economic principles before the *Wealth of Nations*." Its originality, they agree, "all the more remarkable because it [was] entirely devoid of providentialist or teleological content," was first and foremost in its developed notions of the self-organization of economic systems.[73]

Cantillon's self-organizing thinking permeates the *Essai* throughout: in his assumptions about population (prefiguring Malthus), in his theory of value and prices (Hayek identified him as "the originator of the self-adjusting price specie flow mechanism"), in his model for the flow of money and commerce (Hume's midcentury "Of the Balance of Trade" also apparently benefited from advance access to a manuscript copy of Cantillon's essay), and indeed in how all these connect to each other in a macroeconomic self-reproducing equilibrating process.

At one point, for instance, Cantillon put together a thought experiment about a region that suddenly gets more money from mines. In a long detailed narrative, which Jevons found to be "one of the most marvellous things in the book," Cantillon shows how such a region first benefits from an economic boom involving increased consumption, employment, and trade. Then, through multiple self-propelled processes, the boom "gradually" but "necessarily" turns to bust, as labor begins to emigrate, money to flow out to foreign manufactures, local manufactures slowly to decline, and the circulation of money to slow down, until the state ultimately falls into ruin. This model of a self-organizing equilibrium correcting an initial disruption was historically tested, Cantillon suggested, since it was "approximately what has happened to Spain since the discovery of the Indies." But it was also a general theoretical truism, as Cantillon put it elsewhere: "When a State has arrived at the highest point of wealth . . . it will inevitably fall into poverty by the ordinary course of things. The too great abundance of money, which so long as it lasts forms the power of States, throws them back imperceptibly but naturally into poverty." Note Cantillon's insistence that such correctives take place "inevitably" and "by the ordinary cause of things," without external intervention. This self-organizing dynamic occurs on every level, national and local. "If there are too many Hatters in a City or in a street for the number of People who buy hats there," Cantillon offered as another illustration,

> some who are least patronized must become bankrupt: if they be too few it will be a profitable Undertaking which will encourage new Hatters to open shops there and so it is that the entrepreneurs of all kinds adjust themselves

> to accidents in a State (*les Entrepreneurs de toutes espèces se proportionnent au hazard dans un État*).

Adam Smith could hardly have put it better.[74]

The very last sentence in Cantillon's *Essai* put its theory back into its own historical context. It was a parting warning against manipulations of the stock market, for instance through the collusion of a bank and a minister in raising the price of public stock. "But if some panic of unforeseen crisis drove the holders to demand silver from the Bank the bomb would burst (*on en viendroit à crever la bombe*) and it would be seen that these are dangerous operations."[75] The long shadows of Law's System and the Mississippi and South Sea bubbles—or "bombs"—could not be mistaken.

Isaac Gervaise, the anonymous author of *A True State of Publick Credit*, and Richard Cantillon: each one was an isolated voice in the aftermath of 1720, each one experimenting with self-organizing models of economic order, each one prefiguring the political economists of the second half of the eighteenth century, each one having no immediate echo—one delayed by a generation, the others having no known echo at all. It is likely that there was no connection between them. Recall also Sir David Dalrymple, Daniel Defoe (in his 1728 state of mind), the marquis d'Argenson, and John Law himself. And there were other flickers of such ideas, such as the 1721 proposal for the revival of credit in Britain according to which "when more Credit shall be extant than can be employed, it will naturally stop of itself"; or Richard Steele's "own private Opinion," expressed in spring 1720, that "the best Method [is] to let the Annuities run out untouch'd, unsolicited; and that any thing, concerning them, should come from themselves."[76] Even if Montesquieu and Cantillon, or Cantillon and Law, had a chance to discuss their thoughts on self-organization over dinner, or more mundanely read each other's writings, for the most part these early forays into self-organization in economic thinking were probably not influenced by one another or by one common source. Rather, these voices, and more broadly the wave of thought experiments in self-organization that we have seen during this decade, appear to represent multiple cases of spontaneous generation, as it were, of self-organization ideas.

These convergences, as we have argued, draw our attention to a particular historical moment, and to the pressures generated by specific circumstances. This chapter has shown in some detail how the financial havoc of the beginning of this decade, and the subsequent swift resumption of order, provided the opportunity and the motive to come up with precisely such responses. The simultaneous move toward self-organization by disparate individuals, however, also suggests that these were not idiosyncratic breakthroughs but rather reliant on shared intellectual possibilities already available and ready for the taking. Chapters 1 and 2 thus singled out new departures, tools, and frames of mind that had evolved over the previous half century to become key elements in the self-organizing narratives that burst into the open in the 1720s. Now, once the self-organizing idiom crossed a threshold of public presence and articulation, its supple potential for imagining sources of order in so many different circumstances ensured its spread far beyond economic reflection. The following chapters chart the inscription of self-organization onto eighteenth-century culture and Enlightenment thought on a far wider scale.

# * 2 *

PROLOGUE

# An Island of Dreams

We know no time when we were not as now;
Know none before us, self-begot, self-rais'd
By our own quickening power, when fatal course
Had circled his full orb, the birth mature
Of this our native Heaven, ethereal sons.
Our puissance is our own.

JOHN MILTON, *Paradise Lost*, 5.859–64

The waters of Lake Bienne in Switzerland part around the island of St. Pierre, a narrow promontory made famous by its Enlightenment admirer, Jean-Jacques Rousseau. September 1765 found Rousseau there, retreating after humiliating travails in Geneva and Bern to an isolated spot sheltered from the world's attention. On the island he celebrated the "happiest time of his life," several weeks spent on the shores of this little paradise. Afternoons he retreated to a "hidden nook along the beach," where he lost himself:

> There, the noise of the waves and the tossing of the water, captivating my senses and chasing all other disturbance from my soul, plunged it into a delightful reverie. . . . The ebb and flow of this water and its noise, continual but magnified at intervals, striking my ears and eyes without respite, took the place of internal movements which reverie extinguished within me and was enough to make me feel my existence with pleasure and without taking the trouble to think. From time to time some weak and short reflection about the instability of things in this world arose, an image brought on by the surface of the water. But soon these weak impressions were erased by the uniformity of the continual movement.

Where land dissolved into water, Rousseau melted too. In reverie, the edges of his soul blended seamlessly into the world. The "ebb and flow" of water replaced the "internal movements" of thought. Inside and outside, mine and yours, humans and nature: the reverie softened all of these distinctions. The mind of Rousseau rested in an intertidal zone, an infinite present in which "time is nothing," where past and present lose their claims on our souls, and where "the point separating . . . fictions from . . . realities" evaporates. In this place, we discover and enjoy "nothing external to ourselves, nothing if not ourselves and our own existence." "As long as this state exists," Rousseau sighed, "we are sufficient unto ourselves, like God."[1]

It is no wonder that for Rousseau in 1765—harassed for his political and religious heterodoxies and publicly shamed by luminaries like Voltaire for his atrocious treatment of his family—the dream of self-sufficiency might well have appealed. On the shores of Lake Bienne, he dreamed his way across the intertidal zone between self and world, bringing them into the harmonic convergence he always longed for. This method was hardly appropriate for humans tangled in ordinary and mundane affairs, however. Rousseau knew this well. For his whole career, in fact, he was obsessed with the *difficulty* of this move across the intertidal zone, the difficulty of bringing human beings across those mysterious thresholds that stand between them and their perfection. "Man is born free, and everywhere he is in chains," the 1762 *Social Contract* opened, and the threshold between the freedom of an original nature and the bondage of society gave his major works their enduring pathos. It was a pathos of puzzlement: How did it come to pass that we children of nature find ourselves alienated from both it and each other? When did our selves come unhinged from the world? His entry on political economy for the *Encyclopédie* put this dilemma in political terms:

> By what inconceivable art could the means have been found to subjugate men in order to make them free; to use the goods, the labor, even the life of all its members in the service of the state without constraining and without consulting them; to bind their will with their own agreement; to make their consent predominate over their refusal; and to force them to punish themselves when they do what they did not want?

Human beings made their political world to set themselves free, but in doing so, they abandoned their own freedom and alienated themselves from their own autonomy. The *Social Contract* proposed its own inconceivable art as a form of therapy for this—an imagined compact that would preserve our original and natural freedom even in a state of social and political union—but its optimism was always tempered by the magnitude of the task. The Lawgiver needed to accomplish this, Rousseau tells us, is an extraordinary being, someone capable of "changing human nature" and combining "two apparently incompatible things . . . : an undertaking beyond human force and, to execute it, an authority that is nil."[2] This is no task for common men.

Seven years earlier, in his *Discourse on the Origin and Foundations of Inequality*, Rousseau set the problem of thresholds into a historical frame, wondering at the "immense distance there must have been between the pure state of Nature" and the moment that social institutions like language and politics emerged onto the historical scene. Born of nature, humans are her immediate creatures no longer. His conjectural solution to the problem—his effort to recreate a time before society and language by "setting aside all the facts," as he put it—sounds bizarre to a modern reader. But giving free rein to speculation was a magical bridge over the intertidal zone, a way of cognitively connecting things fundamentally different. We know we have a natural origin, but a full understanding of it is lost to us. "We know neither our nature nor our active principle," Rousseau wrote in *Emile*. "Impenetrable mysteries surround us on all sides; they are above the region accessible to the senses." In a world where "imperceptible effects often unite to produce a considerable event," Rousseau chided Voltaire in 1756 after the earthquake at Lisbon shook what little faith in providence he might have had, what looks like chance might be design.[3] We simply do not know the extent of either our freedom or our dependency. At times we feel like Satan, autonomous, "self-made," and knowing "none before us," in Milton's words. Yet we also cannot help but recognize our original ties to a nature and divinity that shapes this self, this all-too-human and limited "puissance." We started in one place and have become something else, but what that might be is hard to say.

In centuries past, the source of this difference between what we were and what we are had a name. The name was "sin," and it named the corruption that could never recover its original goodness. It is not an accident that Milton's Satan thought himself naturally self-sufficient. Diabolical naturalism denies the essential difference between "now" and "then," and proclaims equality between self and the world of nature. By contrast, divine naturalism embraces the original divergence but then projects the ultimate reconciliation between man and a nature redeemed far into the future. In this Christian vision, sin and soteriology go hand in hand—we live in a fallen world, but we can endure it because we know that, however difficult, our fallenness is sensible and just, that we can learn from it to orient ourselves to an ultimately benevolent divinity, and, most of all, because we know that eventually this miserable state of affairs will come to an end.

The practices and disciplines of Christianity were, in short, a way of living and thinking between radically different states. It was a complete answer. By binding the fallen world to a redeemed world-to-come, it gave believers a set of mental tools applicable across the foreseeable range of human experience. Christianity resolves the quotidian and the cosmic into *one* coherent system, and it can fold even moments of apparent incoherence, as we saw, back into a story of God's ultimate concern. The Enlightenment was, by contrast, a land of incomplete and heterogeneous answers. It might be defined as a set of efforts to abide in the fallen state—perhaps both to regret the loss of our original home and to imagine a reconciliation in some distant future, but also to settle for the condition in which human beings find themselves and learn to live in what Pierre Bayle called "the ruins of reason." Like a beach, the ruin marks a threshold, carrying signs of its origin but marking the changes that separate past and present. In the words of Newton's great antagonist, Thomas Burnet, the world is "preternatural, like a Statue made and broken." Its present form is testament to prior earths, one that came directly from God's chaos, one that preexisted the Deluge, and one, our earth, that came after, belated and shattered by the awful power of the Flood. Mountains, cliffs, seas, cinders, stones, islands, and hills collapsed into a stately pile: "And so the Divine Providence, having prepar'd Nature for so great a change, at one stroke dissolv'd the frame of the old World, and made us a new one out of its ruines."[4] While at the end of days God may redeem this ruined world, in the meantime, this is where we live.

Rousseau's dreams dissolved the boundaries of this divided world. But only a God could long live in such a state, happy as it was. For ordinary mortals, the thresholds remain. Thus for eighteenth-century intellectual culture, these thresholds inspired endless creativity, and sustained reflection on the possibilities of bridging those states that so fascinated Rousseau: chance and order, nature and culture, freedom and dependency, contingency and determination, among others. The repeated efforts to understand these gave speculations on nature and mankind a peculiar kind of interchangeability. Seen in disciplinary terms, what was often called natural history in the eighteenth century thus moved freely between domains that we now hold distinct: biology, physics, geology, chemistry and medicine converged with anthropology, theology, history, politics, and more.[5] These convergences were not signs of intellectual confusion. Rather, they were vibrant efforts to think moral and natural orders together, to understand human thought in relation to the world and its God, and to allow human experience—the bridge between humans and the world that surrounds them—to be suffused with natural, ethical, and cognitive value. Eighteenth-century thought took shape, in important ways, around thresholds.

The collective effort to think and describe the movement across these thresholds was, often as not, couched in the language of self-organization. This language played with what Peter Reill has called the "middle element," the interstitial and mysterious space of transition between one incommensurable state and another.[6] Repeatedly it posed questions now familiar to us: How do we get something from nothing? How do contingent singularities coalesce into complex orders? What kind of causality might be at work in a world where the whole has different qualities than the parts?

Characteristic of eighteenth-century intellectual culture—and especially what we will at times call Enlightenment—was both its fascination with these questions and the looseness of its answers to them, its experimental approaches conducted across a variety of genres. This looseness gives Enlightenment thought a kind of heterogeneity that can be troubling for those historians of philosophy who put a premium on clarity of expression and consistency of system. But it was in the context of this looseness that the language of self-organization flourished. Indeed, self-organization was so powerful a response to these questions exactly because it answered them with less than metaphysical certainty. Acknowledged circularities in argument; moments of mysterious transformation; resignation to incomplete understanding: these were integral elements in the eighteenth-century descriptions of emergent orders.[7] As we saw in the last chapter, and will see again in the following section, these uncertainties and mysteries did not restrain the development of the language of self-organization, but rather encouraged it. The bridges between chance and order, freedom and determination, simplicity and complexity, were fragile and in need of constant repair. But their value lay in just this fragility, in their refusal to make final determinations on the nature of order, whether human, natural, or divine.

To see this language of self-organization at work, and to see its power in shaping the imaginative landscape, the following section will follow it through two of the most vibrant areas of eighteenth-century intellectual culture: the sciences of life and the sciences of mind. These were sciences that took center stage in the eighteenth century in ways simply unprecedented. Unlike the celestial mechanics of the scientific revolution, or the metaphysics of the mechanical philosophers, the sciences of life and mind were above all concerned with the mysterious ways that order might grow out of the dynamic and variable operations of the world itself. At the heart of each was the effort to understand how it is possible to move from simple elements (inorganic matter or inchoate experience, say) to the intricacies of biological organisms, natural systems, and human consciousness. As each of these sciences developed in the eighteenth century, they endlessly returned to their central organizing problem, namely, how a concatenation of singular events might produce complexities of operation whose nature and behavior were distinct from any of its parts. In life and in the mind, the eighteenth century discovered, or believed it discovered, that simplicity repeated does not just produce more simplicity. At some mysterious point and for some mysterious reason, repetition becomes something else; quantitative iteration produces qualitative transformation. The structures that result—whether living organisms, natural environments, or the complex operations of human cognition—are new and unforeseen. Life and mind are self-organizing.

CHAPTER 4

# The Order and Organization of Life

The island of dreams was also an island of plants. When Rousseau woke up from his reverie, he had to resign himself to living in the world. The sweetest balm for this sad circumstance, he discovered, was botany. The happiest time of life, he said, was spent only in "the delightful and necessary pursuit of a man who had devoted himself to idleness." Botany was the perfect activity for this "lazy man," a luxurious waste of time. "It is said that a German did a book about a lemon peel," Rousseau remarked, but for his part, he would have happily written a book about *every* plant, "each type of hay . . . each moss . . . each lichen" on St. Pierre. Lounging around on his island, not even unpacking his luggage, Rousseau planned for an endless future, "two years . . . two centuries, and the whole of eternity" of puttering. With no goal nor purpose, these were the pleasures of Jean-Jacques, at least in those short days before he was once more cast adrift by the exigencies of politics and scandal.[1]

In the past, botany often stood as one of the more egregious examples of Enlightenment managerial instrumentalism. The brute force of its ordering impulse took, it has been said, concrete scientific form in the "dogmatic confidence" of Carl Linnaeus, whose taxonomies and tables chopped the floral world into regular squares on a grid of knowledge. No less pernicious, and related in form, were the commercial and imperial exploitations that brought the natural world under Europe's heavy boot. Here the comments of the British urbanite Joseph Addison stand as an epigraph for the period. "The Fruits of Portugal," he wrote,

> are corrected by the Products of Barbadoes: The infusion of a China Plant is sweetened with the pith of an Indian Cane. We repair our Bodies by the Drugs of America, and repose ourselves under Indian Canopies. . . . Our Eyes are refreshed with the green fields of Britain, at the same time that our Palates are feasted with Fruits that rise between the Tropics.

Imperial and domestic dominion converged in the botanical project, which not only categorized all of the world's flowers and grains but also offered techniques for their management at home. Succulent plants fed the appetites of Europeans hungry to expand their historical boundaries into the worlds of Asia, Africa, and the Americas. At home, gardens cultivated and altered these surprisingly labile organisms, and so became ostentatious sites for showcasing the technical skills of scientists and kings. The technocratic desire to regulate the economy equally applied to the world of plants. Like its commercial counterpart, floral cameralism valued self-sufficiency and careful administration of local resources. The Nordic cold of Sweden did not, as Lisbet Koerner has shown, dissuade Linnaeus from cultivating sugar cane, tea, even bananas (!) at home. This bid for a tropical paradise in Uppsala was, needless to say, unsuccessful. Still, it reflects the stupefying confidence in human power to shape the natural world that so often attaches to the term Enlightenment.[2]

As other scholars have been pointing out with increasing urgency in recent years, however, the Enlightenment was not really the "triumphant calamity" imagined by its harshest critics. Certainly the prescriptive literature of the eighteenth century abounded with recommendations for control. Schemes for agricultural improvement were beloved by prescriptive journals and weeklies, which produced boundless schemes for better beekeeping, crop rotation, hybridization, and so on. Yet Rousseau's plants were altogether different. True, when he went botanizing, he tucked Linnaeus under his arm. But his *experience* of botany, the experienced moment of contact between human and inhuman, tested the very limits of instrumentality and purpose altogether. Botany, Rousseau commented in his letters to Madeleine-Catherine Delessert, "has no other real utility than that which a thinking and sensitive being can draw from the observation of nature." The effort to give plants use-value, to make them serve as mere instruments for others, only "denatures" them, makes them monstrous. Languorous curiosity, a "luxurious idleness," best befits the investigator of nature's flowers and grasses.[3] No accident, then, that Jean-Jacques's days on St. Pierre were spent half lazing on the beach, half in botany. Both were a kind of reverie, a dissolution of the self into the world that suspended purposeful activity. Botany was, for him, an intertidal science.

In Rousseau's musings on plants, we can hear echoes of an ancient conversation. We have already overheard Lucretius in chapter 1. Here, in Rousseau, it is the language of Aristotle that sounds, Aristotle for whom plants defined the edges of living nature and were thus particularly interesting for thinking about edges between life and death. This edge was, for Aristotle, marked by the *soul*, the "cause or source of the living body. . . . It is the source of movement, it is the end, it is the essence of the whole living body." Living things are intrinsically purposeful, serving not the ends of others but their own ends. Purpose inheres in the very souls that give them life, and their lives, in turn, are developments of this purpose, a telos that takes them from birth to death. Plants have the most primitive souls, what the Greek called the "nutritive soul," a foundational ability to grow, develop, and reproduce, without any of the higher functions (movement, will, sensation, or thought) found in animals and humans. As such, plants are minimally alive. They have a purpose, but they cannot orient themselves toward it in any willful way. "Things pass one by one gradually from life into non-life," Aristotle wrote, curious about the middle ground "between life and the deprivation of life." The incarnation of this indeterminate space was the plant, "indefinite in form," neither quite animal nor quite mineral.[4] Plants grow right on the border between purposeful life and mute stone. No wonder that Rousseau liked them so much, as they marked that edge between the living land of St. Pierre and the dead waters of Lake Bienne.

The eighteenth-century sciences of life grew up around thresholds like these, restlessly investigating the spaces between the objects and processes of nature.[5] This chapter charts these sciences of life and the self-organizing stories that made them such a powerful part of the intellectual imaginary of the period. It claims to compass neither all the figures involved in the sciences of life nor their many particular debates—these were legion. Nor does the chapter constrain itself to a precise chronology. The ideas of self-organization that emerged so powerfully after 1720 stayed in deep dialogue with the scientific conversations of the previous century, indeed those of antiquity. Nor, finally, does it quite present a story of competing theories (vitalism, epigenicism, preformationism, iatromechanism, and so on) although these terms do appear, and the scholarship that addresses them has been essential to our understanding of the period.[6] However important these stories are, they are not our story.

Instead, the chapter will investigate the language of self-organization, finding it in widely varied places: in the France of the 1750s, in the England and Germany of the 1780s and beyond. Fundamental to this language are issues already seen in this book: interest in purposes and teleology, excitement about dynamic forces, impatience with mechanism, creative uses of causality, fascination with complexity and variability, and interest in emergent properties. These issues were an integral part of the weave of the life sciences. Indeed, the life sciences were so essential to Enlightenment natural philosophy, we believe, because they allowed such wide latitude for speculations about them.

Just as importantly, the chapter pays close attention to the phenomenology of Enlightenment natural history, the ways that the descriptions of life exceeded the possibility of its deductive explanation. Life cannot be deduced, the life scientists discovered, to the delight of some and the chagrin of others. Instead it must be narrated in less-than-analytically-complete ways. Self-organization was a key mode of this narration, a set of stories that repeatedly described the emergence of something from nothing. These stories imagined new possibilities of understanding a world whose complex orders are immanent in the very materials that make it up.

## QUASI-TELEOLOGY AND THE PURPOSES OF LIFE

Integral to these stories and to the eighteenth-century sciences of immanent order were purposes, reasons *why* the simple parts of the world defy inertial heaviness and organize themselves into things higher and more complex. Purposes were not supposed to be part of the scientific endeavor. Final causes and teleology were both excluded from the mechanical hypothesis that helped launch the scientific revolution. Science should focus its attention, the argument went, on "that from which the change . . . first begins"—what Aristotelian metaphysics would call efficient causes—rather than "the end, i.e. that for the sake of which a thing is," or the purposes at which events aim.[7] The restriction of the field of inquiry was deliberate. It was intended, among other things, to segregate the domain of God's purposeful activity from the world of natural law. Natural laws could be offered as descriptions of observed regularities, without offering any hypothesis about *why* nature is organized in this particular way, and not some other one. When Newton famously threw up his hands and declared that he had no idea what caused gravity, or why it worked, he conjured up a science purged of purposes.

This restriction of the causal concept comes naturally to us. It is quite difficult for moderns to understand what a cause *is*, except in terms of efficient causes. Insofar as causes are events that make something happen, it seems, they are necessarily about mechanisms of change. Even in antiquity, as Michael Frede has noted, it was not easy to conceive of how a "final cause . . . should be called a cause" since it doesn't seem to include a means for changing the world. What, after all, did Aristotle *mean* when he said that final causes "come first," that it is the purpose that somehow causes the process?[8] It seems simply peculiar.

It is less peculiar, however, when we realize that, for Aristotle, the final cause was not interesting because it gives us an account of how things change. Rather, it talked about *why* they changed, and how human knowers might understand this change in the first place. It was interesting because of the question that it answered: Why are things like this, and not otherwise? Even if we have a hard time embracing Aristotle's teleology, the question does not just vanish because it is ruled improper for scientists to consider. Newton's great critic Leibniz, for example, resurrected Aristotelian problems with increasing urgency in the later 1670s, as he sought to develop a scientific home for both divine wisdom and teleology.[9] Even for Newton, and certainly for his admirers, the question of "occult qualities"—the *why* of gravity—never ceased to press.

Indeed, it is one of the (apparent) paradoxes of the period—and one of its aspects that gives it an especially ludicrous appearance from the perspective of modern science—that the very prohibition on purposes generated such excessive attention to purposeful nature. The design theory that was so powerful in natural history from the early eighteenth century onward was nothing if not an effort to convert the natural world into one replete with purposes and, by extension, moral and theological edification for mankind at large. It was this theory that, as we have seen in chapter 1, found moral stories in the most mundane or discouraging corners of nature, in the shells of snails, the blindness of moles, the vomit of dogs, and so on.

Easy pieties, no doubt, but they concealed a question that perplexed many of the less pious too: Can we even have a meaningful science devoid of purposes? In 1765, Rousseau thought not. Or at least, he found any science like that dispiriting and dull. The pleasure of botany disappears "as soon as we mingle a motive of interest or vanity with it," for then plants become "instruments of our passions" rather than self-standing and free. If instrumentalism subjects all things, living and dead, to the claims of human needs and art—the value of things rests in their value *to us*—Rousseau affirmed instead the existence of purposes inherent in the materials themselves. "These forms, these colors, this symmetry were not put here for nothing," he remarked in his *Fragment on Botany*, and it is the particular job of the botanist to appreciate the latent purposes in things:

> To the eyes of the botanist plants are only organic beings, [since] as soon as the vegetal is dead and ceases to vegetate, its parts no longer have the mutual

> correspondence which . . . makes it a unity, it is a simple substance, matter . . . a dead earth, which no longer belongs to the plant realm.

A botanist is a botanist insofar as he treats what Aristotle called the *entelechy*, the unity and "being-for" something that distinguishes the living from the dead.[10]

The eighteenth-century turn to the life sciences, in short, put the question of purposes squarely at the center of natural philosophy. After all, already for Aristotle, purposes were most important for the study of living organisms. "It is not enough," the Stagirite said to Democritus and Empedocles, "to say what are the stuffs out of which an animal is formed," since a dead animal is made of the same things. The study of the living has to begin with the substance or essence (*ousia*) for the sake of which organic processes unfold.[11]

How was this to be done in the eighteenth century? Aristotelian problems and categories were on the table, but clearly no one could simply resurrect Aristotelian arguments. Yet just as Lucretius was retooled and put into the service of a new concept of providence, so too could the ancient doctrine of final causes be given a new orientation in the eighteenth century. In 1728, the term "teleology" found its first Latin usages in the philosophy of Christian Wolff, the German metaphysician and pupil of Leibniz. His *Philosophia rationalis, sive Logica* developed the concept from his earlier work on natural organisms, and it made teleology into the science of God's intentions in the natural world. In the life sciences, however, teleology quickly departed from natural theology and became something both more powerful and ambiguous. What John H. Zammito has called the "decisive eighteenth-century retrieval and reformulation of Aristotle's notion of entelechy, or intrinsic purposiveness," entailed purposes immanent in living organisms. There are purposes, but they don't necessarily come from *outside* the organism. Rather, their origin is unspecified, their ontological status undetermined, and their mechanisms grounded in the very operations of the living systems themselves. The result was what we will call "quasi-teleology," namely, the appearance not so much of a consistent *theory* of final causes as a heuristic. The life sciences—at least a significant sector of them—embraced explanations that worked *as if* some kind of purposive agent were in play. They described their object of study *as if* it were a purposive thing and put these purposes into the service of empirical scientific research, even as they (unlike Aristotle) often refused to specify how these purposes might align with other explanatory commitments. In doing so, they told stories of animate nature that filled it with self-organizing potential.[12]

That purposes haunted the study of life doubtless had to do with the perplexing challenge that life posed for scientific and philosophical inquiry. Rousseau once remarked how "the generation of living and organized bodies is . . . an abyss for the human mind." The Scottish surgeon and scientist John Hunter—someone who spent far more time than Rousseau on these questions, and a towering figure in late-eighteenth-century medicine—could not have agreed more. "Life," he wrote, "is a property we do not understand: we can only see the necessary steps leading toward it."[13] Life is puzzling because it seems to stand above and beyond the parts that make it up; it is something that leaps into being out of nonliving nature, and then has the ability to shape its own parts, to reproduce them, replace them, and put them to work for its own purposes.

For Hunter, puzzlement about the emergence and nature of animation stimulated both empirical and metaphysical research. His 1794 *Treatise on the Blood* was replete with technical observations and experiments on the behavior of mammalian blood. These experiments did not, however, satisfy his curiosity about the function of man's only liquid organ. This function, he had begun to suspect already in the 1760s, was far more mysterious than anyone had supposed. Blood was not merely a fluid pumping in the body, he finally decided, but a real and self-standing living thing. "To conceive that blood is endowed with life, while circulating, is perhaps carrying the imagination as far as it well can go," he admitted, but given what he knew about the properties of it, he could not imagine how "we should think it to be otherwise."[14]

Why should Hunter think the blood was alive? Superficially, because we die without it. But that could be said about almost all of the organs. Moreover, it was, Hunter was the first to admit, hardly intuitive to imagine that a fluid body, "whose parts are in constant motion" and which could be spilled without "affecting itself or the body," could possibly be alive.[15] The vision of living blood was not incidental to Hunter's thought, however. Rather, it made the complexity of life concrete and familiar.

The blood lent material form, for example, to the necessary (in his view) supramechanical qualities involved in all living beings. As Hunter insisted, life "can never rise out of, or depend on organization . . . Mere organization can do nothing, even in mechanics, it must still have something corresponding to a living principle: namely, some power." What the Dutch physician Herman Boerhaave—leading luminary of early-eighteenth-century mechanist medicine—called the "supports, columns, rafters, cantilevers . . . pulleys, ropes, push-buttons, bellows, sieves, filters, canals, troughs, and reservoirs" of the body are never enough to explain life.[16] Doubtless there are mechanical operations without which life does not exist. Breathing works like a bellows, the heart like a pump. But neither a bellows nor a pump is alive. The *difference* between the two, the phase shift that happens when, say, a machine (human or otherwise) becomes animate, is what Hunter wanted to put his finger on. For this remarkable qualitative shift, a ghost in the machine was necessary.

But for Hunter, this was a *materialist* ghost, incarnate in blood. A "materia vitae diffusa" permeated the entire animal body, surrounding and enabling all of the organs to stay in constant communication with every other part, and the blood gave this physical form. The blood is "a perfect whole of itself, having no parts dissimilar." Divide the blood in two, and still you have blood. Remove a pint of blood from the body, and yet the blood continues to flow. Take that same pint, and put it in *another* person—Hunter was fascinated by transfusion—and, if the types match (we now know though Hunter did not), it still works as blood. It is an organ that acts across the entire organism with perfect unity, indeed acts across organisms as an infinitely divisible yet always complete substance. As such, blood captured the mysterious zone of transition from death to life, enabling "reciprocal action" between the parts, yet diffuse in its effects, always moving but in such a way that its motion "neither . . . can affect it or the body."[17] Like the water that toppled Rousseau into reverie, the blood's fluid state carried the body out of death into life.

These speculations unfolded inside the framework of quasi-teleology. Hunter was a materialist, but his living matter was a purposive thing. The fluidity of the blood is "intended for its motion and its motion is only to convey life" to the body, he wrote, while the coagulation of the blood serves what he calls the "union of the first intention," that purpose of healing that clotting and scabbing supply. More generally, an organ he defines as a machine that serves "some other purpose than itself," and ultimately *all* the organs orient toward the purpose of living. If the body were just a machine, these purposes would have no place since the total function of each part would lie within the nexus of effective causes. But in a self-organizing body—an animate body—the parts serve a purpose whose ends transcend local organization, whose ends are *invisible* to mechanical explanation.[18]

Hunter was hardly the first to see either blood or purposes at the heart of life. Already in 1651, William Harvey had described the blood as "spirit," something "celestial" that animates all living forms. The great Aristotelian mechanist of the mid-seventeenth century, the discoverer of the heart's pump-like nature, as well as a metaphysician of the blood, Harvey pursued his inquiries with a robust toolbox of teleological ideas.[19] By the turn of the eighteenth century, the German doctor Georg Ernst Stahl—often credited with founding modern "vitalism," but more aptly an "animist" in his insistence on the *soul* as a life-giving thing—sharply distinguished between the mechanisms that "constitut[e] the material for any body" and the "higher and external force [that] puts it into motion and directs it toward an effect as its final purpose."[20] A body becomes an organism, in this view, precisely insofar as it obeys a teleological imperative.

Aristotle this was not. The distinction between matter and form so crucial to the Greek's understanding of an organism, for example, vanished in the new sciences of life. In its place was something altogether more obscure. Both an effect observed in nature and a cause of the differences between animate and inanimate things, life was something that had to be invoked for a science to develop, but its nature was shadowy. Life, Ralph Cudworth insisted in the 1670s, cannot "properly Intend those Ends which it acts for, nor indeed is it Expressly Conscious of what it doth." Life is a thing driven by purposes, but these purposes are always opaque to itself. The "Vital Sympathy, by which our Soul is united and tied fast, as it were with a Knot, to the Body, is a thing that we have no direct Consciousness of, but only in its Effects," Cudworth went on.[21] Life has a purpose beyond mere mechanism. But the nature of this purpose—Aristotle's final cause—is ultimately unknown, and enacted without volition. Useful, certainly, but for whom and for what, it is unclear.

The eighteenth-century sciences of life were replete with quasi-teleology. At times it was explicit, with defined ends on account of which processes happened. Much more often it was highly attenuated, simply a way of acknowledging the distinct ways in which living systems depart from dead matter, the ways they reproduce, regenerate, and operate in the service of their own persistence. So it was with Hunter's blood: purposes were both present and absent. The blood bears life, but its own form is amorphous. A substance whose goal was stipulated but opaque, bearing a "property we do not understand," as Hunter put it, blood incarnated the quasi-teleological.

To step back, the languages of life performed a crucial set of services. They afforded a descriptive palette to writers eager to understand the leap from nothing to something, from death to life. They let living creatures organize themselves. They let these creatures carry in themselves the forces of their own creation, without granting them full independence from the world around them. Plants, animals, and mankind are self-organizing, that is, alive, and yet their autonomy from nature or, in many cases, God is necessarily incomplete, for their purposes are opaque to themselves. The languages of life performed heuristic functions as well. They filled the gap between a necessarily metaphysical search for causes and an avowedly empirical description of effects.

And they allowed the complexity of living systems both to coordinate and to surpass the individual parts that made them up.

## THE METAPHYSICS OF THE SEED

The complexity of the eighteenth-century life sciences was bound with their very object of study. Little wonder, since unlike mechanics, where the processes studied can be modeled with simple machines, animate nature is intrinsically complex.[22] Explanatory reduction in biology is fiendishly hard even today. It was no less difficult three hundred years ago, when scientists could only dream of the techniques we use today. Nor was the "biology" of the eighteenth century any less loaded with questions of metaphysical (e.g., philosophical and theological) import than it is now.[23] The order of things that creates mankind itself cannot, we suppose, be freed of controversies about matters of ultimate concern. These controversies were energetically pursued in this first age of self-organization, when *the* central problem of the life sciences—the origins of life—became so culturally and intellectually urgent.

One solution to life's evident complexity was ancient, elegant, and apparently simple: the seed. Already in antiquity, scientists knew how essential seeds were for the reproduction of living organisms. From life comes life, in this theory, whether in the plant or animal world. Even in antiquity, however, the theory had its dilemmas. Where, for example, do seeds come from? Why do seeds produce just one kind of creature? What role does sexual activity play in making seeds? Why do offspring look like their parents? Both simple and complex answers were given. In Democritus, for example, we can find the fairly mechanical notion that sex knocks loose an already formed seed, while in other ancient writers, the seed emerges in some way from the brain, or is composed of some blend of the body's humors, or perhaps is secreted from the entire body and all its parts.[24]

After 1650, when the sciences of life emerged as an expansive area of inquiry, the seed was rediscovered, an apparent panacea for the paradoxes of living nature. Seeds answered questions that this book has repeatedly discovered at the center of the period's intellectual culture, namely: How do you account for both the variety and the order of the world's things? How do you get something from nothing? They did so with increasing persuasiveness as technologies of observation grew in both sophistication and availability, and as researchers found in their microscopes seeds never seen before. Regnier de Graaf, for example, confirmed in 1672 that there were structures in the mammalian ovary—so-called Graafian follicles—that seemed to release *something* like an egg, even if he could not discover this egg itself. Once the analogy between oviparous and viviparous reproduction was (at least in theory) confirmed, suddenly it began to seem plausible that the entire complex order of living nature could be mapped onto discrete quanta of biological data. As Elizabeth Gasking points out, the "major problem . . . [of] how differentiated parts could ever form from unorganized matter" was solved by recourse to the seed.[25] It was solved by declaring, in effect, that *there was no problem*, since differentiated parts were always there, ready to unfold.

The doctrine of preformationism, as it came to be called, was extraordinarily powerful in the later seventeenth century. Leading microscopists like Jan Swammerdam (Netherlands) and Marcello Malpighi (Italy) made satisfying empirical discoveries as they studied the developmental stages of both insects and embryos, and different versions were elaborated, some stressing the sperm, some the egg, as the original seed. Metaphysicians were quick to follow suit, since here they could discover the wise organization of the world by God's provident hand. Complex versions of this appear already in Leibniz, who declared that "living bodies are still machines in their smallest parts *ad infinitum*" and insisted on the ability of these machines to determine their own ends, in much the same way that every monad "is big with its future." Seeds were, for him, a limit case for organic order, infinitely numerous, diverse, and small, much like monads themselves. Preformationism tended, however, to be far less sophisticated than this. Thus our Oratorian Nicolas Malebranche, who marveled in 1674 how "in a single apple seed there are apple trees, apples, and apple seeds, standing in the proportion of a full grown tree to the tree in its seed, for an infinite, or nearly infinite number of centuries."[26] Creation of living things on the fifth day was, once and for all, complexity and unity merging in the idea of the seed.

It is important to recognize that the seed fetish was not, *pace* Reill and Jacques Roger, just an ideological product of late-seventeenth-century Christian piety. As providential materialism solved real intellectual puzzles, so too did preformationism. Its questions and its answers were integrally linked to the larger issues of order and complexity that structured so much of the eighteenth-century intellectual inquiry. For Leibniz, seeds allowed teleology and mechanism to converge—infinitely small machines whose purposes were built into their very fabric. For Malebranche, the seed let him argue that "nature's role is only to unfold these tiny trees."[27] By converting the problem of generation into one of growth—that is, no new living things are created, but rather they develop from the potential of the seed—preformationism seemed to solve the "something from nothing" puzzle that so perplexed generations of early modern scientists and philosophers. Aristotle's question for Empedocles was, after all, a deep one. How do you transform matter into living forms? The seed took the place of the Aristotelian soul, distinguishing living from dead by a principle of animation implanted at the very dawn of things by God himself. Pious, yes, but also useful in solving real dilemmas.

Furthermore, the seed extended, as Gasking has also noted, a highly productive scientific principle, namely, as Aristotle wrote, that the natural body is something that is "organized [ὀργανικόν]", that is, a body with organs. Preformationism extended this principle ad infinitum. At the very smallest scale, bodies are already complete, possessing all "the parts essential for life" even before the organism fully emerges into view.[28] Interestingly enough, the technology of the microscope seemed to point in just this direction. As observers discovered in pond or pepper water organisms never seen before by the human eye, it seemed to grow *more* plausible, not less, that there were living things all the way down.

This empiricism was metaphysically loaded. Take, for example, the case of the eighteenth century's most public microscopist, Albrecht von Haller. Haller was a Swiss student in Germany when he discovered botany and began a lifelong interest in the life sciences. A renowned poet, he was an accomplished anatomist too and, for some years, professor of medicine at Germany's foremost eighteenth-century research institute, the University of Göttingen. He also underwent one of the most dramatic scientific conversions in the period. While at Göttingen, he confidently declared his faith in some form of spontaneous generation (see below). It would soon be discovered, he remarked in 1746, "that Animals and Vegetables . . . generate themselves from a fluid that thickens and that organizes itself little by little according to laws unknown to us."[29] Ten years later—and now back in his native Berne and running the local saltworks—he famously flipped, professing faith instead in the theory of preformed seeds.

The shift from one to the other was dramatic, but it also signaled something about the empirical ambiguities at play. As Maria Theresa Monti has noted, Haller's shift to preformation depended on a commitment to "the mechanisms of organogenesis before the structures had passed the threshold of perceptibility." As much as the microscope revealed, it could never finally confirm the claims researchers wanted to make, since all of the important work of development happened well beyond the range of their (even enhanced) vision. For this reason, scientists from the dawn of the microscope onward had to make a metaphysical commitment to what lay on the other side of the veil. Is the visible world merely the top layer of ever-smaller ordered structures? Or does order just come to an end at some point, and become chaos? Haller was clearly committed to the former proposition, but this was at best an inference. Thus, for example, Haller treated nascent embryos with vinegar. What a joy when invisible structures magically appeared, revealing to the eye, he thought, those elements always already present in the egg itself. His critics quickly argued (rightly, as it happens) that the vinegar actually created the structures he was hoping to find.[30] But in a sense, it did not matter, since there was in any case no way of knowing what was happening beyond the threshold of sight.

The movement from nothing to something (whether by God's creation or by some other natural process) thus happened in an enigmatic place, beyond the threshold of sight and into the land of conjecture. These processes put paid to what the philosophe Denis Diderot called our "prejudice that nothing happens beyond the range of our senses."[31] Virtually *everything* concerning life happened just beyond our sight, right where the empirical became the metaphysical. As a result, eighteenth-century biologies have a deeply paradoxical air to them. They profess to describe the phenomena of living nature in empirical terms. Yet the phenomena they study demand commitment to a metaphysics beyond empirical confirmation. What is life? Do we just know it when we see it? What is the difference between an oak tree and the tree of Diana, produced by the crystallization of silver nitrate and beloved by alchemists and natural philosophers alike? Determination of the object of study required a theory of animate nature impossible to confirm by means of the senses.

The microscope simply cannot tell us, in other words, when organic complexity emerges. Even the committed mechanist Marcello Malpighi—someone who hoped that life "could be understood as a machine, by taking it apart and examining its components one by one"—confessed in 1673 to an intractable bewilderment when he watched an embryo develop in a chicken egg. "The bourns of mortal life, I fear, are too uncertainly defined" to be seen with the naked eye, he wrote. So "while . . . we are studying attentively the genesis of animals from the egg, lo! in the egg itself we behold the animal almost already formed, and our labor thus is rendered fruitless." For other preformationists, however, this was a *virtue*, not a problem. For the later Swiss natural philosopher Charles Bonnet, for example, this proved a great truth, namely, "that there is no true generation in nature." "Generation" is a malapropism, a name improperly applied to "a development which renders to us visible what we could not see before."[32] The lens demonstrates experimentally, Bonnet thought, that the physical processes leading from death to life are beyond the empirical. "Lo!" is the only response to the sudden appearance of the organism.

For Bonnet, Haller, and their ilk, these limits on empirical knowledge were a source of pleasure. Theories of the seed, as Jacques Roger has shown, nearly always included a sentiment like this one:

> If one attempts to derive clear ideas of the first formation of some organized bodies, one soon realizes that the force of our reasoning, and the extent of knowledge we are permitted to have, cannot possibly bring us there; we must begin with the development and growth of beings already formed, without attempting to go further back.

The French entomologist René de Réaumur's prohibition on "going further back" echoed the prohibition on teleology so fundamental to the early scientific revolution. Because reason itself cannot penetrate to the final causes of things, science has to start with things already in motion. In this sense, and others too, preformationism was a direct heir of seventeenth-century mechanism.[33] But just as mechanism never successfully excluded teleology—at least not for long—neither did its eighteenth-century biological variants. Indeed it was precisely the revelation of purpose that made seeds so metaphysically satisfying.

The limits of human knowledge, in fact, more expanded than restricted the purposiveness of natural organisms. Réaumur, Haller, Bonnet, and all those who saw the world brimming with miniaturized beings did confine the dynamics of development within the husks of seeds. But these husks, already bursting with the beings to come, sheltered a powerful space for divine purpose. Every seed was pregnant with the future; every being now engaged in unfolding purposes mysteriously implanted by God's hand. Behind the husk was a metaphysical preserve for the divine, a bounded yet infinite space for him and his works.

In the seed, the finite and the infinite converged. We might imagine this convergence using a convenient geometrical figure. Discovered in the mid-seventeenth century, the "truly marvelous and extraordinary" trumpet of Evangelista Torricelli lets us sense—if only momentarily—why the seed might have exerted such power over the early modern scientific imagination.[34] Start with a hyperbolic curve, and rotate it around an axis to generate a three-dimensional shape that narrows to a point on one side, and opens to a funnel on the other (fig. 4). If you do this, the Italian mathematician showed, you generate something that has an *infinite* surface area but a *finite* volume. As edges of the curve stretch to infinity, the volume enclosed by the shape converges on a definite number, but the surface area on the outside grows endlessly. You can, in other words, fill the inside with paint but *never* have enough paint to

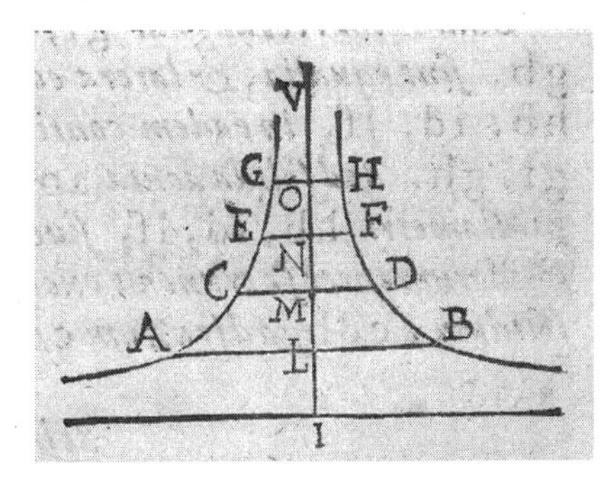


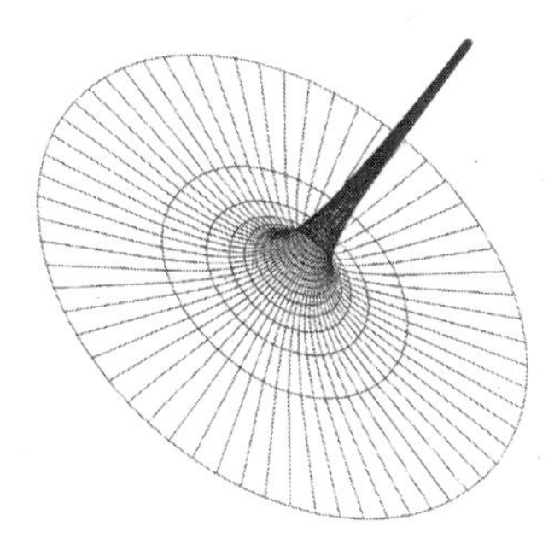

FIGURE 4. An acute hyperbolic solid ("Torricelli's Trumpet"). From Evangelista Torricelli, *Opera geometrica* (1644). Courtesy of The Bancroft Library, University of California, Berkeley; graphic representation (right) used with the permission of Richard Stauffer.

cover the outside! Philosophers (Hobbes among them) were flummoxed at this deeply counterintuitive idea, and mathematically speaking, this paradox turns out to be only apparent.

Yet the same astonished pleasure of discovering a concrete relationship between finite and infinite attended the metaphysics of the seed. While appearing to restrict the play of purposes, it actually enhanced them, turning all living creatures into the telos of the very tiny. Concealed in all of us, they discovered, is a trace of the divine, working its way through the generations. This trace is protected by the seed's husk, which incarnates the resistance of the natural world to the human intellect, and confirms the distance that lies between the merely mortal and God himself.

The seeds of Haller and Bonnet did not, in short, merely recapitulate the ancient sciences of life. Instead, they offered resources to confront the most urgent questions of the age. The preformationism of Haller and Bonnet was so compelling precisely because it offered an entire set of answers to the problems of order and emergence. These answers preserved purposes without abandoning empirical science. They did not so much defend an older Christian theology as renew how this theology might find a home in the world. The science of seeds supersaturated nature with purposes, filling the immanent world with traces of the divine made imaginable by the technology of the microscope.

## THE FLOWS OF NATURE

Seeds were not self-organizing. No dynamism; no development; no unpredictability; no powers not already contained in their husks. They answered questions of order--of how life comes to be from the materials of the world—but by *stopping* any final flow between nature and the human. But this was not the only way the question was answered, as Rousseau and Hunter indicate. The other way—the way that installed stories of self-organization at its center—was no less involved with purposes. But it also involved an endless reflection on flow, on the softness of the line between nature and the human. "Epigenesis" was a term poached from the antique thesaurus by a character we've already met, the English anatomist William Harvey. For Harvey, epigenesis names the emergence of life by "the superaddition of parts . . . out of the power or potentiality of the pre-existent matter." Preformationists, in his view, make the "common mistake of seeking for the cause of the diversity of parts in diversity of matter." In other words, they deflect the problem of life's orders by supposing that diverse living parts merely develop from parts already extant in the seed. Epigenesis, by contrast, takes living parts to be emergent. They differentiate themselves out of a common substance: albumen, for example, or (recall Hunter) the blood, a substance that "acts above the forces of the elements," one that "*is spirit* by reason of its admirable properties and powers."[35] Life is self-perpetuating, self-organizing.

Spontaneous generation—maggots born from meat, and so forth—featured in our first biology textbooks as an original superstition of the life sciences. What Marc Ratcliff has called "spontaneism" was powerfully attacked by the mechanist natural philosophers of the seventeenth century. Their targets were both Catholic neo-Aristotelianism, with its insistence on the distinction between inferior and superior beings and their respective biological origins (the former from putrefaction, the latter from sexual reproduction), as well as the more authoritative work of Harvey.[36] Only in the eighteenth century do we see, by contrast, the first really sustained efforts to make epigenesis a scientifically credible way to imagine the emergence of life.

As with seed theory, epigenesis took root in a matrix rich in both metaphysics and empiricism. By the third decade of the eighteenth century, as Ratcliff has shown, attention to the details of the microworld became ever more exacting. With it came increased focus on subjects more developmentally ambiguous than birds and mammals: insects, microbes, fungi, coral, and the variety of species whose generation was often less than straightforward. Few used this microworld to more creative effect than John Turberville Needham, the first Catholic priest admitted to the Royal Society, and a man who lent epigenesis observational credibility in a set of papers published from the early 1740s.[37] His work began while at the English College in Lisbon, and later he moved to Paris, where he met and collaborated with the famous French naturalist and *intendant* of the Jardin du Roi, Georges-Louis Leclerc de Buffon. In conversations with him, and with the president of the Prussian Royal Academy of Sciences, Pierre-Louis Moreau de Maupertuis, he began research that cast him into the European-wide conversation about the nature and origins of life.

In his 1749 *Observations upon the Generation, Composition, and Decomposition of Animal and Vegetable Substances*, Needham described a set of remarkable experiments. Among the more dramatic of these was one centered on an infusion of wheat, mashed and suspended in water. What Needham saw through the microscope was, for him, nothing short of miraculous:

> To the naked Eye, or to the Touch, it appear'd a gelatinous Matter, but in the Microscope was seen to consist of innumerable filaments; and then it was that the Substance was in its highest Point of Exaltation, just breaking . . . into Life. These Filaments would swell from an interior Force so active, and so productive, that . . . they were perfect Zoophytes teeming with Life, and Self-moving.

Dead vegetable had become living animal, extending tendrils that would, on their own, grow into new "wheat islands" (fig. 5) and budding small animals capable of moving on their own. These were the "eels" that Voltaire satirized in the 1760s, when the two men confronted each other over mutual charges of atheism, but in the 1740s, their discovery gave Needham the proof he wanted that life was a force saturating the world.[38] This force explained what, in his view, seed theory could *never* explain, namely, those living phenomena that failed to conform to any neat rule. Wheat blight—he likely meant *Fusarium graminearum*, a fungus that produces masses of microscopic filaments—was his example, an "extraordinary Exuberance of matter" resulting from a "Want of Resistance" to the extraordinary powers of vegetable nature. Blight was, moreover, just one of the natural forms that (he believed) disproved seed theory. Monsters, mules, and self-regenerating starfish were some of the organic forms that made seed theory nearly inconceivable to Needham. How, he asked, can it "be agreeable to reason" that these creatures too came from seeds?[39] Here was life bursting right through the preformationists' husk.

Liminal natural phenomena like these threatened, among other things, the neat distinctions between species embraced by Enlightenment taxonomists. In an important way, a strong concept of the species gave a "goal" to all development. The telos of an embryonic foal, for example, was to manifest the characteristics of the species *Equus caballus*. Its development was "complete" once these characteristics were achieved. Deviations from "proper" development

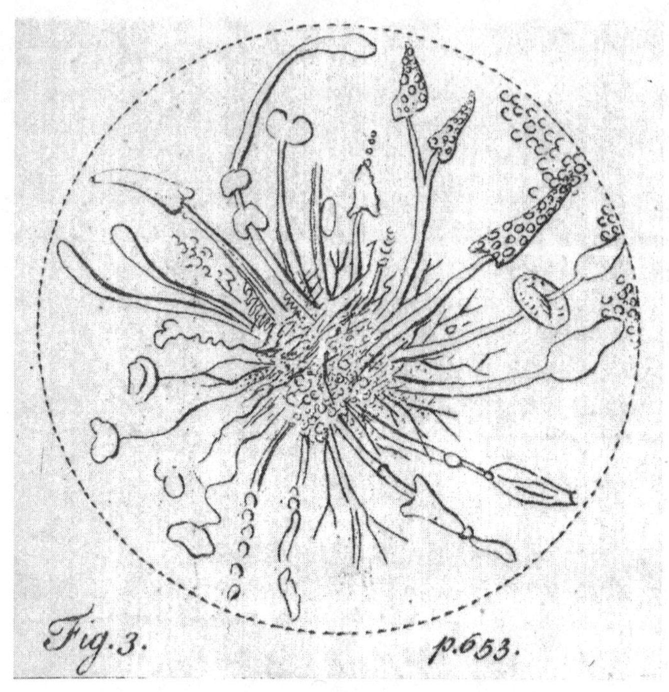


FIGURE 5. "Wheat islands." Original image from John Turberville Needham, "A Summary of Some Late Observations . . ." *Philosophical Transactions of the Royal Society of London* (1748). Courtesy of The Bancroft Library, University of California, Berkeley.

could, in turn, be judged by a species norm. There is an obvious tautology here: the purpose of a horse is to be a horse. But this was not really an issue, so long as animal generation only involved, as Bonnet put it in 1762, the "development of the germs" already underlying each species. Seeds solved this problem, by directing each organism into the niche waiting for it on a taxonomic chart. In the 1740 edition of his *Systema naturae*, Linnaeus had already made the connection explicit. "If we regard the works of God, it is obvious enough to everyone that single living things are born of an egg, and that every egg produces an offspring similar to the parents," he wrote in his first general observation on the study of nature, and "thus there are no new species being produced today."[40] In this sense, seeds were the causal substratum of taxonomy, just as the latter gave a descriptive architecture to the former.

Of course, the issue of species was never as settled as all that, as historians have long known. The two giants of eighteenth-century natural history, Linnaeus and Buffon, for example, violently disagreed whether or not species have ontological status or are mere conveniences for organizing nature's varieties. Yet both men wavered on the point. Foucault to the contrary, the Swedish scientist was hardly content with an "arbitrary" or artificial set of classification practices, or with merely arranging a descriptive grid for natural objects. Rather, he struggled mightily with the species concept, especially after discovering in the 1740s exactly the phenomenon that interested Needham, namely, the plasticity of natural forms. The discovery by a Swedish student of a "mule species in the vegetable kingdom"—one plant having produced offspring of an apparently different species—made it quite unclear what natural logic underlay species formation. By 1767, in his twelfth edition of the *Systema naturae*, Linnaeus would add new language dealing with plants, language that would speak of *two* logics of vegetable speciation, one by the divine creation of "generic plants," and the second by "redoubled generations (*per generations ambigenas*)" as plants cross and mix to produce "all possible existing species." For his part, Buffon excoriated Linnaeus for failing to understand that species were both real and immortal natural objects, "the sole beings of Nature, perpetual beings, as ancient and as permanent as it is." And yet he too was puzzled by hybrid forms, moments when it becomes difficult to "fix the line between what Naturalists call *species* and *variety*."[41]

No wonder, then, that threshold creatures, ones that cracked the husk not just of the species but even of the individual organism itself, would have been so intriguing to the epigeneticists. These creatures tended to sit, in fact, right at the threshold where the nature of the organism became obscure. Indeed, Linnaeus himself, in that twelfth edition of the *Systema naturae*, came up with an entire genus called *Chaos*, whose species included those microscopic eels cherished by Needham and other "living molecules" whose study, Linnaeus remarked, he would leave to others.[42] The most startling of these threshold creatures in the eighteenth century—the one that riveted both scientific and popular audiences—was the polyp.

This curious organism began to assume an iconic status in 1741, when the Genevan scientist Abraham Trembley presented a set of shocking discoveries to the French Royal Academy of Sciences and its leading entomologist, Réaumur. Over the next few years, natural philosophers in Paris, Leiden, Berlin, and England reproduced Trembley's findings. The story was sensational. The creature itself was a marvel, a freshwater multicellular organism related to sea anemones, with animal characteristics (especially locomotion) yet a structure so simple that Leeuwenhoek had already classified it as a plant. Not only did the polyp manifest that ancient saw *natura non facit saltus*—"nature makes no leaps," providing a bridge from vegetable to animal kingdoms—but it also did so in a remarkable manner. The first reports from the French Royal Academy were awestruck:

> The story of the Phoenix reborn from its ashes, as fabulous as it might be, offers nothing more marvelous. . . . The chimerical ideas of palingenesis or regeneration of plants and animals, which some alchemists have thought possible by joining and reuniting their essential parts, only aimed at re-establishing *one* plant or *one* animal after its destruction . . . but witness Nature which goes far further than our chimeras. From each portion of an animal cut in 2, 3, 4, 10, 20, 30, 40 parts, and, so to speak, chopped up, just as many complete animals are reborn, similar to the first. Each of these is ready to undergo the same division, & to be reborn the same from its remains, without it being known at what point this astonishing multiplication will cease.

With no mother or father, the phoenix was "self-begot, self-rais'd," in Milton's terms. It was uncanny enough to see a starfish regenerate a leg, or watch a lobster regrow a claw. But here was the marvelous rebirth *of the complete organism itself*. If seed theory reduced generation to development, the epigenetic polyp erased the line between them. It inserted a complete ambiguity between parent and child, germ and complete life-form. "With light steps she strolls, the plastic goddess Nature!" exclaimed Johann Christian Reil at century's end, and her first organism is the polyp, "an organic tube . . . itself means and end, an enclosed whole, the first and rawest sketch of individuality."[43] Like John Hunter's blood, the polyp blended part and whole, means and ends. How can we even speak of a seed in an organism for which "birth" itself is an uncertain state?

The polyp seemed Nature's rebuke to preformationist enthusiasm. It also offered resources to address a question confronted by all theories of natural order (in the eighteenth century and afterwards): How is information encoded and carried from generation to generation? Albrecht von Haller put it like this: If you believe in epigenesis, he wondered, "why should the material coming from a hen always give rise to a chicken, and that from a peacock give rise to a peacock?"[44] How could it be that, from an undifferentiated material (blood or albumen or whatever), one thing should develop and not another? In order to answer this—to explain the empirical facts revealed in the microscope, to resolve the philosophical conundrum, *and* to preserve the space of divine activity—preformationists like Haller allowed only God to carry data across the generational boundary. Their seed was a reservoir of information whose expression not only revealed divine purposes but was also given force by them.

It was also a reservoir impermeable to the effects of the exterior world, developing only immanently from its own blueprint.

Trembley's regenerating polyp embodied other informational possibilities. It indicated that enough information—both about structure (a divided polyp only ever makes another polyp) and about expression (the polyp needs no direction from beyond itself)—can inhere in the material to accomplish its own purposes. Haller's question still stood. Getting from chaos to the form of the polyp was presumably a big step. But still the polyp showed that organic material is plastic, and capable of ordering itself without any seeds involved. Moreover, it linked this ordering impulse to its environment and the world at large. The observer's knife began a process of ordering that would not have otherwise happened. The knife did not determine the final outcome, however, since obviously only some organisms perform such miracles of rebirth. Rather, the polyp and its environment were involved in an intricate feedback loop, inputs becoming outputs in ways both circular and flowing.

The French mathematician Pierre-Louis Moreau de Maupertuis, someone fascinated with Trembley's discoveries, spent much time trying to understand these complex informational processes. The man who famously "flattened the earth" from its ideal spherical perfection lamented "that we do not know at all how . . . an animal can fashion itself." Nonetheless, he believed, only some complex system of autogenesis could explain how information is transmitted contingently across the generations. Nothing expressed this contingency more than the variety of individuals. Granted that species exist, he noted, they open up a real problem: What to do with the *mistakes*? Why, in other words, don't *all* individuals conform neatly to the species norm? His beloved example of variability was "a family known in Berlin where the children are regularly born with six fingers." This phenomenon, he noted,

> is inexplicable in any of the most commonly accepted systems of generation. . . . In ours there is no difficulty whatsoever: the first monstrosity was an accidental effect. . . . The habit [*l'habitude*] of the location of the parts in the first individual makes them replace themselves in the same manner in the second, in the third, etc., so long as this habit is not destroyed by another, more powerful, whether it be on the side of the father, or the mother, or by some accident.[45]

Seed theory explained continuity perfectly well. But it could account for neither the appearance of the mistake nor its conservation over time.

The first—the moment of change—Maupertuis attributed to pure chance, a force he saw at work throughout the natural world. As we have seen in previous chapters, chance is often important to self-organization, a force of differentiation that opens up the possibility of complex and unpredictable order. Among life scientists, this was particularly true for Maupertuis. Chance, he remarked in an essay written in the 1740s, produced "an innumerable multitude of individuals," of which only a "small number turned out to be constructed in such fashion that the parts of the animal could satisfy its needs." The rest, an "infinitely greater number," he drily said, simply died off.[46] Once you subject organic development (as in the polyp) to the whim of an observer with a knife—or more generally, to the confining and encouraging powers of the environment—then the world's living forms are cut loose from wise design and flung into the arms of chance.

Nor was chance the only issue. Just as important was the stipulated ability of matter to preserve its own variability and to transmit the changes down through time. What we call heredity was deeply mysterious for a seed theorist. Children resemble both parents, after all, and it seems to beggar the imagination that every mother is filled with the seeds of all future couplings, however furtive and unanticipated. Organic matter must, Maupertuis insisted, have some capacity for responding to the contingent conditions of its production, and for moving information from one being to another. For this strange ability, he invented a host of names: *habitude*, *désir*, *aversion*, and *mémoire*. All of these named "some kind of principle of intelligence," a purposive activity in matter itself, and were rooted specifically in what Charles T. Wolfe has aptly called Maupertuis's "endowed molecules," that is, material particles with a disposition to complex order and organization.[47] These purposes were not entirely self-generated. The world too had its place, the world whose effects reverberated through living organisms in ways both causal and diffuse. If "memory" was the mechanism by which information was transmitted across generations, it was also the way the world made itself felt in every single organism. The barrier between living being and world is soft, a beach rather than a cliff.

## FROM MONSTERS TO THE FORCES OF NATURE

As it turns out, the chance operations of nature—monsters and other teratological organisms—were magnets for scientists interested in these soft thresholds. In the 1720s, for example, French natural historians debated whether monsters threatened the very concept of God's wise design. The physician Louis Lémery asked the money question: "When one considers . . . what is virtually always found in monsters . . . [the] disorder, confusion, deformity, the degeneration and absence of certain functions . . . is one to say that a design has given rise to such works?" A question that has not lost its relevance even today. But in the eighteenth century, what Lémery called the "infinite number of oddities" of the world confounded any simple or mechanical account of organic development.[48]

Few had as ready access to these oddities as Haller's great rival, the young German biologist Caspar Friedrich Wolff. In its early days, Wolff's itinerant career was thwarted by his heterodox natural philosophy and his lack of access to the oddities and corpses he so longed to dissect. For years after he published his famous *Theoria generationis* (1759), he could only look on in envy as Haller discovered yet more confirmation of his organic theory in the entrails of conjoined twins and hundreds of other specimens. In 1766, however, and at the recommendation of the mathematician Leonhard Euler, Wolff became the curator of the natural history collection at the St. Petersburg Imperial Museum, which gathered more anatomical curiosities than virtually anywhere else in eighteenth-century Europe. The museum was rooted in imperial mania. Its founder, Peter I, was compulsively fascinated with the abnormal, and went so far as to issue royal edicts (in 1704 and 1718) that forbade the killing or hiding of monstrous children and ordered that all teratological samples be sent directly to his curiosity cabinet. The punishments for disobedience and the rewards for finding new and strange samples were both significant. His collections quickly grew enormous. Years later, the Russian museum offered Wolff what he could not find in Germany, a huge data bank of dissection materials. Monsters, Wolff hoped, would point a way beyond the "lifeless mass" of preformationist nature—which merely "cast[s] off one piece after another, until the affair comes to an end"—and toward a nature that "destroy[s] itself and that create[s] itself again anew, in order to produce endless changes." Using them, he planned a masterwork, a *Theoria monstrorum* that would show how nature's *vis essentialis*, its "essential force," produced the varieties of life that we see all around us.[49]

Although Wolff fell well short of his ambitions (some thousand pages of manuscript are still held in St. Petersburg), his efforts were significant. His interest in monsters testified, for example, to a Lucretian sensitivity to the importance of chance and fortune in the making of complex natural systems. Indeed, unsurprisingly, aspects of Lucretius were integral to the understanding of living systems as immanently ordered. Even the priest John Needham found himself approvingly cited in La Grange's 1768 French translation of our heretical ancient. The lines in question were these:

> Do atoms of soul remain, or not,
> in a dead body? But if they remain, still present,
> we may not properly deem the soul immortal,
> since it has gone but left some parts behind.
> But if it has fled with all its parts intact,
> leaving no single particle in the body,
> How can a corpse all rotting come alive
> with worms? (3.713–20)

Built out of the same things of the world—those atoms driven by chance that we saw in chapter 1—the soul emerges from the body. It is an accidental product. The thing that *we* are, "*that* soul, *that* body, in *that* conjunction," will come to pass just once, at one intersection point of nature's "myriad ways" (3.846, 856). It is this conjunctural happenstance that makes everything unique and gives the world its endless variability. At the same time, too, the soul is a material product. The "worms" that grow from our bodies are extensions of the living principle of the soul. Here, according to the eighteenth-century translator, Needham confirmed what Lucretius had intuited, the foolishness of believing that "all the accidental generations that one might offer are caused by eggs" and the importance of material principles, the "development of juices," to the birth of the living.[50]

No less important—and no less Lucretian in resonance—was the sense that the dynamics of organic development were not governed by merely mechanical operations but depended on some *force* inherent in materials themselves. Earlier in that same book, in fact, the ancient pagan insisted how life depends on a "force (*vis*) . . . hidden deep, deep down . . . the inmost element of the body . . . the soul of all the soul" (3.273–75). This was not quite the "secret Lucretian system," however, that someone like Haller suspected underlay the epigenetic position.[51] There was, in fact, very little of system here. Rather, Lucretius—like Aristotle—was resurrected just enough to be put to work for the eighteenth-century language of self-organization.

Already in Newton's time, for example, the question of force was in the air. The physician (and early enthusiast of vegetarianism) George Cheyne thought he was following in the footsteps of the master, for example, when he declared in 1715 that motion was "not *essential* to Matter." But just as quickly he recanted, wondering if energy was not, in fact, one of "the *primary* Qualities of Matter." For Cheyne, however, self-moving matter did not entail Lucretian impiety any more than it did for Georg Stahl, pious animist of precisely the same period. Stories of forces, energies, and powers instead spoke an Aristotelian language, imagining a physical world where matter was not precisely identical to itself, where the *real* substance of things is distinct from their material accidents. Most importantly, they spoke a Christian language, in which God himself is the name for the "active Principle which animates, as it were, the dead Mass of Bodies."[52]

Pieties about active matter did not need to take an orthodox form. God served all sorts of Enlightenment ends. One of the more idiosyncratic was the German philosopher and theologian Johann Gottfried Herder, someone who experimented freely with the language of self-organization. Later in the century, Herder would imagine a world emerging from chaos under the tutelage of the divine hand:

> While once the material of future worlds floated around in the infinite, it pleased the creator of this world to let its materials form themselves according to their inherent inner powers . . . [and] thus a harmonic world system formed from a floating, coalescing chaos . . . eternal proof of the law of nature, that order emerges out of the state of confusion by virtue of implanted divine powers.[53]

Herder's divinity was hardly the architect-in-chief, nor the immediate crafter of the world's forms. Instead, Herder's divine was a principle of self-organization, guaranteeing that things trend toward complexity and organization rather than dispersion and confusion. Its forces worked through the materials, indeed were inherent in matter itself. The grammar of this organization was, therefore, a reflexive one: materials "form themselves"; they "emerge" from chaos, occupying once more that middle space between activity and passivity.

Needless to say, though, the language of forces also could be scandalous, nowhere more so than in the hands of the French physician and literary provocateur Julien Offray de la Mettrie. The mid-1740s—the same decade when Haller, Buffon, Needham, and others were so intensely engaged in their biological debates—were also years marked by La Mettrie's publications and the authorities' violent responses to them. The Paris Parlement, for example, sentenced his *Natural History of the Soul* (1745) to be publicly burned and forbade its publication, while even the rather more permissive Dutch authorities suppressed *Machine Man* shortly after its 1747 release. These reactions were conditioned by a host of fears, Spinozism, atheism, and libertinism among them.[54] Yet La Mettrie (despite his evident relish for scandal) was not far outside the major currents in eighteenth-century natural history.

From the title, *Machine Man* looked like a brief for pure mechanism. "Matter is in itself a passive principle; it has only a force of inertia," La Mettrie wrote in his *Natural History of the Soul*, or, again, "the human body is a clock," in *Machine Man*. In fact, La Mettrie never failed to recognize the limits of his ostensible mechanism. There was, as Charles T. Wolfe has commented, more Epicurus than Descartes in his approach to material things. In particular, La Mettrie attributed unique powers to organized matter, namely, the power to shape itself. If the body is a machine, it is one that "winds itself up, a living picture of perpetual motion." The living organism "contains the motive force which animates it and which is the immediate cause of all the laws of movement." "Matter is mobile. . . . It has the power to move by itself," he continued in distinctly unmechanist terms. La Mettrie was no closer than anyone else to discovering exactly what this motive force was. Indeed, he declared it "folly to waste one's time trying to discover its mechanism." Even this sentiment played a materialist role for La Mettrie. Since we are ourselves participants in the world of matter, we cannot get behind it to see how it operates. We cannot open up the clock and check out its works. Instead, we must become what Natania Meeker calls "voluptuous philosophers," embodied participants in a material world filled with unpredictable forces and agents.[55]

Interest in the "self-creating power" of nature that Yves Citton has attributed to "Spinozist vitalism" was not limited to the libertine faction. The sense that in nature there *must be something more* than dead matter was absolutely pervasive in the Enlightenment. Already in the early eighteenth century, for example, chemists were discovering what they would call "affinity," which the Frenchman Etienne-François Geoffroy described as mysterious "rapports between different bodies which allow them to unite easily with one another." Before the age of any real atomic theory, and certainly before any knowledge of chemical bonding, scientists were stuck observing the various tendencies matter had to compound with itself. So Geoffroy published a "*table des rapports*," a table that organized basic chemical substances into categories of reactivity and mutual affiliation.[56]

But the question of *why* certain substances affiliate, and others do not, was simply unanswerable in any empirical way. As a result, some (like the self-taught chemist and, later, anti-Revolutionary balloonist Louis-Bernard Guyton) believed the laws of affinity must mirror the causal laws governing celestial mechanics, seeing in the microworld a precise reflection of the macro. These, however, were in the minority. More common were the sentiments of the Swede Torbern Bergman, that affinity "seems to be regulated by *very different laws*" than gravity and attraction. In a 1748 manuscript, the Scottish scientist

William Cullen tersely remarked that "chemistry not only considers a set of properties different from these which the mechanical philosophy considers, but a set of properties also that may depend on different circumstances."[57] The idea that there might be *different* regimes of causal law in different natural domains is heresy if you believe, as a mechanist should, that causal law is uniform across the range of physical phenomena.

But, as this book has been trying to argue, few people actually did believe this. Or, put better, few people were able to describe the world as if this were actually true. They paid attention to the "extraordinary," "singular," "surprising," "strange," "marvelous," "paradoxical," and "obscure" phenomena that made up the natural world. Different domains of nature, many thought, were subject to different causal laws. As Elizabeth Williams has noted, for example, a crucial characteristic of midcentury chemistry was a sense of the "irreducibility of chemical phenomena to the laws of physics." The entry on "chemistry" for the *Encyclopédie* noted, first, that "the phenomena of organized bodies have to be made the object of a science essentially distinct from all the other parts of physics" and, second, that even this latter physics needed subdivision. The laws that governed "masses," the French doctor Gabriel-François Venel wrote, were "absolutely different from those that governed the mutual relations of corpuscles," the proper subject of chemistry. "We dare defy anyone to present us an explanation of a chemical phenomenon founded on known mechanical laws that we cannot show to be false," Venel boldly continued. Explanatory registers *differ*, in the same way that living tissue differs from the elements that make it up: a living thing is a "machine that confounds all of our mechanical ideas." Moving between realms requires leaps of the experimental and philosophical imagination. In his entry on death in the *Encyclopédie*, the physician Louis de Jaucourt described the passage from person to corpse in terms that Rousseau would have recognized, as a "*reverie determinée*," a settled reverie as life leaves the body. Just as Jean-Jacques found his very identity thinning out into the world as he reclined on the banks of Lake Bienne, the passage from life to death is a mysterious one. "By degrees we begin to live," Jaucourt wrote, and so too we die, passing through that intertidal zone almost imperceptibly.[58]

"Reverie" was just one entry in a thesaurus of terms used to describe the movement between physical matter and its living incarnation. In his *History of Animals* (1749), Buffon spoke, for example, of a "procreative force that exerts itself perpetually . . . a mystery whose depth it seems we are not permitted to plumb." His idea that organisms orient themselves, and are oriented, by an "internal mould"—what Reill aptly calls a "physical symbol of the 'extended middle'"—gave theoretical formulation to the circularities of cause and effect involved in living systems. The internal mould was a "hidden structure" that preexisted the final form of the creature, yet was itself transformed by its development. Buffon did not stop there. He also came up with the idea of the living organic molecule, an "intermediate stage" between the living and the dead. Organic molecules are "always existing, always active," and their free play along the threshold lets them explain the widest variety of organic phenomena. "There are perhaps as many beings, both animal and vegetable, that produce themselves by the accidental combination of organic molecules," Buffon wrote, echoing first Lucretius and then Aristotle. "It is to the generation of these species of being that one can apply the axioms of the ancients: *corruption unius, generatio alterius.*" The threshold between mere matter and life was haunted by an enigmatic activity, an energy or a substance aimed at increased complexity.[59]

Others developed different idioms, more or less exact. There must be a "something," Herder complained, inside a plant that produces the effects it manifests, but "we have no name for this inner state." "Developing," "driving," "sensing": these were just a few of the attributes of the organic "something" that Herder found useful. The circle of French doctors in the famous medical institute of Montpellier talked about this "something" with the language of "vital principles" or "vital faculties" that were inherent in matter in ways that acknowledged their enormous difficulty in understanding. The doctor Paul-Joseph Barthez spoke of "tonic forces," powers that "effect movement whose progression is not perceptible and which are all the better hidden the more that they form among the diverse organs a perpetual condition of oppositions that are extremely varied." For the chemist and later theologian Franz von Baader, one name for Herder's "something" was "love," that "universal striving of all part of matter to mutually combine" without which the world would be "a desolate, eternally dead chaos . . . a non-thing."[60] Love was as good a name as any for that tendency in matter to organize itself, to grow more complex, to develop its own potential.

The most enduring name for this "something" was the German neologism *Bildungstrieb* (the "developmental drive") of the comparative anatomist Johann Friedrich Blumenbach, another of the Göttingen doctors whose interests in the science of living complexity shaped the later eighteenth century, not least the work of people like Herder. His contemporary notoriety depends on his interest in craniology, and specifically his efforts to discriminate between the different cranial capacities of the human races. But Blumenbach was far more than just a pioneer of what later became scientific racism. He was an important and rare conduit between German and English scientists, a pathway encouraged by the Hanoverian succession in England. Blumenbach was one of the few eighteenth-century Germans whose works were translated quickly into English, for example. His connections to the London scientific establishment also brought Göttingen (founded by George II in 1734) many of the artifacts collected by Captain Cook on his famous South Sea voyages, and indeed resulted in Blumenbach's appointment as court physician to the British king in 1788.

Like his predecessor at the University of Göttingen, Albrecht von Haller, Blumenbach wondered about the powers of matter. Haller had made famous that curious power of movement possessed by dead muscle tissue. A freshly killed frog's heart will contract when pricked, as we know from high school biology, and this power Haller called "irritability." Blumenbach went much further, uncovering a total of *five* inherent forces in matter: irritability, sensibility, and contractibility, as well as the "*vita propria*" (the specific life particular to a specific organ or animal), and finally the *nisus formativus*, or *Bildungstrieb*. There are, he said, "throughout all of nature the most unmistakable traces of an almost universal drive," a drive "to give matter a defined form [*Bildung*]," a drive that is inherent in the organic material itself, and lasts from its generation to its death. Only a drive like this could possibly account for the immense lability of natural forms, from the tobacco plants so transformed by his colleague Joseph Gottlieb Kölreuter's hybridization that Blumenbach thought them a new species, to the entirety of the natural world itself. "A whole creation of organized bodies has, in all likelihood, perished, and new one replaced it," Blumenbach impiously declared, seeing the world as a place of endless "changeability" and "impermanence." From an original "web of slime" (*tela mucosa*), the first organic materials emerge; species appear, they die, and then are "reborn" in new forms. The earth has seen, as he dramatically said, a "total revolution" in its life forms.[61]

The "developmental drive"—like all of these various forces invoked in the later eighteenth century—performed a curious work for the life sciences. Here is Blumenbach, making a confession:

> The term *Bildungstrieb* . . . explains nothing itself, rather it is intended to designate a particular force whose constant effect is to be recognized from the phenomena of experience, but whose cause, just like the causes of all other universally recognized natural forces, remains for us an occult quality.

On the face of it, this is a peculiar statement. Even if we stipulate that the *cause* of the drive is unknown, and agree that a "term" is not a force, how can Blumenbach claim that the drive "explains nothing"? Especially when, for him, it explains not only something but virtually *everything* about the natural world? Blumenbach referred readers to Newton here, but Newton, at least in the *Principia*, never needed to explain gravity to make his mathematics come out right. Blumenbach's project—to explain the origins of the enormous variety, diversity, and complexity of the earth's living forms—was never a merely descriptive enterprise, by contrast, but always also a causal one. His objections to the preformationists were entirely focused on their accounts of the *causes* of diversity. Yet, for Blumenbach as for Hunter before him, the nature of the causality at work was unclear. At times, Blumenbach condemned teleology: to "say that an eye is designed for seeing" would be just as ridiculous as saying that "stones are designed to hit someone on the head." Yet, the developmental drive, as Timothy Lenoir has perceptively noted, was always a "teleological agent," since it suggested an innate direction in natural processes toward complexity itself.[62] Both necessary and impossible, quasi-teleology pushed the paradoxes of nature into full view.

Some of these paradoxes found poetic form in a work written by the Swiss Pietist Georg Christoph Tobler, edited by (and for a long time attributed to) none other than the young Johann Wolfgang von Goethe. "Die Natur" was written in 1782 and "published" in the handwritten and distributed *Tiefurter Journal*. The brainchild of the patron of Weimar classicism, the Duchess Anna Amalia, its contributors included Goethe, Herder, and others. Tobler's poem contained a series of countervailing aspects of the natural world. Here are a few of them, in list form:

> Everything is new, and yet always ancient.
> We live in her midst and are alien to her.
> She speaks ceaselessly to us and never discloses her secret.
> An eternal life emerges and moves in her, but she never advances.
> We work with her even when we want to work against her.
> She isolated all things in order to bring them together.
> She rewards herself and punishes herself; pleasures and tortures herself.

The paradoxes were a bit facile, but they sought to preserve nature in a middle state, between revelation and concealment, familiarity and difference, purpose and accident. In that sense, they make explicit what Herder called the "oscillations" between states of both nature and human knowledge of her.[63] They leave us stuck in the center of the paradox, without apparent exit.

### TIME AND THE EXTENSION OF PARADOX

In his *Physics*, Aristotle also discovered an irreducible paradox, this one in the nature of time. Time is particularly hard to understand, he mused, because "not only do we measure the movement by the time, but also the time by the movement. . . . They define each other."[64] When the hands of a clock roll around its face, are they registering time? Or motion? The loop cannot be untangled. Time played its role in the self-organization stories of the eighteenth century too, less as paradox itself than as a way to keep causal paradoxes in suspended animation.

The doctrine of seeds, for example, compressed time into a virtual singularity. "All mankind are exactly the same age, the great grandfather not a second older than the youngest of his great grand children," Blumenbach laughed, but his point was a serious one. The seed was an antitemporal structure. Inside of it was squeezed all times, past, present and future. His own quasi-teleological processes, in contrast, unfolded in a highly differentiated timescape. They were drawn out along a temporal axis, whose contours affected the processes themselves in yet another recursive way. We might echo Peter Reill and call this temporality history. It is certainly true, as Reill notes, that the presumption that natural organization tends more toward complexity than toward chaos does capture something fundamental about eighteenth-century teleology. Yet we want to emphasize something a bit different here, something Jacques Roger calls "duration," the extension of time so essential to the Enlightenment sciences of life.[65] Extension in time did not always entail a direction of historical progress. It could become rather a temporal space in which causal law was both preserved and suspended.

Look, for example, at the idiosyncratic biological theory of someone like Diderot. Few were so enthusiastic about the polyp. "On Jupiter or Saturn, perhaps human polyps!" he wrote, and reveled in the thought of "men dissolving into an infinity of atomic men." But the real force of the polyp, in his view, was the indeterminacy it cast on any future state:

> Even if man does not dissolve into an infinity of men, he resolves at least into an infinity of tiny animals whose metamorphoses and organization, past and future, are impossible to predict. Who knows if ours isn't the nursery for a second

> generation of beings, separated from this one by an incomprehensible interval of centuries and successive developments?

Given time, all things have the potential to become something other than what they are, and to do so in unpredictable ways. He marveled at Needham's microscopic work, observing that the main difference between his miniature wheat islands and the macroscopic development of life was time alone. "What is our human lifetime [*durée*] in comparison to the eternity of time?" he asked. "Everything changes, everything vanishes." The "duration and vicissitudes of millions of centuries" thin and soften the very concept of the living being. Time, he wrote in his *Thoughts on the Interpretation of Nature*, "must in the end introduce the greatest difference between the forms that existed in the ancient past, those that now exist, and those that will in the remote future."[66]

For Diderot, time thus served as a force of differentiation, not a captain steering toward a particular future. Things grow, separate, and decay. What is a being but a "sum of a certain number of tendencies," a movement toward "a limit," Diderot wrote, imagining less a determinate universe than one filled with inclinations and yearnings. We live, therefore, as creatures of time. Diderot's blind geometer reflected on this at the very moment of his own death. His last words meditated on time and yearning: "How many worlds . . . have dissipated themselves, re-formed themselves, and dissipated again, perhaps at every instant. . . . What is a world? . . . A passing symmetry; a momentary order." "We shall all pass away, unable to fix . . . the precise time that we have lasted [*duré*]," he ended.[67] Pondering on human finitude revealed the truth that the world's orders are sustained in time, a truth that for Diderot left open both possibilities of development toward a future, and diffusion by the exigencies of chance and change.

Time was everywhere in Enlightenment biological theory. It softened the chains of causality, inserting elements of unpredictability and chance back into the larger order of things. When Maupertuis imagined the world, he saw it poised in a contingent temporal arc. It is "not impossible," he wrote, that after a massive flood or fire, "a new combination of elements, new animals, new plants, or even entirely new things would reproduce themselves." On a less global scale, when he described the production of new species through the transmission of "errors" through the generations, he leaned on time's extension as a crucial precondition of natural diversity. In another implicitly Lucretian moment, Maupertuis commented that from "the force of repeated deviations [*écarts*] would arise the infinite diversity of organisms that we see today."[68] Over time, he went on, things imperceptibly shift, always on the move.

In fact, duration was the precondition for *any* theory that sought to connect self-organization to the chance influences of an external environment. The transmission of traits over generations was not just a problem but also an opportunity, since it opened up the space of transformation enabled by temporal extension. The earth is a stunningly complex thing, the German biologist Carl Friedrich Kielmeyer remarked in 1793: nearly "seven million different organisms," each with tens of thousands of individuals, each of these filled with thousands of organs.

> And if this wasn't enough . . . we should note as well *how saturated time is* with all of these things: each organ is so adapted to the changes in all the other organs, and so unified in a system of concurrent and consequent alteration, that each of them is . . . alternately cause and effect of the others. Each individual, stimulated by the organs, lasts for a greater or lesser amount of time; and at each moment of this duration, the system of effects that we call its life, and the system of organs that composes its organism, changes itself.

Time's saturation with life is a striking image. Notable too, because the saturation goes both ways. Living organisms are beings in time. They are saturated by inputs but their outputs are structured by the state of things *at a particular time*. Complexity and simplicity go hand in hand here, as usual in self-organizing systems. The "division of the powers among the organs at different times" still follows, Kielmeyer insisted, the "same law," and yet the same law can have different effects, depending on when things happen.[69]

Time worked to conserve both uniformity and heterogeneity. Insofar as it is uniform, as Herder pointed out, nature "has put real limits on its species and would rather let a creation die than essentially disturb or debase its structure." The heterogeneity of nature, moreover, ensures that the "dog can take on a wolf-like aspect," not from the immediate impact of external events, but from their subtle incorporation into the very fabric of the species itself. "Climate is a chaos of different causes, which work unequally, slowly and in different ways, until they perhaps at last penetrate to the interior, and there, through custom and creation, change it," Herder wrote.[70] He too preserved something of the Epicurean here, the sense that this process of change, extended over time, is nonetheless unpredictable. The swerve only "perhaps" happens: how and when is fully unclear.

Far more than Diderot, of course, Herder had an eye toward progressive development, the divine telos that produced what he called "the crown of organization on our earth, humanity." But he was too sensitive to the oddness of change to find plausible a reduction of the natural world to some evident design. Change happens in time, and in ways difficult for anyone to imagine. We are, in his metaphor, like the caterpillar:

> See there creeps the ugly . . . caterpillar, its hour comes and the exhaustion of death afflicts her; she walls herself in . . . she has the cocoon for her shroud . . . now the inner organic powers begin to work. Slowly at first works the metamorphosis and seems like destruction . . . the new creation is still unformed in its parts. Gradually these form themselves and come to order . . . a few minutes . . . [and] the entire structure is changed. Who would imagine the future butterfly in the figure of the caterpillar?[71]

Who indeed could imagine this? Progress, perhaps, but also change that works along temporal registers—slowly, gradually, a few minutes, a few millennia—and then things are different in ways that no one could have envisioned.

It is duration too that returns us back to the image of the phoenix, that potent myth of self-creation that we already discovered in the polyp. Things must die that things might live. Nature needs death, but death does not touch her essence, "which rises over all ruins, always as a phoenix grows from her ashes, and blooms with rejuvenated force." Herder may have been repeating a line from his old teacher, Immanuel Kant. It was in his 1755 *Universal Natural History and Theory of the Heavens*, a very early work, that Kant first evoked the "phoenix of nature, which burns itself only in order to revive from its ashes, once more rejuvenated, through all the infinity of times and spaces." The phoenix not only embodies the suspended temporal paradox; it is also a figure for the spontaneous emergence of living forms from the dead ashes of things. It is the bird whose death signals life, whose nature turns inward and becomes the source of its own demise and resurrection. Insofar as natural organisms themselves embody this phoenix principle, one might—as in fact the French doctor Xavier Bichat did—imagine life as itself *no more* than a resistance to death. But Herder and, for that matter, Kant were interested in the extended figure of circulation, where time is both essential and transitory. "Many plants must have preceded and died before an animal appeared," Herder insisted, and it was only "finally" that the "crown of the organization of our earth, Man" made his entrance onto the stage of being.[72]

Not everyone embraced the ringing changes of nature with such enthusiasm. Buffon, for one, looked for a fixed point in the concept of a species. They are "the sole beings of Nature, perpetual beings, as ancient and as permanent as it is." Like his "interior mould," the enduring "constant form" that underlies the material changes in living organisms, the species apparently stands firm against time. Yet just as the interior mould develops itself into something we can recognize by "intussusception," by the absorption of "foreign material" into itself over time, so too are the species cast into the temporal flow. They are, he proposed, visible as fixed forms "only when we consider Nature in the succession of time, and in the constant destruction and renewal of creatures."[73] Their objective immortality can only be measured against the movement of nature through time.

The flow of time, then, made self-organization possible. It offered a space for chance and environment to work their effects in the production of living systems, and a space for these systems to develop immanently into something greater than the sum of their parts. Time also helped to control the *mysteries* of this development. Lending duration to a process was a way to elongate the distance between cause and effect and to open a path for the curious reversals of causal flow that self-organization constantly entailed. Hard constraints softened as the dynamics of order extended themselves further and further into some unknowable future.

## CAUSALITY AND THE OBSERVATION OF LIFE

Time, in short, managed the paradoxes of self-organizing causality. The attentive reader will have already noticed many of these, since they were endemic to the languages of biological self-organization. Forces emanated from within materials, yet were also implanted there (for some) by a divine hand. Or perhaps they emanated solely by virtue of material organization (as for someone like La Mettrie). In that case, however, they depended on some quantum shift in the material itself, from inanimate to animate. The very idea of quasi-teleology—in which purposes are immanent in organic development, but never in a way actually *intended* by that development—left the causal nexus ambiguous.

This was an ambiguity insoluble by the force of reason. It could not be analyzed or deduced away. But it *could* be managed descriptively, by pulling imaginative resources from nature's own mysterious forms. Herder, for one, looked to a physical substance whose nature captured the circularities of self-organization: the blood. Years before Hunter, the German exclaimed how the blood is a "compendium of the world: lime and earth, salt and acid, oil and water, the powers of vegetation, forces, sensibilities are organically unified in it, and mingle together." In this heterogeneous mixture of material and force so fundamental to living matter, "we do not know, where this force begins, and where it ends." So the blood offered an imaginative clue to an essential truth about the world:

1. Where there is an effect in nature, there must be an effective power; where stimulus shows itself in striving or even in convulsion, there too a stimulus must be felt from within. . . .
2. No one can draw a line, where an apparent effect can serve as proof of an inner power and where it can no longer serve as such.[74]

On the one hand, we must approach nature as if causality were constantly at work, as if every effect were the expression of some power. Yet the *relation* between the effect and that inner power is ambiguous. It cannot be fully described, nor can we feel certain in our causal taxonomies. The blood did not *prove* anything about causality, but it did *show* the complexities of nature's immanent order.

In the *Dream of d'Alembert*, Diderot used a different natural analogue to imagine these complexities, one borrowed from the Montpellier doctor Théophile de Bordeu:

> The world, or the general stock of all matter, is a giant swarm . . . Have you seen how it forms itself at the end of a tree branch, a long cluster of little winged animals, each clinging one to the other by their feet? . . . This cluster is a being, an individual, some kind of animal. . . . If a single one of these bees is minded to pinch in some way the bee to which it is clinging, what do you think will happen? . . . [It will] excite throughout the cluster as many sensations as there are animals; and the whole group will stir itself, rouse itself, and change its location and form.

Diderot found the bees irresistible. He described the polyp, for example, as a "cluster of infinitely small bees" each with a "living memory of one position." Bees, he was pleased to note, confused the very concept of an organism. Maupertuis, with whom Diderot had significant disagreements, had shared this pleasure. In considering the variability of the earth over time, and our difficulty in determining how each element relates to the generation of the whole, he likened both to "a swarm of bees . . . a body that has no resemblance whatsoever to the individuals that make it up."[75] The bees captured both a problem in the objects we study and the conditions under which we study them: Is the proper object of science (in this case) the individual bee or the swarm? And how should we relate the actions of one to the actions of the other?

The swarm proved an especially rich image of the causal challenges of self-organization. What in Bernard Mandeville had embodied the paradoxes of complex *social* systems, in Diderot embodied those of *natural* ones. A bee "is minded" to pinch his neighbor (an unpredictable action, with a mechanical result), and the result is the movement of the whole swarm (a mechanical action, with an unpredictable result). Between the actions of an individual bee and those of a swarm, there is an incommensurability that cannot be reconciled. The bee certainly did not *intend* to cause this movement, nor could she have predicted it from the outset. Her actions are, like those of life itself, opaque to her. The problem is not dissolved by moving up to the level of the swarm. It, of course, has no idea *what* started the ball rolling in the first place. The observer is left in quandary. Do the uncertain causal relations belong to himself, because he simply cannot trace all the various pinches? Or do they belong to the entire system, which moves along two apparently separate tracks, that of the bee and the swarm? It is not clear.

The Enlightenment sciences of life—at least a significant sector thereof—embraced, in short, a kind of constitutive perplexity. Their ambiguities were constitutional, unavoidable. Even so, they were productive of new insights about the world and its organisms—and not least because of the perplexity itself. "It is owing to their wonder that men . . . at first began to philosophize," Aristotle famously said in his *Metaphysics*. This is no less true in the Enlightenment. Buffon's *History of Animals* began, remember, with a tribute to the mysteries of nature, Rousseau's "abyss for the human mind" that drew researchers onward, into lives filled with speculation, discovery, and scientific production. This sense of mystery—already familiar to us from earlier chapters—was ubiquitous. Our surgeon John Hunter also spoke of the "wonder, admiration, and curiosity" that attends the "production of animals out of themselves." No surprise there, for such wonder, he wrote,

> is commonly the case in effects whose immediate causes are so obscure, more especially when we are ourselves both effects and causes of the same. . . . The first process set on foot in the formation of an animal is so small . . . and its situation so obscure, that its operation cannot be traced by taking it up at stated times.[76]

Time plays its role again—we are, Hunter makes clear, as men on the way, forced to see the world in time. More than that, too, we are the very same products of this world that we are observing, "both causes and effects of the same." In spite of, in fact, *because of*, this obscurity and mystery, we are compelled to push ahead, carrying our knots with us. Cause and effect, world and observer: these flow together in a parable of Jean-Jacques's reveries.

## A BRIEF HISTORIOGRAPHICAL POSTSCRIPT

In 1966, Michel Foucault's *The Order of Things* made a famous claim about the difference between the life sciences of the eighteenth and the nineteenth centuries. In the eighteenth century, life was "spread out over an immense table." Its seminal figure was Linnaeus, and its chief project the building of equivalences between natural beings. Classification was the center of Enlightenment natural history, and as a result, life "became the province of an ontology which dealt in the same way with all material beings." But by the time of Cuvier, life had become "autonomous," irreducible to any other phenomenon. It was, he insists, only at this moment that biology became truly possible, when life had "become wild once more." The order of things, he insists, fundamentally changed between 1750 and 1800, not so much in what he calls "opinions" but in the "internal conditions of possibility" of studying the living world.[77]

These claims are wrong. They are wrong in their content—life in the age of Linnaeus was *not* the province of an ontology that dealt the same way with all material beings, for example, but was rather a site for sustained reflection on the thresholds between living and dead and on the plural ways that living creatures organize. Taxonomy was only *one* means of managing the variety of nature's productions, and certainly never managed to contain the explosive challenge of living organisms to the sciences. And Foucault's claims are wrong in their method. To look only for "deep structures"; to insist on hidden unities or so-called epistemes; to restrict ourselves to speaking of "internal conditions of possibility": these are to miss what animated Enlightenment natural history (and much of the rest of its intellectual production too). The Enlightenment "order of things" does deserve attention, but it deserves attention in its own terms, sensitive to the modalities and motivations that lent it its peculiar energy.

What lent it this energy was *not* some comfortable reliance on coherence, uniformity, and symmetry. Enlightenment natural history projects did not, as should be clear by now, form anything like coherent systems. Their idioms of self-organization were, to echo Foucault, much too wild for that, both in their content and in their method. To use the language of Aristotle, natural history was diaporetic, a science that began in the experience of perplexity. In some sense, for the Greek, all sciences start in perplexity, with a sense of a knotty problem that needs to be unraveled. The successful philosopher surveys the problem and then resolves it with a leap into a higher domain of truth. Sometimes, however, such a leap into truth is not possible, for the knot cannot always be unraveled in a satisfying way. Sometimes, as Aristotle put it, the "knot is in the object" itself, and this is revealed by its *refusing* to collapse into one or another facile resolution. In that case, good philosophy does not simply stop—it continues to "state the difficulties," to remain in that state of perplexity that itself propels scientific advance.[78]

The eighteenth-century sciences of life were diaporetic in just this way. Since the knot was in the object itself—the living organism and the circularity of its self-organization—their stories of emergence were, as many well knew, impossible to square with basic metaphysical commitments, say, to cause and effect. Yet the sciences of life soldiered on, writing natural histories that were replete with the language of forces, of circular causality, of error, chaos, variety, and diversity. Their creativity did not flourish despite their contradictions but *because* of them.

Finally, then, these sciences were untidy, since their concepts often oscillated between what we might call the moral and natural, between the human experience of the world and the human explanation of it. The project of teleological renovation that began in the science of life extended, in other words, right into the human being itself. Because whatever else biology was in the Enlightenment, it was almost always a self-conscious science, one reflexively interested in the human ability to understand life at all. The problem of life was not just its enigmatic origin but also our own participation in it. We are living things, and even our tools of observation are given us by those same living principles. For this reason, among others, the problem of biology and the problem of the human understanding were always intertwined. The human mind, like the human body, needed its own organization, an emergent process that would take it from the acids, salts, and oils of the world into the complexities of thought, consciousness, and morality.

CHAPTER 5

# The Emergence of Mind

What is a dream? The German man of letters, scientist, and satirist Georg Lichtenberg was not sure, but it did not stop him from wondering. His 1801 collected writings, for example, tell the story of a young duchess who died in childbirth. Her infant son died too, and afterward their bodies were neatly set into a coffin and driven to the churchyard for burial. Along the way, however, the corpse wagon hit potholes and the coffin was severely jarred. Worried about their dead, the mourners opened the coffin and discovered inside an indecorous and shocking tangle of limbs. Later, Lichtenberg dreamed of the episode. In his dream, he told the story of the duchess to a friend. But he "forgot" the presence of the dead child until a dream-character, a third person, reminded him of it. Who, Lichtenberg later asked in some amazement, "reminded me in the dream about the child"? Surely it was I, he wrote. But if so, why didn't I remember it? "Why did my fantasy create a third [person], who surely must both surprise and shame me" for forgetting this detail? How do we "know" something that we don't remember knowing? This doubling and splitting, Lichtenberg elsewhere observed, is typical of the dream state. It happens "more often, I say, than can be merely accidental. . . . It likely stems from the fact that our brain is doubled, symmetrical." This, he remarked, "deserves our attention."[1]

The German physiologist and psychiatrist Johann Christian Reil did pay attention. For much of his career, he studied brains and their altered states. In 1803, he published a curious book on the human mind that tried to take both into account. The *Rhapsodies on the Application of the Psychic Method of Cure for Madness* was an enthusiastic reflection on mental peculiarities and their cure by means psychological rather than physical. Reil read Lichtenberg with interest. The story, Reil reported, indicated some real truths about the dream state. It is, he wrote,

> the product of a partial wakening of the nervous system. . . . Either fantasy alone awakes, or, along with it, some of the sensory organs, the capacity for motion, and so on. Thus the difference between dream, talking in one's sleep, and sleep walking. The self-consciousness wavers in its collective relations. Fantasy ebbs and flows by itself, no sensory impressions bridle it any longer. The dreamer . . . takes its stories for real objects, and plays each different role as its own. . . . The actors come out, the roles are divided; and of them, the dreamer takes only one, which he binds with his personality. All the other actors are as other to him as other men, though they are also creations of his own.[2]

What intrigued both men was the complex nature of thought, its "collective relations" as a physical and psycho-spiritual phenomenon. In dreams, thought is autonomous from the world. Since sensory data is cut off by sleep, dreams witness the imagination operating on itself. Yet the dream is often in dialogue with the world, the memories and stories of experience shaping the dreamscape. And finally, the brain and nervous systems do not disappear in sleep. As material substrates, they are no less important for the dreamer than the faculties of memory and fantasy. World, body, and imagination, all acting in tandem.

All of these were already at play on that island of St. Pierre. With his senses so captivated by the restless motion of the waves and water, Rousseau entered a waking dream on the sparkling shoreline. Thought stopped its usual activities. The "ebb and flow" of water, he wrote, "took the place of [the] internal movements" of his mind.[3] The imagination took wing, image following image, apparently free from their sensory surround. Yet it was just this surround—his particular location on the edge of the water, lulled by waves and light—that set the reverie into motion. Inputs from the world liberate the soul from the shackles of worldly causality. And dreams double up on themselves, freely creating a world from the material world, the experience of which has made the dreamer who he is.

The Enlightenment, it has often been said, was the age of reason. Indeed, there is no doubt that reason was a preeminent human faculty as well as a powerful norm in the eighteenth century. Calls to be reasonable in religion, ethics, philosophy, and politics sounded from around the European landscape, inside and outside of churches, universities, and coffeehouses. But bigger game was afoot than reason, as our opening stories hint. For the first time in the eighteenth century, *mind itself* became a subject of most urgent interest, and with it arose a host of questions about its origins, operations, and ultimate ends. Mind was more than just reason. As Lichtenberg, Reil, and Rousseau show us, mind bridged rationality and fantasy, thought and world, memory and imagination. Dreams were to mind what monsters were to life, evidence of inner complexity, variability, dynamic formation, and the limits of reductive schemata for understanding its operations.

This mind took root in the languages and stories of self-organization. As in the sciences of life, self-organization made its appearance in Enlightenment cognitive theory to manage this basic question: How do we get something from nothing? How, from the chaotic sensations that beset us from birth, does something like a mind emerge? These were new questions in the eighteenth century, and they spurred intensely creative meditations on the nature of human thought. Thought was not something that could be taken for granted anymore, something merely stipulated as foundational to human beings. Rather, its story needed to be told from the bottom up.

This chapter explores this story, and in so doing explores a moment when the mind became, for many, an emergent thing. Not something already inscribed in the human being, but dynamic and self-creating, an ever-surprising feature of the human engagement with the world around it. As in the previous chapter, stories of the mind's self-organization were rife with causal enigmas, appreciation for variability and complexity, moments of mysterious change, and curiosity about accident. The search for mind, however, crystallized other aspects of self-organization much more sharply. The mind is, after all, the very tool for pursuing ideas of self-organization. Thinking about thinking: the reflexivity of the process invited speculation on how such thoughts might be thought in the first place. The practices of philosophical inquiry—the move of metaphysics inward, the interest in analogical over deductive methods, the embrace of uncertainty as a cognitive fact—all emerge in and through the language of the self-organizing mind. As a result, this chapter ranges widely, from the neural monism of midcentury materialists to the debates about the origins of language to new directions in probability research.

We are not the first to notice how obsessed the eighteenth century was with the problem of mind. This was the century when psychology first became a discipline in a recognizably modern sense, and when epistemology—theories of where knowledge comes from—colonized virtually every sector of an older philosophical corpus. Our goal here is neither to survey this entire terrain nor to organize its dominant schools of thought. Self-organization, we have already argued, was not so much a school of thought as it was a common language for narrating stories of order in a world whose design could not be taken for granted. As a result, it is to be found less in normative statements about how rationality, reason, and cognition *should* behave than in the richly elaborated (if often incoherent) set of descriptions of how these actually *do* behave.[4] Given the logic of this book, we would stress again the historically specific nature of these descriptions, grounded as they were in the experience of new forms of complexity that helped make self-organization such a compelling way to make sense of the world. Self-organization is more phenomenology than metaphysics, a way of approaching a world whose orders have become troublingly opaque.

## MIND, BODY, AND BRAIN: THE EMERGENCE OF A QUESTION

The complexities of the eighteenth-century mind began in the dreams of Descartes. How can the senses be trusted, his 1641 *Meditations on First Philosophy* asked, when they take for real the "visions which come in sleep"? Sleep, like madness, deceives the mind. When the French king Charles VI decided, circa 1400, that he was made of glass, all of his senses confirmed this delusion as real. What if a malicious demon convinced you of the same? What if it should deceive the philosopher into thinking that "the sky, the air, the earth . . . and all external things are merely the delusions of dreams which he has devised to ensnare my judgment"? Cartesian thought departed from this point, moving with confidence toward the radical idea that *no* trustworthy knowledge can begin with the outside world and its possible illusions. Only the truths of the mind, the "thing that thinks," provide any certainty at all. With the dream, Descartes collapsed the variety of things Aristotle had once treated together in *De anima*—bodies and the three types of souls, vegetative, sensitive, and rational—into two things: *res extensa* and *res cogitans*.[5] Extended things and the mind.

Dreams solved one problem, only to create others. If the soul is a thinking thing, thought is both before and beyond the body's experience of the world. But if *this* is true, John Locke later asked, what about when we sleep? In sleep, we don't know that we are thinking at all. And how can we be thinking, if we don't think we are? As Locke put it:

> Characters drawn on Dust, that the first breath of wind effaces; or Impressions made on a heap of Atoms, or animal Spirits, are altogether as useful, and render

> the Subject as noble, as the Thoughts of a Soul that perish in thinking; that once out of sight, are gone for ever.

If the soul switches off every night, just when the body itself shuts down, how can it be a thinking thing? And if it is still thinking during that evening hiatus, then what kind of thought is it? For the dream *ought* to be the purest form of thought, perfectly untouched by the effects of the world. In sleep, we should find the mind's "native *Ideas* . . . those more natural and congenial ones which it had in it self, underived from the Body." But what we find is instead a mess, a congeries of thoughts "all *made up of the waking Man's* Ideas . . . oddly put together." Disordered, wild, fantastic, uncontrolled, and bizarre, how could dreams possibly be the native thought of the soul? "Methinks, every drowsy Nod shakes their Doctrine, who teach, that the soul is always thinking," Locke wryly concluded.[6]

Instead, Locke described the mind emergent. Touching the world is not a barrier but a beginning for thought. Children are not born speaking Socratic wisdom. Until their "Understanding turns inward upon itself," their thoughts are "like floating Visions." This turn inward is dynamic and recursive. It begins with a perception, which "furnishes the mind with a distinct *Idea*, which we call *Sensation*." Then comes a moment of *attention*, when "the *Ideas* that offer themselves . . . are taken notice of, and, as it were, registred." Later on, when that idea needs to be revisited, it recurs through the operation of "*Remembrance*."[7] Through this recursive behavior, extended over time and in experience, thought becomes conscious of itself. And only thinking that is conscious of itself, insisted Locke, can truly be called *thought*.

This move from nothing (disorganized sensations more or less randomly experienced) to something (a sophisticated and self-conscious mind) was a mysterious business. If we begin as blank slates, where do those features like memory and attention, so essential to our mental organization, come from? At one point, Locke calls them "faculties" and makes them into a "Power" that inheres in all human minds. At another point, he calls them "modes," that is, ideas about thinking that only occur *after* the "Mind turns its view inwards upon itself."[8] Never does he quite explain them, or derive their properties from the inputs that generate them. Instead, they leap like Athena from Zeus's head. They emerge unpredictably, as effects of the mind's self-organization in its encounter with the world.

Already in the late seventeenth century, then, the mind was a fascinatingly complex object. A complete theory of mind—tracing the operations of mental life along smooth causal highways—was elusive. Even for an apparent rationalist like Leibniz, it was not the clarity of the mind's operations that was so striking but its opacity. Can the soul, Leibniz tellingly asked, ever fully be conscious of all its own operations? And his answer was no. "There is in us an infinity of perceptions unaccompanied by awareness or reflection . . . , alterations in the soul itself, of which we are unaware because [they] are either too minute or too numerous," Leibniz insisted. The world of sensation, as we saw in chapter 1, is like "the roaring noise of the sea," which we hear as a jumble of confused particulars, and whose elements we are unable to separate out. We walk through our lives accumulating sensations without ever being aware of them, only to have them unexpectedly return to our minds. "Dreams," Leibniz said, "often revive former thoughts for us in this way," recalling things that we didn't know we had lost. All of these minute perceptions, Leibniz went on, "constitute that *je ne sais quoi*, those flavours, those images of sensible qualities, vivid in the aggregate but confused in the parts."[9] The mind is not a tabula rasa, in Leibniz's view, but a *tabula complexissima*.

This complexity only intensifies when we attend to the mind's correlate, the body to which it inevitably attaches. As Fernando Vidal has noted, Descartes may have been a dualist—famously denying souls to animals, and insisting on the mechanical operations of the body—but he was also adamant that human beings must be treated as a *union* of body and soul. "I and the body form a unit," he remarked, and dedicated the last work published in his lifetime, the *Passions of the Soul* (1649), to understanding how this unity might work. It was in this treatise that he discovered a home for this unity in the pineal gland, what Nima Bassiri has called a "virtual reduplication of the organizational totality of the body . . . to which the soul must necessarily be joined." From the pineal, the soul "radiate[s] through the rest of the body," much as the nerves themselves—those "little threads or tubes coming from the brain and containing . . . a certain very fine air or wind which is called the 'animal spirits'"—radiate to every part of the corporeal machine.[10]

How the Cartesian unity should be understood (even whether it *can* be understood) is a famously difficult problem, and one we cannot hope to resolve. More important than an answer, however, is the state of the question. From this point forward, as Vidal argues, the "science of the rational soul in its union with its body" would become a sine qua non of the study of mind. Nerves would serve as a physical correlate of this union, pathways of communication between the mind and the world around. Given that we experience external objects but that these "be not united to Minds," then it is evident, Locke wrote, that "some motion must thence be continued by our Nerves, or animal Spirits, by some parts of our Bodies, to the Brains or the seat of Sensation, there to *produce in our Minds the particular* Ideas *we have of them*."[11] Not just mind and world, then, but also matter and motion, and specifically, the motion communicated along the nerves to the center of our mental processes, the brain.

The problem for Locke, Descartes, and everyone else was determining what mechanical function these nerves actually had. Nerves channel motion, experience confirms, but the mechanism is frustratingly unclear. Since antiquity, for example, anatomists had never unambiguously revealed any air or liquid passing through them. Galen had reported that he could insert a hog bristle into the optic nerve, but even this finding was made uncertain when the Italian anatomist Andreas Vesalius looked more closely at the brain in his 1555 *De humani corporis fabrica*. "I never found any passage of that sort" despite looking for it, he reported, and though he did not have a better explanation for nerve function, the question remained an open one.[12] In the absence of anatomical evidence, the mechanism of the nervous system was, to say the least, obscure.

As they took shape in the later seventeenth century, then, theories of mind faced a perplexing causal knot, another Aristotelian diaporesis. Mechanism worked well in theory but also tended to falter in the face of the world's complexities. In this case, it faltered in the face of both experimental limitations and intricate neural phenomena. Few denied the important role material causes played in affecting mental events and functions.[13] But it was nigh impossible to correlate these events with a determinate set of inputs. Again, phenomenological *descriptions* of cognition did what a theory could not: namely, capture the circular relations between soul and body and conjure up, as if by magic, the spontaneous operations of mind. As in the world of the life sciences, the result was a profusion of explanations characterized by a potent mixture of the material and the spiritual.

This promiscuity in description began at the very origins of modern brain science, with Thomas Willis, the most important late-seventeenth-century cranial anatomist and an early member of the Royal Society. In some senses, Willis was a mechanist. Mental functions, he reported in his *Anatomy of the Brain* (1664), depended on the motion of animal spirits around the brain. Imagination is the "undulation or wavering" of the spirits outward in the brain, and memory, the "regurgitation or flowing back of the Spirits" back toward the center. Memories are stored in "Cells or Store-houses" in the brain, and so on. Yet Willis was also an exceptionally observant fellow, and he could not fail to notice that the nerves do not have a central cavity, like a vein.[14] They could not work like a hydraulic system, therefore, pumping fluids or air from place to place. So how did they work?

Willis was not sure. Rather than commit to one mechanism, then, he invented many. Animal spirit acted—for example, and following Gassendi and, interestingly, Epicurus—like light. The animal spirits are like "rays diffused from the light it self," he wrote, since "light is scarcely carried swifter through a diaphanous Medium, than the communication of the Spirits is made from one end of the nervous System to the other." At the same time, too, they are like fire, a "heap of most subtil Contiguous particles, and existing in a swift motion, and with a continued generation of some, renewed by the falling off of others." These were not the only metaphors. Juices, ropes, and fibers also served to describe neural activity, a baroque profusion less the consequence of the obscurity of Willis's thought than of the object he was studying. The sensitive soul, as he put it, "cannot be perceived by our Senses, but is only known by its Effects, and Operations."[15] The eyes open and sensation begins, but what animates this soul cannot analytically be isolated. It can only be seen, and described, obliquely.

He did have one consistent name, however, for the relationship between animal spirits and soul. He called it a *hypostasis*. This technical term goes back to ancient medicine. Hippocrates uses it to name the precipitates from urine, for example, and Willis too used it in this way in his diagnostic essays. But hypostasis meant more than just what he called a "laudable sediment." By the late seventeenth century, it was a richly metaphysical term as well, one with roots in both classical and Christian traditions. In Stoicism, it was the "becoming real" of the primal material, the assumption of physical form by the reality that lurks behind the appearance, what might be called being. Neoplatonists like Plotinus or Porphyry made the hypostases into the fundamental reality underneath everything that exists. In some sense, the entire universe is the bodying-forth of a soul, alive and self-actualizing in time, not a "work of art . . . [made] by an artisan or artist" but "produced from within, by [an] organizing power of Nature." No surprise, then, that Christianity would find use for this nexus as well, in its doctrine of the Trinity. The hypostatic union of Christ is the joining in *one* hypostasis of two natures, divine and human. Early Christian heresies split on these points, but by the fifth century, the doctrine was fairly stable, and had room to grow as well: Aquinas would use the hypostasis, more generally, to refer to the "union of soul and body," which gives rise to the person, or the human being.[16]

For Willis, the hypostasis carried a residue of both traditions, material and metaphysical. It helped him to name that mysterious union of matter and spirit that make up the soul. The "companies or throngs" of the "moving animal Spirit . . . constitute the Hypostasis of the bodily Soul," Willis wrote, its real essence. Under a "beamy Systasis and contexture," Willis announced elsewhere, the animal spirits "effect or cause the whole Hypostasis or subsistency of the sensitive Soul." So the sensitive soul, the perceiving soul, (now in a full metaphorical lather) consists of "choyce, most subtle" particles that,

> as a flower arising out of the grosser mass, do mutually come together, and do constitute fit passages, which they produce thorow [*sic*] the whole frame of the Body, having got one continued Hypostasis, to wit, very thin, and as it were Spirituous, and equal, and extended to the whole.

The sensitive soul is a "shadowy subsistence" of the animal spirits, which "being chained and abhering [*sic*] mutually one to another, are figured together in a certain Species." These souls together are a dynamic production of these mobile spirits, which "heap themselves together," and eventually "frame the Body" itself. In yet another descriptive frenzy, the soul is "as it were the Specter, or the shadowy hag of the Body"; it is "intimately united with it . . . as its prop or stay," but "being made of a most subtil texture . . . , it cannot be perceived by our Senses."[17]

Clearly the hypostasis did little to settle Willis's language down. Instead, it provoked his poetical gifts of description, while leaving the mind-body problem a mystery. But this may well have suited his temperament. The hypostasis captured the curiosities of causality in the mental realm. We cannot see the soul, so we can know it only by its effects in the body. And yet we know too that these effects in the body—the operations of animal spirits—recursively constitute the soul. "By a Circular necessity," Willis puts it, body and soul emerge together.[18]

## NEURAL MONISM: THE MATERIALS OF THE ENLIGHTENMENT MIND

From the beginning, then, the sciences of mind spoke the language of self-organization. This was not the only language they spoke, of course. When the great rationalizer of metaphysics in the early eighteenth century, the German philosopher Christian Wolff, put psychology on the disciplinary map in the 1730s, he had little interest in the dynamic formation of mind.[19] But he was unusual in this regard, and by the 1740s, interest in just this dynamism was widespread. Along with it came the promiscuous descriptions of emergent mind—those fascinations with causal loops, mutual exchanges between mind and body, and so on—that we have already seen in Willis.

These descriptions were particularly rich in that area pioneered by Willis (but tellingly also ignored by Wolff), the physiology and neurology of the material mind. This was an enormous growth area in the eighteenth century. "Brainomania," one commenter has characterized the period's interests, and with some truth.[20] Take the example of the doctor Robert Whytt. A Scotsman—educated at Edinburgh in the 1730s and later a student of the great Dutch doctor Herman Boerhaave in Leiden—Whytt began his career as a researcher into the terrors of bladder stones. In the 1740s, he began to reject Boerhaave and the hydraulic model of the body that we saw him offer in the previous chapter. Whytt then turned to more general works on physiology, neurology, and the nervous system, and by the end of his life he was at the top of the Scottish medical establishment: member of the Royal Society, elected president of the Royal College of Physicians in London, and finally royal physician when the king traveled to Scotland.

For Whytt, nervous physiology made clear just how limited the mechanical hypothesis was for explaining the operations of mind and life more generally. Common experience, he insisted, consistently proved these limitations:

> When a candle is placed before the eyes, if one of them is covered with the hand, or any other opaque body, the pupil of the other will be observed immediately to become wider. . . . This *consent* in their motions must be inexplicable upon mechanical principles alone.

Eye dilation is autonomic. Pupils widen in the dark regardless of our wishes. But their dilation does not depend solely on external stimuli. As the candle experiment shows, you can easily produce an effect (dilation) that happens in spite of an adverse mechanical cause (more light). In a sense, the eye has to *choose* between the candle and the hand, and in doing so creates effects that run against simple input-output causality. The body is filled with similar causal knots. Yawning is involuntary, for example, and yet it is "so very catching, as frequently to go round a whole company." Even the *idea* of a yawn, he exclaimed, can do the same. Or take the human heart. "The contraction of the heart is indeed the cause of the motion of the blood," Whytt reported, and yet without the blood, "this action of the heart cannot be performed: these two causes, therefore, truly act in a circle, and may be considered mutually as cause and effect." In the case of the candle, involuntary motion happens without external cause, but with no act of will. In the case of the yawn, involuntary motion happens not just without consent, but without any apparent functional *need* for a yawn. And in the case of the heart, the system is fully circular, each a cause of the other. In a human body, Whytt wrote, "there is no mover that can properly be called FIRST."[21]

These causal puzzles did not trouble Whytt at all. On the contrary, such easily observed phenomena proved the existence of what Whytt called a sentient principle or being "which in a peculiar manner displays its powers in the brain, and, by means of the nerves, moves, actuates, and enlivens the whole machine." Like "plastic nature," this sentient principle was not precisely a soul in the traditional sense, although Whytt often used this language to discuss it. Nor was it exactly matter, since matter was, for Whytt, a dead thing. Nor, finally, was the sentient principle easily locatable in one place. It is, as he put it, "present, at least, wherever the nerves have their origin," minimally in the "large space of the brain," but also wherever else nerves travel in the body: the eyes, ears, heart, stomach, bladder, and more.[22] Only via the operations of such a principle could the real and empirical behavior of a body be imagined.

What we will call his neural monism—his refusal to distinguish between what Gary Hatfield calls "mental phenomena . . . [and] mental substance"—underlay the conflict that erupted between Whytt and central Europe's greatest doctor, Albrecht von Haller, in the early 1750s. When Haller discovered the ability of muscles to contract even after death, he took this as a sign that their actions were distinct from that of sensibility and enervation. But Whytt insisted on a simple equation: no sentient principle = no motion. Even the cut nerve must be sentient, in other words. The sentient soul looks much like our polyp of the last chapter, something that "may be cut up into as many pieces as the anatomist pleases." This made the Scottish doctor uncomfortable, and he denied believing it. Nevertheless, if "the soul . . . is not confined to the brain . . . but is present everywhere in the body," it is hard to see how he could be understood in any other way.[23] The body is so soaked by the soul that, after death, the soul-body continues its activity, at least for a time.

As should be clear, there was no trace in Whytt of a traditionally Christian or Aristotelian or Platonist soul. He neither divided the soul into parts, nor spoke of its immortality, nor ascribed to it *any* real qualities besides sentience and activity. Indeed, it is hard to say what kind of thing he thought the soul really was. As with Blumenbach's *Bildungstrieb*, however, this was not a weakness but a strength in Whytt's system. There is "no need of understanding the nature of the soul," he declared, for "it is sufficient, if we know from experience, that it feels . . . and has a power of moving the body." Or again:

> How, or in what manner, the will acts upon the voluntary muscles, so as to bring them into contraction, is a question beyond the reach of our faculties;

> and indeed, were it otherwise, *the answer would be here of no great importance*, it being sufficient that experience convinces us that the will is really possessed of that power.

As Vidal has noted, eighteenth-century psychology (like biology, as we have seen) was a resolutely antimetaphysical practice.[24] And we can see how little metaphysical commitment Whytt had to the doctrine of the soul and its powers. Instead, his commitments were phenomenological. Like agency in an economic system, the soul did not have to be explained but merely stipulated. Only its *activity*—the soul captured in actions through the body—has any relevance for a real cognitive science.

Even someone committed to the Christian soul, however, could find attractive the self-organization of the eighteenth-century mind. The Swedish nobleman Emanuel Swedenborg is better known for his religious creativity and inspiration—there are still active Swedenborgian churches in the United Kingdom, the United States, and other places—but his early writings were all on the natural sciences, and the longest of these was his treatise on the brain. Written in the early 1740s, shortly before he began to experience visions and prophetic dreams, the manuscript was lost until the nineteenth century, when the English Swedenborg society translated and published it (the only extant edition even now). Here, in a Christian register, we can find another version of eighteenth-century neural monism, another language of self-organization applied to the soul and its emergence.

Like Whytt, Swedenborg bound the soul irrevocably to embodied and phenomenal life. Indeed, Swedenborg could not conceive the soul absent a physical brain. The "soul designs and forms the whole of its body . . . by means of the cerebrum," but it does this immanently, in dynamic exchange with its own embodiment: "*The cerebrum . . . first represents in itself the idea or ideal of its soul, in the form and nature of which* . . . the body as its image is formed."[25] It is not that the soul makes the body as a carpenter might, according to some preexisting design. Rather, the body and soul are engaged in a reciprocal activity, the brain imagining a soul in whose image a body is formed.

This mutual construction he calls the "*circle of life*," a circularity that applies to cognition as well. Thinking happens because the nerves are "able to undergo infinite states of mutation, and . . . each mutation corresponds to an idea." As for Locke, so for Swedenborg is experience a precondition of thought:

> The phenomena of the surrounding visible world penetrate . . . from the external organs of sense to the inmost sensorium, or to the soul itself, and from this

> by voluntary determinations they flow out into actions; by a perpetual circle, according to the flux of the fibres . . . , they thus flow from the surrounding physical world back again into the soul. . . . This perpetual circle . . . resembles a spiral line, which perpetually bends upward and thus moves in a transcending line.

Leaving aside Swedenborg's prophetic moment—his vision of transcendent spiritual progress—what we find is an empiricist theory of cognition. At its center is the organ of the cerebrum, the "perpetual nexus of the lowest with the highest things," the anatomical medium across which the "transcendent circle" flows between the things of the world and the things of the mind.[26] The things of the world enter via the senses, precipitating a fluxing cascade of mutations along the neural fibers (in other words, "ideas"). In a sense, then, thought does not occur just in the *brain* but in the entire body. Or, to put it differently, the entire body itself becomes a brain.

Swedenborg's vision is that of a soul operating "*like a certain Deity of its microcosm or universe*."[27] Unlike God, however, the soul cannot make something from nothing. So its self-formation is a dynamic and material one—the development of its full potential is only achieved in its efforts to become embodied. Cognition too must therefore retain elements of both the spiritual and the material. Neurology is a divine science, a form of phenomenal theology.

We have seen before how the language of self-organization can bridge the intellectual chasms between the pious and the impious. Whytt found his sentient principle fairly agreeable to Christianity, and Swedenborg was a deeply if idiosyncratically pious man. Our third character from the 1740s was of an altogether different attitude, yet he too found self-organization a useful way of imagining the mind. When we last met the scandalous French doctor Julien Offray de la Mettrie, he was imputing self-organizing forces to organic matter. His 1747 "machine man" was a strange machine indeed, one whose "springs all wind each other up without us being able to tell at which point on the human circle nature began." Three years later, in his *Treatise on the Soul*, he again looked for an active principle, not above and beyond matter, but one bound into matter itself. "Matter is mobile. . . . It has the power to move by itself, and . . . is susceptible of sensation and feeling," La Mettrie declared, a deliberate slap in the face of both strict mechanists and the overly pious alike.[28]

As in his writings on life, on cognition La Mettrie seems at first of the party of the machine. The soul or mind is a product of bodily experience, sense organs exciting "diverse movements of the spirits" in the body's nerves. These spirits hurry through the nerves as little "globules" that push one another, throwing the last one "at the soul, which wakes up at this hammer blow and receives more or less vivid ideas, according the movement impressed on it." But as we have so often found, the apparent clarity of this mechanism grows cloudier the closer you look. In *principle*, the outside world forms our thoughts, but in *practice*, the process of this formation cannot be specified. In part this has to do with the nature of the world, whose objects bury many "properties . . . in [their] smallest elements." In part, though, this has to do with the nature of our nervous systems:

> Each sense has its own little department in the cerebral matter and . . . the seat of the soul is therefore made up of as many parts as there are different corresponding sensations. Who could enumerate them? And how many reasons there are to multiply and modify feeling to an infinite degree: the material of the nerve cases, which can be more or less solid, their softer or harder pulp, their more or less lax position, the difference in construction at one end and the other, etc. . . . Each nerve differs from the others at its origin.

In an important, even Swedenborgian sense, each nerve is unique. If this is mechanism, it is a Leibnizian mechanism, with each gear of the nervous machine itself a monad, irreducible to any of the others. The soul is made of innumerable sensations, none of which can be converted into any of the others. Experience is, la Mettrie wrote, the "true physical cause of all our ideas, but how extraordinarily small this cause is!"[29] Each minuscule cause vibrates along a unique nerve, multiplying feelings "to an infinite degree," as La Mettrie put it. Here again, in the world of self-organization, there is a qualitative difference between the aggregate and the individual, between the system of order and the components that mysteriously generate it.

In this system of order, every experience is unique, every node in the system independent. Inputs from the world cannot be regularly correlated with outputs from the brain and nervous system, since these are in a process of what La Mettrie calls "permanent modification." The causal nexus between mind and world becomes rather opaque:

> Our ideas come neither from a knowledge of the properties of bodies, nor from what the change experienced by our organs consists in. They are produced by that change alone. According to its nature and its degree, ideas arise in our souls which have no connection with their occasional or efficient causes.[30]

La Mettrie's point seems to be this: cognition is determined not by the world but by the "changes" that our organism produces in response to the world, changes that derive from the reactions of infinitely diverse and sensitive neural agents. In this sense, ideas "have no connection" to experience itself but instead are our own ideas, generated by a mind that emerges from and operates on its own mental events.

For all of these neural monists, these fundamental facts turn the brain into a zone of causal loops. In fact, the function of the brain is to host these loops. It is a medium for reciprocal acts of creation, what Georges Canguilhem described more generally as reflex. Sensations are, in his words, like "explosion[s] where the effects multiply the cause conformable to rules which are neither those of geometry nor of arithmetic." Causality is characterized by "flux and reflux," motions that turn back upon themselves in forms of circular doubling. In a system where, as with La Mettrie, each neural experience is unique, unpredictability is built in. Cognitive reactions to identical stimuli may well be different, given changing circumstances in the mind. The "spontaneous Actions" of the animal spirits, in Willis's terms, find parallel in that French doctor's vision of matter that "has the power to move by itself."[31] Self-organization indeed.

## COMPLEXITY AND A MECHANICAL THEORY OF MIND

In the context of our neural monists, David Hartley's *Observations on Man* (1749)—a book that, the American doctor Benjamin Rush claimed, proved "the indissoluble union between physiology, metaphysics, and Christianity"—was an apparent outlier. With geometry and God in equal measures, Hartley pursued a pious mechanism to the bitter end. The sacred cow of human freedom, he felt, must be sacrificed on the altar of the basic truth that "every Action . . . arises from previous Circumstances." If we fail to commit to this truth—insisting, for example, that "Action A, or its contrary a, can equally follow previous Circumstances"—then we effectively "destroy the Foundation of all general abstract reasoning." Even more troubling, we destroy piety and salvation itself. The amelioration of sin, "correcting what is amiss, and improving what is right," demands that the cognitive process be deducible in a metaphysically secure way. Without linear causality, moreover, there would be no way to "reduce the State of those who have eaten of the Tree of Knowledge of Good and Evil, back again to a paradisiacal one."[32]

Arid ground for the lush complexities of our neural monists. Yet even here, perhaps despite Hartley's best intentions, we can find the language of self-organization. Start at the beginning, Hartley advised, with the simplicity of a perception. With it, A causes B: vibrations are triggered in the "Aether residing in the pores of [the] Nerves." These vibrations, in turn, trigger other vibrations in the "medullary Substance" that coexists with this aether in the nerve, and in turn, this last vibration wiggles into the brain. Once there, something mysterious happens. A sensation happens. This effect, however, he supposes to be "immechanical and to arise from the immediate Agency of God," since it is "impossible . . . to discover in what Way Vibrations cause, or are connected with Sensations, or Ideas."[33]

We can hear an echo of an earlier pious mechanist here. Unlike Malebranche, however, Hartley felt compelled to explore the nature of mind further. Vibrations, it turns out, are not nearly as simple as you would think. Repeated vibrations, for example, produce what he calls a "Disposition to diminutive Vibrations . . . Vibratiuncles and Miniatures." Some nerves seem more likely to vibrate one way than another: they have a "texture" that somehow preserves the traces of their sensation histories. These traces he calls, with a telling tip of the hat to the world of accident and superstition, "preternatural" vibrations, since they seem to exist beyond the reach of natural causality. If the same object is "impressed again and again, for a sufficient number of times," the medullary substance "will not return to its natural original State . . . but remain in the preternatural State."[34] The preternatural state (as the name indicates) violates the "A causes B" calculus. In some circumstances, A no longer causes B, as when we grow so accustomed to a sensation that we stop noticing it. And in some circumstances, B happens regardless of A, as when one thought triggers another one, without the original impulse anywhere to be found.

Preternatural activity was not an exception to Hartley's nervous theory. It was crucial to it. Without it, he had no account for how thoughts can be provoked by other thoughts, or how memory works, or how ideas relate to one another. Yet this power of association is fundamentally nonmechanical:

> The Power of Association is founded upon, and necessarily requires, the previous Power of forming Ideas, and miniature Vibrations. . . . But then (which is very remarkable) this Power of forming Ideas, and their corresponding miniature Vibrations, does equally presuppose the Power of Association.

In other words, the mind's fundamental parts are not linked together in a line, but rather rotate around each other in a "mutual indefinite Implication." The mind organizes itself, as associations create ideas that create associations and so on. Nor do the complexities stop here. Once the associative engines of mind start chugging, they produce not just complex ideas that seem to have no relation to their parts, but even "decomplex" or hypercomplex ideas whose relations to sensation are so attenuated as to be wholly unavailable to human scrutiny.[35] Metaphysical commitments, however deeply felt, were no barrier for self-organization, no barrier for the eighteenth-century appreciation of the complexities of mind.

## HUME AND THE SCIENCE OF THE *JE NE SAIS QUOI*

> People don't like to take ticket #1 in a lottery. Take it, reason loudly says, it can win the 12,000 dollars as well as any other one; don't take it under any circumstances, whispers a *je ne sçai quoi* . . . and it will not be taken.
>
> GEORG LICHTENBERG, *F814*

Eighteenth-century interest in cognition was not limited to material, physiological, and neural aspects. The new Enlightenment mind also demanded inquiry into the nature of its own highest operations: reason and the science that it pursues, philosophy. Already in Descartes, understanding the nature of mind was intimately connected to the practice of philosophical inquiry. But if the philosophy of the seventeenth century formed around the desire for deductive certainty, that of the Enlightenment dreamed of discovering the "nature and potentiality of thought itself," as Ernst Cassirer once wrote, seeing reason as a "kind of energy, a force which is fully comprehensible only in its agency and effects."[36] Not a metaphysical first principle, reason is here, we discover, understood as emergent.

Few were as suspicious of metaphysical first principles as David Hume, a figure who will take center stage in the next chapter. The Scottish bon vivant had no interest in brain science, but he had no doubt that the mind was a product of embodied experience in the world. Indeed, the job of philosophy was to figure out exactly how this product came to be. Philosophy must understand what minds are by looking at what they *do* in the world—it must tell stories of how a mind and its qualities are generated in and through its encounter with a world.

Take causality. We have spent a lot of time in this book arguing that metaphysically satisfying understandings of causality—that, e.g., every effect proceeds from some necessary cause—failed in the self-organizing systems that the eighteenth century was so eager to describe. Hume was the first philosopher we know who discarded the metaphysics of causality altogether, insisting in his *Treatise on Human Nature* (1739–40) that it is not a cause but a *consequence* of the human experience. Causality is not a necessary feature of the world, a first principle that starts an infinite regress back to some original Unmoved Mover. It is a feature of our *minds* as they undergo a set of experiences that repeat themselves over time, *this* happening after *that* over and over again in a process he called a "constant conjunction."[37]

Needless to say, the sense of causal relation that conjunction creates in our minds is no product of reason. Rather, it is "produc'd by a number of past impressions and conjunctions," which "arises immediately, without any new operation of the reason or imagination." Truths that we take as fundamental are contingent, accidents of circumstance and habit. In fact, it is *because* of its contingent nature—because, as Hume puts it, "I *never am conscious*" of reasoning my way to causality; because causality seems so self-evidently true; because I never notice where my certainty comes from; because causality arises seemingly of itself—that causality feels so fundamental to our mental constitution.[38]

The deep nature of an embodied mind is thus shrouded in mystery. "We cannot penetrate into the reason of the conjunction," Hume wrote. Instead, "we only observe the thing itself, and always find that from the constant conjunction the objects acquire a union in the imagination." Nor is this process of habituation available to objective scrutiny. Experience, as Hume says, "may operate in our mind in such an insensible manner as never to be taken notice of." It may "produce a belief . . . by a secret operation." This is even true in cases of self-reflection, when we consider our own active minds and detect a "certain *je ne sçai quoi*, of which 'tis impossible to give any definition or description, but which everyone sufficiently understands."[39]

Echoing Leibniz above, Hume's cognitive theory was a science of the *je ne sais quoi*, what Richard Scholar has called the "experiences that elude explanation." Already Daniel Defoe had found commercial credit to be an "invisible *je ne scay quoi*," and Jean Le Clerc had spoken of a similar *je ne sais quoi* in the fortunes of lottery winners, since the operations of both were opaque to rational calculation.[40] Hume's analysis of mind, however, made *all* the things of the world opaque to human knowers, for a thing does not reveal "either the causes which produced it, or the effects that will arise from it." Suppose a man finds "a watch or any other machine on a desert island," Hume asks, what could he conclude? Here he recalled the traces on the Rhodian shore—the shapes drawn in beach sand whose deliberate structure testifies to the presence of an intelligent design—only to wipe them away. For Hume, the watch disclosed nothing beyond what we bring to its discovery. It is mute. No "mere operation of our reason" can find anything in objects themselves, since the "ultimate springs and principles" of things "are totally shut up from human curiosity and enquiry." The "scenes of the universe are continually shifting . . . but the power or force, which actuates the whole, is entirely concealed from us," Hume declared in his *Enquiry concerning Human Understanding* (1748).[41]

In this work, he spoke of the "secret springs and principles" of the mind and hoped to discover its "laws and forces," much as Newton had done for astronomy. Despite such hopes, he had to admit that it was not just objects that resist explanation. Even *our own minds* are mystifying. First, error is always a possibility. Thought can "excit[e] any idea it pleases"; it is not fully in control of its own functions, since its "motion is seldom direct, and naturally turns a little to the one side or the other." The effects of this mental Lucretian swerve can be corrected (by, say, good philosophy), but the source is never eliminated. Second, we cannot explain our own mental processes. We make inferences from experience—B followed A, hence A caused B, hence all As cause Bs, etc.—but why does the inferential chain happen? "There is required a medium, which may enable my mind to draw such an inference . . . [but] what that medium is, I must confess, passes my comprehension." The human mind and its capacity to *have* ideas at all is, he concludes, a "real creation; a production of something out of nothing."[42] The mind that emerges over time, through experience, without necessary direction—the self-organizing mind—can never be fully present to itself. Its final springs remain hidden.

There is something dispiriting, "deflationary," as one Hume scholar calls it, about this whole story, especially as it might apply to our understanding of ourselves as unique and enduring persons. The more you reach down toward that thing called "identity," some mental constancy, the more it recedes into obscurity. But hope is not altogether lost. The science of a self-organizing mind must labor in the field of contingency—mysteries and swerves again—yet even in "our wildest and most wandering reveries, nay in our very dreams," we discover something like an order. The order is not a deductive order, derived from necessary first principles, but a *descriptive* order that starts in medias res. And it is a provisional order, one that does not try to penetrate *behind* phenomena to some inner truths. Instead it proceeds, as Hume puts it, by a "species of ANALOGY," playing along the surfaces of phenomena and their appearances in time.[43]

## THE METHOD OF ANALOGY

There are analogies all over this book, from the world of living organisms to the operations of economic systems, as thinkers feverishly used one kind of order to understand another. But the philosophy of the mind-emergent elevated analogy from unspoken practice to deliberate method. Analogy insists that things be understood comparatively; it is a *correlative* rather than deductive method. And, for the Enlightenment anyway, it insists on what we might think of as a fictional or at least constructive element in the cognitive process. Analogies do not inhere in the world—as perhaps they did in the Middle Ages and Renaissance—but instead are the mental tools for world-ordering, the means by which the mind makes the mysterious leap from raw experience to complex thought.[44]

"I strongly favour inquiry into analogies," Leibniz wrote circa 1704, but it was Leibniz's grudging admirer, the French philosophe Étienne Bonnot de Condillac, who in the 1740s really began to pursue analogy as an engine of mental development. If Hume was a critic, Condillac was a constructivist, building models of mental order in service of a sensationalist theory of mind. In his *Essay on the Origin of Human Knowledge* (1746), analogy glues together what is, in the beginning, a mere series of disparate things. From inchoate yet iterated experience, one thing begins to recall another. As it does so, it recalls too all those things that are associated with it. Simple perceptions grow into larger complex ideas, which then can themselves be the subjects of more analogical comparisons. Cognitive systems are thus built of "chains whose strength will lie entirely in the analogy of the signs," things bound together in relations of resemblance, rather than identity. Over time, repeated experiences and comparisons allow us to recognize that we are "a being that is constantly the same 'self.'"[45] We are, in a sense, creatures of analogy.

Condillac was far more confident than Hume about the powers of philosophical analysis. He spoke of "a sort of calculus" by which ideas might be "compos[ed] and decompos[ed]" to discover their various combinations and components. Even so, he recognized the curious difficulty that we face when we try to translate between our knowledge of things and our experience of them:

> When I see a bas-relief, I know, without any doubt, that it is painted on a flat surface. . . . yet this knowledge, repeated experience, and all the judgments I can make do not prevent me from seeing convex figures. Why does this appearance persist?[46]

Here the calculus will not do. No matter how much we compose or decompose our ideas, we cannot help but see the bas-relief *as if* it had depth. We know it is flat, but we see it otherwise. Perception cannot be transmuted by any simple additive process into cognition. Nor can the process be reversed. Instead, experience and knowledge coexist as parallel yet connected elements of the human mind.

Analogy allowed for just these sorts of loose, nonhierarchical connections between things. As such, it was a particularly apt way of approaching the mind. When Leibniz spoke of the "harmony between souls and bodies, such that each perfectly follows its own laws without being disturbed by the other," this might profitably be understood as a species of analogical thinking: the mental and the material move in a synchronized rather than a causally determined way. More generally, analogy connects things (like souls and bodies) across different realms. Thus Maupertuis, in his *System of Nature*, comments that because it is "impossible to know by experience" species that are different from us, we have to judge them "by analogy." This constructive process bridges the gap between "elementary perceptions" and the real order of things. It accomplishes that alchemical transformation of aggregate data into a system more complex than the sum of its parts. This can be problematic, since the elements then lose their particularity, their "*sentiment particulier du soi*," as Maupertuis puts it.[47] But it is also necessary for comprehension at all, letting us see self-organization without having to explain it away.

Analogies abounded in self-organizing systems, we suggest, because such systems require techniques of description free from the rigors of deductive certainty. Analogies assume a principle of nonidentity between things. Indeed, they assume the *impossibility* of reducing one thing to another. Their order instead emerges from the process of comparison itself and the extended description of the symmetries and dissonances between things. Observations are made. Then observations about the observations are made. The two processes are related, but not deductively. The process of relating is contingent. It requires an active observer. Discovering similarities demands time and attention, and is bound by circumstance. The method of analogy is, in short, a generative principle that *creates* order from within the process of description itself.

## THE CO-CONSTITUTION OF MIND AND LANGUAGE

Analogy models the emergence of mind as a dynamic process, responsive to the environment around it. This emergence is unpredictable, since nothing in raw experience necessitates any particular form of consciousness. And sometimes it might fail altogether. We can imagine a person, for example, whose experiences were so randomly distributed that no personality, no *mind* at all emerged. Condillac actually knew of such a person, a Russian child raised by bears whose mind remained, despite years of biological existence, a blank. No reason; no memory; no reflection; no sense of self; no human cognition at all. And why not? Because bears cannot provide any connective tissue between the child's mind and the world: "Their roar does not have sufficient analogy with the human voice."[48]

This story turned out to be more powerful than one might suppose. When Condillac connected mind to language, he started a ball rolling that has not stopped even today. In his own context, his *Essay* not only inspired Rousseau to pen a brilliant work on the origins of language (circa 1761). It also inspired the Berlin Academy of Sciences to challenge Europe's scholars in 1769 to determine whether "men abandoned to their natural faculties are capable of inventing language."[49] Thirty-one essays were submitted, and the prize went to Johann Gottfried Herder, whose *Essay on the Origin of Language* would go on to much success. The connection of cognition to language began here, in eighteenth-century efforts to understand the emergence of mind.

Condillac's insight was as powerful as it was (from a linear perspective) paradoxical. Language, his *Essay* announced, was both cause *and* consequence of human mental operations. It neither arises from some extant natural faculty nor is a gift from God. Its origin lies elsewhere, in the mysterious passage from nothing to something that characterizes self-organizing systems. The first humans had no language, Condillac's story goes, and so "connections did not last long" between perceptions and consciousness. But one day they found themselves together, and began to observe the "cries of passion" each made when suffering, and to feel "suffused with sentiments" at them. Pity and mutual company conspired

> to make it habitual for them to connect the cries of the passions and the different motions of the body to the perceptions which they expressed . . . [and so this] use of signs gradually extended the exercise of the operations of the soul, and they in turn . . . improved the signs and made them more familiar.

Any single word cannot arise "except by chance," but over long generations—duration again playing its role—signs and mind "mutually assist each other" in growing complexity.[50] Language is an emergent thing, co-constituted with human cognition, each advance reinforcing the advance of the other, until finally the mind and language together leap free of their physical matrix. From simple nature, humans bootstrap themselves into complex culture.

Contemporary critics of Condillac could not fail to notice how metaphysically unsatisfying this account was, how the various parts took each other for granted. Rousseau, for example, charged that Condillac "assumed what I question," namely, the existence of higher mental processes, language, and society in primitive man. He assumed, in other words, the presence of *something* when there should rightfully be *nothing*. But Rousseau's effort to get back to the "nothing" only made the puzzle more perplexing. The mind boggles at the problem of origins, he admitted:

> Let us for a moment cross the immense distance that must have separated the pure state of nature from the need for Languages. . . . New difficulty, even worse than the preceding one; for if Men needed speech in order to learn how to think, they needed even more to know how to think in order to find the art of speech.

"The events I have to describe could have occurred in several ways," he wrote. "I can choose between them only on the basis of conjectures." The problem, as he saw it, was that these higher-order faculties had such tiny seeds. The "astonishing power of very slight causes"—and in Rousseau's case, this meant accidental encounters in the woods, the building of the first huts, and so on—always made deductive arguments tentative and incomplete.[51]

Herder was no less irritated yet no less reliant on less-than-deductive logics of emergence. "Words arose because words were there before they were there," he sniffed at Condillac. His own 1772 essay, however, was filled with similarly miraculous moments of self-production. Its beginning described how human cries are like chords "whose sound and strain are not ruled by will and slow deliberation," but by a "sympathy" that resonates among all who hear. And its moral was told wholly in the language of self-organization—that man is by nature a creature who reflects, that from "the entire floating dream of images" a miraculous moment of care (*Besonnenheit*) allows us to select one "mark," and with this mark "eureka! . . . human language is invented." Language is a "self-made sense of the mind," emerging out of the very activities that make humans human.[52]

Aha, it is tempting to say, caught in your own trap. But to criticize Rousseau and Herder (much less Condillac, as much later Jacques Derrida was wont to do) for metaphysical incompletion misses the point.[53] Co-constitution; circularity; mysterious emergences: these were not liabilities but assets in the language of self-organization. Once the cognitive faculties were cut free of their a priori status, their origin became contingent and their development dynamically related to their settings.

The body and brain, as we have seen, was one such setting. Just as important were the other complex systems into which minds were integrated.

Condillac's interest in the sociability of mind—where the observation of others, and the observation of oneself observing others, was made crucial to self-formation—was widely shared in the eighteenth century. Indeed, the entire language of sympathy (to be discussed further in chapter 6) migrated from medical to social analysis in just this period, in an effort to show how complex mental attributes like virtue might arise unexpectedly from one mind observing another. In these systems, minds grew out of their interactions with other minds. When philosophy migrated out of the study and into the complexities of the world, theories of mind followed. No longer autonomous, mind was yoked to other complex orders of the world, *collectivized* in ways unimaginable for someone like Descartes.

## OBSERVING OBSERVATION: THE MATHEMATICS OF MIND

This collectivization can be seen in other ways too. The seventeenth century saw the invention of "apperception," for example, Leibniz's term for the mind's special awareness of itself. By the middle of the eighteenth century, the observation of observation had become a stock issue in the analysis of mind. The German translator of Charles Bonnet's *Analytical Essay on the Faculties of the Soul* (1760)—the Jena philosopher Christian Schütz—named this as the first difficulty of any psychology, that the mind must "learn its own essence and its powers out of its [own] operations." "Sustained self-observation," Jean Trembley later announced, is the foundation of cognitive research.[54] The observation of observation could be pursued with ever more focused self-scrutiny (on attention, see below). It could also be pursued in the *aggregate*, the observation of observation made scientific through the application of large numbers.

Earlier in the book, we discussed some features of early probability theory, especially its efforts to apply mathematical and statistical methods to the observation of natural phenomena. These quantitative methods in the seventeenth century revealed, as we saw, regularities and orders invisible at the level of an individual event. By the mid-eighteenth century, probability theory would become ever more recursive, applied not only to phenomena outside the mind but to those inside the mind as well. If knowledge grows with the mind through a concatenation of experiences, then there is an irreducibly contingent aspect to it, a contingency that can be managed by the same tools used to manage other contingencies. Beyond a science of things, probability theory in the eighteenth century became a cognitive science, a science addressed to our *certainties* about things.

The question was this: How certain should I be about the certainties that my mind feels about things? Jacob Bernoulli had already asked the question but had not come up with the answer. In his *Essay towards Solving a Problem in the Doctrine of Chances* (written in the 1740s but only published posthumously in 1764), however, the clergyman Thomas Bayes attacked this so-called inverse problem in probability again. Specifically he resolved, as his publisher and friend, the preacher Richard Price, put it, to "give a clear account of the strength of *analogical* or *inductive reasoning*." Although Price used the term "analogy" loosely, the meaning was clear. What was needed was a second-order operation, the determination of the probability of a probability. We must, Price wrote, figure out how to determine "in what degree repeated experiments confirm a conclusion." That is, what degree of certainty should we feel about our own beliefs given any given amount of confirming (or disconfirming) data? Only by knowing this can we know how much faith to put in the proposition that there are "in the constitution of things fixt laws according to which events happen."[55] In his essay on chances, Bayes solved this inverse problem, thereby quantifying certainty in a more-or-less precise way. If the late-seventeenth-century probability project taught that we could discover an order to apparently random things in the *world*, then, the late-eighteenth-century project taught that we could discover this order in the *mind* as well.

For optimists, the quantification of certainty seemed like the tool needed to let our self-organizing minds transcend the vagaries of experience and attain the reality of a determinate natural order. The French mathematician the marquis de Condorcet was delighted that our knowledge "is founded on probabilities, the value of which it is possible to determine with a kind of precision." His optimism about quantification extended from the human mind to political deliberation and social planning. As far as cognition went, he looked forward to using probabilistic techniques to understand "the technical or even mechanical means of executing intellectual operations" and thus to developing a science of rational decision making.[56]

But this optimism was never unguarded. To be sure, Condorcet wrote, "the events which one attributes to chance are determined by invariable and necessary laws." And yet, he continued, "the expression of these laws is too complicated for us to discover." Outside of mathematics, "there is not one sort of proposition that is certain. . . . Everything is probable." In practice, this means that we cannot reason about most things "as if, in calculating the operations of a great hydraulic machine, one had restricted oneself to the simple application of the general principles of mechanics." Even the most powerful truths are "susceptible of exceptions and even of modifications." And this is because these truths are themselves products of the human mind. Probability is, as Condorcet often insists, a "purely intellectual consideration."[57] As a result, we are liable to constant error, our minds crying out for correction.

Was correction even possible? Even as Bayes was writing his *Essay*, Hume raised his eyebrows at the idea. He doubted, for example, that either better knowledge of statistical outcomes or a better mathematics of certainty could do anything to sway our own cognitive commitments: "What effect [can] a superior combination of chances . . . have on the mind"? Belief and assent can be produced "neither by *demonstration* nor *probability*." In the 1760s, this discussion became concrete, when the great mathematicians Jean le Rond d'Alembert and Daniel Bernoulli quarreled over an issue already familiar to us. As we saw in chapter 1, inoculation for smallpox made concrete the consequences of statistical thinking. Earlier writers had come up with mathematical rationales for inoculation. Only with d'Alembert do we see an effort to *challenge* inoculation on mathematical terms. It was not just, for d'Alembert, that the mathematics of inoculation ran aground on the reality of moral experience—in Daston's words, the "psychology of risk-taking."[58] It was that these mathematics also assumed what they needed to prove, namely, that expectations are evenly distributed, and the value of one life equivalent to that of another.

The problem is that they are *not*. An ancient grandfather would seem to have less to lose than a teenager; a terminally ill child less than a middle-aged adult; a dullard less than a genius. Moreover, the "value" we are discussing depends on our perceptions of it. If the teenager is suicidal, then perhaps that grandfather's worth is *higher*; perhaps the dullard will live an honest life but the genius one of depravity. Even the strongest probabilistic pedagogy—teaching people, for example, about the objective risks of contracting smallpox from inoculation—can do little to stabilize the value of a life, since this is grounded not in the aggregate but in the very personal domain of individual experience and our feelings about it.

The cognitive application of probability, moreover, again highlighted the asymmetry between the truths of experience and the truths of the world. Real things in the world do not behave according to chance or hazard, virtually all of these scientists would have agreed, but we humans need to describe them as if they did. As a result, as in analogies, there is a necessarily constructed, even fictional aspect to our knowledge. Like analogies, too, probabilistic truths are both approximate and retroactive, determined only after a series of observations are made. Like analogies, finally, they are dynamic in their nature, necessarily changing when new perturbances, unlikely events, and prior errors are discovered. They may approach a thorough description of the world, but only asymptotically, over infinite time and with infinite resources.

And so, to return to Bayes's *Essay*, we have to notice that its quantification of uncertainty only redoubled the presence of the probable. Price saw this already. Suppose we have a die whose properties are unknown but whose roll produces the same side a million times. Intuition might tell us that there is just one side to this die. But actually all we can say is that there is a better than even chance that the die has 1.4 million more of these sides than any other. And, he calculated, there is a *less* than even chance that it has 1.6 million more of these sides. "It should not . . . be imagined that any number of such experiments can give sufficient reason for thinking that it would *never* turn any other side," Price insisted.[59] Indeed, he was eager to point out, we can *never* end the process of experimentation, never arrive at certainty.

The die is an analogy for events in the world. The falling of a stone, the burning of wood in the fire, even the rising of the sun tomorrow morning: these are outcomes we can reasonably expect, but "no finite number of returns would be sufficient to produce absolute or physical certainty" about them. We can have "no reason for thinking that there are no causes in nature" that might not change these outcomes. Price's convoluted prose was significant. We may have no reason for thinking that there *are* causes in nature that will prevent the sun from rising tomorrow. But with no reason *not* to think they exist, either, the mind that emerges from uncertain experiences—even the reasonable mind—is in limbo between certainty and ignorance. While inverse probabilities do "make common sense more exact," as Daston points out, they never convert probability into certainty.[60] Instead, they add another layer of probabilistic reasoning between the mind and its experiences in the world, an order whose claims to correspondence with reality are suspended.

Here, as elsewhere, we see again how the language of self-organization bridged secular and religious interests. "Vanity of vanities; all is vanity," said the Ecclesiastes author, and the pious Price felt similarly relieved to discover that the world hides its ultimate secrets from our prying eyes:

> The greatest uniformity and frequency of experience will not afford a proper *proof*, that an event will happen in a future trial, or even render it so much as probable, that it will *always* happen in all future trials. . . . For aught we know, there may be occasions on which it will fail, and secret causes in the frame of things which *sometimes* may counteract those by which it is produced.[61]

By "occasions," Price meant miracles, the sudden eruption of divine contingency into the natural and expected order of things. The probabilistic natures of cognition happily left the human mind open to an unanticipated future, to the secret operations of God.

This combination of materialism and piety shadowed the Bayesian transposition of probability inward. Long before Price, in fact, Bayes was picked up by a thinker already familiar to us. "An ingenious Friend," our cognitive mechanist David Hartley wrote,

> has communicated to me a Solution of the inverse Problem, in which he has shewn what the Expectation is, when an Event has happened $p$ times, and failed $q$ times, that the original Ratio of the Causes for the Happening or Failing of an Event should deviate in any given Degree from that of $p$ to $q$. . . . From this Solution . . . we may hope to determine the Proportions, and, by degrees, the whole Nature, of unknown Causes, by a sufficient Observation of their Effects.

The Bayesian turn arrived late in his 1749 book, and it helped him outline norms for the cognitive investigation of the world. Only the observation of observation, what he called the "Method of Induction and Analogy," could make sense of human experience and thought. Induction was a special case of analogy, Bayes showed, where we have enough data to feel certain about our certainties. Where data are fewer, then analogy proper takes over. At this point, we circulate back and forth between the familiar and the unfamiliar, drawing distinctions and similarities that are less secured by things themselves than in our process of comparison itself. "The analogous Natures of all the Things about us, are a great Assistance in decyphering their Properties, Powers, Laws, &c," Hartley wrote, and "thus all Things become Comment on each other in an endless Reciprocation."[62]

In important ways, then, eighteenth-century probability theory was a cognitive science, developed with an appreciation of the mind's contingent and complex relationship to the world. This was true for Hartley, for Condorcet, for Price, and it was also true for the century's greatest statistical thinker, Pierre Laplace. Laplace was a giant in statistical mathematics, someone who solved the inverse problem in 1774 independently from Bayes, one of the earliest developers of a rigorous mathematics of error, and also someone deeply committed to the regularity and predictability of the world. "The present state of the universe," he famously wrote in his *Philosophical Essay on Probabilities* (1814), is "the effect of its previous state and . . . the cause of that which is to follow," and so for an infinite intelligence (God?), "nothing would be uncertain, and the future, like the past, would be open to its eyes."[63]

Such determinism—so antithetical to the strange causalities of self-organizing systems—was tempered, however, by a real recognition of our human limits. The "actions of the ocean, of the atmosphere and of the meteors, earthquakes and volcanic eruptions": *for us*, the world can be quite the contingent place, one whose "order is perturbed by various causes," causes "impossible to submit to the calculus." Even more puzzling than the world, for Laplace, was the human mind. His *Philosophical Essay on Probabilities* thus moved from his thoughts about world to his thoughts about decision making, and finally to his considerations of psychology. This final domain was the most mysterious and complex. Objects leave impressions, Laplace wrote,

> which *in an unknown manner*, modify the *sensorium* or seat of feeling and thought. The external senses *can learn nothing* of the nature of these modifications, astonished by their infinite variety and the distinction and the order that they maintain in the small space that includes them.[64]

Constrained though we are, however, we are not hopeless. As with nature, we can observe our observations. We can create systems of explanation that are open and dynamic, systems responsive to our own ignorance and to the surprises that inevitably characterize human experience.

Again, and finally now, these systems will proceed not deductively but in chains of analogies. From the natural world to the inner world, and back again, probability and analogy go hand in hand. "Analogy is based on the probability that similar things have the same kinds of causes and produce the same effects," Laplace wrote. Where does he see this? In consciousness:

> The probability of the existence of consciousness . . . decreases as the similarity of the organs to ours decreases, but it is always very great, even in the case of insects. On seeing those of one and the same species carrying out very complicated tasks in exactly the same manner for generation upon generation and without any teaching, one is led to believe that they behave by a kind of affinity, analogous to that which draws crystalline molecules together, but which, blending itself with a collective animal consciousness, produces many most singular combinations with the regularity of chemical combination.

Affinity and analogy, nondeductive relations again, preserve at once the regularities of natural law and the singularities of its observed effects. This complex of linkages articulates various explanatory levels without subordination. In the case of human cognition, Laplace writes, we can call this "sympathy," the "tendency of all similarly constituted beings to live together in harmony." Just as "in a system of resonant strings, the vibrations of one cause all its harmonics to resonate," so too can we see resonances in the human soul. Socially this is true: "Sympathetic feelings, quickened at the same time in a large number of people, are increased by their reaction on one another." And individually too, since the brain connects "all things that have had a . . . regularly successive existence of the sensorium . . . such that the recurrence of one of them calls the others to mind."[65]

Like turtles beneath turtles, it is analogies all the way down. Associations breed associations; these produce cognitive effects that are inaccessible to any deductive schema. The sensorium, Laplace remarks, "is able to receive impressions too weak to be experienced, but sufficient to give rise to actions of whose causes we are ignorant."[66] First-order theoretical knowledge obtains in the world of mathematics. But for phenomenal beings like us, whose habits of mind organize themselves from the materials of their unpredictable experiences, the observation of observation must suffice. This distant viewing opens the space for the self-organizing mind. It allows an aggregated order to stand in distinction from the operations of its parts, and in turn allows these parts to articulate freely of a (humanly discernable) determinate order. Regularity and contingency, complexity and simplicity, causal law and experiential freedom: these contradictory elements cohere inside self-organization.

## PAYING ATTENTION

Stories of self-organization, we have seen, typically tacked back and forth between two analytical perspectives. One described order in the aggregate and investigated the behavior of complex systems *distantly*. The other took a much more intimate view of order and organization, curious about the phenomenology of the individual element. Exactly because there are no easily deduced connections between the two perspectives, stories of self-organization had to be told from various, sometimes conflicting angles. Interests in the collective nature of mind described in the past few sections were thus balanced by a new interest in the mind's experience. It is one thing to say that a mind is formed in dialogue with language or that we need mathematical tools of aggregation for making sense of its outputs. It is quite another to *inhabit* that mind, to imagine its self-creating dynamics from within.

Entire genres of writing emerged in the eighteenth century in order to re-create exactly this—the novel, for instance. Take *Tristram Shandy* (1759), which might be profitably understood as Sterne's puzzled and satirical effort to understand how a mind might constitute itself out of the accidents of human experience. If only his parents, Tristram exclaims at the story's opening, had recalled that they were engaged in the "production of a rational Being"! Perhaps then they might have been more careful about his experiences, since

> nine parts in ten of a man's sense or his nonsense, his successes and miscarriages in this world depend upon [the] motions and activity [of the animal spirits], and the different tracks and trains you put them into; so that when they are once set a-going, whether right or wrong, 'tis not a half-penny matter,—away they go cluttering like hey-go mad.[67]

The entire novel ruminates on this "cluttering," the endless pile of accidents that distract the storyteller, the reader, and our hero from an easily reconstructed story of self-formation. Not that the "rational Being" doesn't finally emerge—Tristram as the narrator at least assures us of this—but the being that he becomes is a creature of distraction.

The comedy of distraction pivoted off a subject fundamental to Enlightenment cognitive theory: attention. Already we saw in Locke a curiosity about the mind's ability to construct its experiences and feed them back into its own cognitive architecture. A few decades later, attention became a standard entry in psychology textbooks. In 1720, for example, the German philosopher Christian Wolff grew interested in quantifying degrees of attention in his early "psychometrics."[68] By the 1740s, sensationalist theorists like Condillac developed a rich phenomenology of attention in an effort to understand how a mind's *own* energies shape its emergence from the busy chaos of the world.

> Imagine someone at a theater performance where a multitude of objects seem to fight for his attention—his soul will be attacked by a large number of perceptions of which he certainly takes notice. But little by little some will please and interest him more than others. . . . From then on he will be less affected by the others; his consciousness of them will insensibly diminish. . . . This operation by which the consciousness in response to certain perceptions becomes so lively that they seem to be the only ones of which we take notice, I call "attention."[69]

As a model of human experience, the theater revealed both our vulnerability to the world and our ability to reorient ourselves cognitively inside it. We enter the world as slaves of perceptions. Only as we attend to particular ones, selecting between those that interest and those that do not, do we become their masters. As we do so, however, we build a zone of cognitive comfort—we become insensible, as he put it, to that which does not please. By paying attention, we distance ourselves from the world and make its perplexing variety manageable.

In this matter of attention, Condillac was more or less a Leibnizian. That oceanic feeling that we have already seen several times was tied, in fact, to an interest in attention. In Leibniz's terms:

> While our senses respond to everything, our soul cannot pay attention to every particular. That is why our confused sensations are the result of a variety of perceptions. This variety is infinite. It is almost like the confused murmuring which is heard by those who approach the shore of a sea. It comes from the continual beatings of innumerable waves.[70]

The soul, for Leibniz, is caught in the space between attention and perception. It cannot attend to what it does not perceive. Nor, however, can it attend to *all* that it perceives. Attention is exactly the name for what discriminates between nothing and all. To pay attention is to select, from the "continual beatings of innumerable waves," *one* wave to consider above all others, knowing full well that this one wave is only part of an ocean of movement and change. The fact that we can pay attention at all marks the difference between a mind and a world. Every time we choose to think of this, and not that, we affirm this difference.

In a sense, attention is what distinguishes *me* from the world. This was Condillac's point as well. Although he said (like a good empiricist) that "perception and consciousness are different names for the same operation," he does not seem to have believed it. The difference between them, after all, is not just quantitative but qualitative. "When objects attract our attention," these particular perceptions make us "recognize them . . . as affecting a being that is constantly the same 'self.'" In this way, "consciousness is a new operation which is . . . the foundation of our experience." Without attention, no consciousness. Life becomes a series of disjointed perceptions, none related to the next. With attention, though, the mind sets itself free of the world. It becomes able to "direct its attention" to particular things that it wants to engage. As Diderot wrote in the 1751 *Encyclopédie*, attention is a "kind of microscope" that shows us a "thousand attributes" of the world that "escape from a distracted view."[71] Distraction is a Tristram Shandy–mind, bouncing from sensation to sensation and slave to the world. Through attention, a mind becomes its own mind.

Attention served as both proof and agent of the dynamic constitution of mind. Later in the century, Herder said just this. In his 1772 *Essay on the Origin of Language*, for example, a human becomes human only when

> the power of his mind acts so freely that it can distinguish *one* wave from the entire ocean of sensations that roar through all the senses, can hold onto it, concentrate its attention to it, and be conscious that it is being attentive.

More oceans and waves, here, but it was a thought that Herder sustained over long periods of thought. His *Ideas toward a Philosophy of the History of Mankind* (1784–91) repeated the idea nearly verbatim, that the mind "summons a form out of the chaos of things that surround it, on which it fastens itself with attention; thus it creates through its inner power a unity from the many."[72] Attention is a form of both world- and self-creation. From a concatenation of things in the world, it selects some few for significance, and in this moment of discrimination produces the entire complex that we call our own mind. Given an initially minor discrepancy between self and world—one thing matters more than another and so we pay attention to it—the higher mental faculties bootstrap themselves into existence.

As the study of psychology developed in the second half of the century, then, attention came to guarantee the mind's freedom vis-à-vis the world. Like agency in an economic system, attention stood for that moment of choice from which cascade a host of unpredictable effects and larger ordering dynamics. For a pious investigator like Charles Bonnet, attention was the "mother of genius," and he called for a "history of attention," which he thought more consequential for understanding human thought than even logic. The "mobile force" of attention proved that the soul is "blessed with activity . . . that [it] can determine for itself the sensation that pleases it more." When we attend to an object, our perceptions of it "fortify themselves and multiply themselves," growing immanently at the intersection of mind and experience.[73]

For Bonnet, the activity of attention—and the freedom of the human soul—was reconcilable with mechanism. Others disagreed. The Scottish physician Alexander Crichton—who translated Johann Blumenbach's *Essay on Generation* in 1792 and whose 1798 *Inquiry into the Nature and Origin of Mental Derangement* was a founding text in British psychiatry—called the mechanical theory of mind "completely absurd." Of course, the observer must understand "the mechanical and chemical powers which operate in the human body." But at a certain point, we "meet with many phenomena which cannot be account[ed] for" by these agents. Impressions travel along the nerves, but when they reach the brain, "new phenomena occur, which are totally dissimilar" from the nervous impulse: "A thought arises!" All efforts to cognize this process run aground on the mind's peculiar powers, not least its ability to alter its own physical substrate. "A thought alters the whole of [the brain's] action," he noted. These reflections on mental plasticity immediately led Crichton to the faculty of attention, the "parent of all our knowledge," for it gives enough data to cognize both self and world.[74] More than this, it is the *observation* of one's own attention that proves (phenomenally, not metaphysically) the importance of our own volition in mental life. A criminal can resist torture by focusing fiercely on the gallows that await him should he confess. Proof enough, in Crichton's view, that even the most powerful physical motives can be counteracted by mental acts.

In his university town of Halle, an early character in this chapter—the German psychiatrist Johann Christian Reil—had read Crichton's book, and he too was fascinated with the attentive mind. Like Crichton, he believed that brain and mind were plastic and adaptive organs. Their qualities are "expanded and modified . . . by the independent capacities that they themselves develop." As the mind moves through the world, it charts its path with an analog of attention, that same faculty of *Besonnenheit*, or "care," that Herder saw at the origin of language. Care is the "compass on the sea of sensation"; it is the "persistence of the perceptive faculties of the soul" even in moments of distraction.

> Thanks to care, the soul must retain, even in the middle of its exertions . . . a so-gentle sense, against the impressions of the world and its body, . . . that those objects that are connected to its present interests . . . remain in focus to consciousness.

In essence, care is a precursor to attention. It is the capacity of a mind to focus on an object without closing out the rest of the world. Its initial acts are "arbitrary . . . because it seizes what accident brings along." But if it begins in accident, it is also crucial to mental volition, since the ability to notice something lets us attend to it more carefully. Accident becomes order, as the mind incorporates its experiences into itself. This ability is not static. Rather, it is a form of dynamic equilibration between different perils: on the one side, distraction, the urge to "to notice all things"; on the other side, depression, the compulsion to take "such a firm grip . . . on one object" that we cannot notice anything else.[75] By noticing just enough, the mind's self-organizing powers flourish.

## MADNESS AND MENTAL LIFE

Observing how the mind creates itself did not mean the end of science. Rather it shifted the scientific perspective from the outside to the inside, from *norms* to *case histories* of cognitive development. In doing so, it made visible a range of human cognitive diversity hidden in earlier ruminations on the mind. Distraction, depression: just as monsters revealed the secrets of biological life, madness became less an exception to mental life than a guide to its most intimate dynamics. From Vincenzo Chiarugi to Benjamin Rush, from Immanuel Kant to Philippe Pinel, from Thomas Arnold to Alexander Crichton: the later eighteenth century routed discussions of cognitive function through the fields of its dysfunction.[76]

Reil pioneered these new directions, so we wind down our chapter with him. In Reil, in fact, we can find a beautiful convergence of the elements of the Enlightenment mind. A careful anatomist of the brain, he both developed new techniques for dissecting an organ notoriously difficult to investigate ("without preparation, the brain is too soupy and fluid" to cut, he commented) and published a series of detailed plates in his journal, the *Archiv für Physiologie* (1795–). He was an important advocate for the reform of the then barbaric treatment of mental patients. He dabbled in the theory of automata, although he believed that to be truly interesting machines would have to build *themselves* in response to "the accidental relations of external influences."[77] And finally he was a keen observer of the varieties of mental phenomena, and eager to understand the mind emergent in all of its dynamism.

Just the dynamism of cognition, after all, made mental illness so interesting. A simple mechanical mind admitted no variety of behavior. Human beings, however, were filled with mental tics, aberrations, and curiosities. Large symptoms start with tiny causes but "grow in size as they roll on, like avalanches, that disrupt the entire microcosm" of the mind. Mental illnesses, Reil wrote, "tend above all to propagate themselves," to take over a mind and reorganize its operating system. They disrupt the mind's causal equilibrium. Instead of a positive loop between the mind and the world, the insane create a negative loop. A cataleptic mind might become fixed on the world, for example, staring "steadfastly at a single object," and thus lose its own self-creating powers. Just as troubled is a mind in which "fantasy runs amok and with wild speed from one object to the next," where the mind is unable to moderate its own powers, leaving the victim helpless to distinguish between the self and the world.[78]

In either case, mental illnesses reproduce like viruses, parasitizing the self-transforming capacities of healthy cognition. For this reason, illness exactly analogizes mental activity. It "changes unceasingly, grows, recedes, shifts its form." In a sense, mental illness *is* cognition in another guise. The modern reevaluation of insanity as a special form of creativity seems to have one origin here. So, for example, the engineer and mystic Franz von Baader described madness as a "means by which our inner, always restless power" dealt with ideas that an "undisturbed" brain could not. Even Plato, von Baader commented, "recommended a certain sickliness as a disposition for more lively insight." Herder was even more emphatic, seeing madness as exemplary of *all* thought. The lunatic makes "his own world" out of his thoughts. But that's true of us sane ones as well: "In the same way move *all* the associations of our thoughts."[79]

The mind's restless and recursive activity makes mental illness so pernicious. Yet it is also what opens it up to intervention from the outside. The "psychic method of cure," as Reil called it, used the nervous system to operate on itself. The nervous system operates via "circular actions," which "announce themselves as the soul's effects" but then recursively produce "simultaneous modifications of its inner state." Stimulating it will "modify its powers . . . in such a way that the dynamic relations of the soul . . . are rectified." Body and soul are thus linked. The neurosis—a term originally coined by the Scottish doctor William Cullen—is both a mental state and a physical one. As a result, new cures suggested themselves. A cure for depression, for example, might be to show the sufferer an object that "through its size and majesty awakens the faculty of care." What to do with "world-numb Platonists, who sense the sparks of a higher being in the virtues of the female sex and so become fools"? Give him a "bordello nymph" to cavort with, and so transform his entire mental order.[80] The psychic cure used the resources of the mind to cure the mind, working to convince the mind to reconstruct for itself a healthy personality.

To come full circle, then, we return to Lichtenberg's dream of the dead duchess. Dreaming is the place where, as Reil puts it, "self-consciousness wavers," where fantasy "ebbs and flows by itself" because "no sensory impressions bridle it any longer." Dreams are not just the broken remnants of the day's thoughts. They are instead witnesses of the mind's self-generating capacities, the same capacities that build attention and reflection, and that produce madness and lunacy. The ever-perceptive Leibniz, in an early fragment, remarked on just this same amazing power of dreams, how their "spontaneous organization carried out in a moment . . . more elegant than any which we can attain . . . while awake." Such a power points to some "harmonious principle, I know not what, in our mind, which, when freed from separating ideas by judgment, turns to compounding them." In the dream, the world disappears, and the mind can operate on itself. But this operation is not without harmony and order. "An entirely different sympathy between the organs of the [mind's] microcosmos generates itself" in sleep and in waking life, Reil wrote. Sleeping and awake, the mind is hard at work, transforming the very things that we most call our own: personality, identity, memory. The "different epochs" of our moral existence, indeed of our person, are products of all of this psychic life, in dreams and awake, experiencing the world in our idiosyncratic ways.[81]

## CONCLUSIONS: ON PANORAMAS, VERTIGO, AND REVERIES

In the early nineteenth century, the French mathematician Laplace marveled at the psychological phenomenon of attention, which both "increases the acuteness, at the same time as it weakens concomitant impressions." The power of attention he found especially noticeable in one of the late-eighteenth-century media spectacles, the panorama. Invented in the 1780s, the panorama was a full-surround circular painting, often of riotous battles, complicated cityscapes, or mountain ranges. It was, in fact, in the Alps that the idea for the panorama seems to have been conceived. Looking at the mountains in 1776, the Swiss natural philosopher Horace-Benedict de Saussure realized that to portray the "infinite variety of the scene" and to "convey an impression of the whole chain," he might need a circular drawing. This panoramic drawing would abbreviate the usual number of individual pieces, the synthesizing of which "requires an enormous amount of attention." The panorama puts the pieces together for the spectator. But for the illusion to become real—and this is what interested Laplace—the spectator must pay very close attention. Panoramas put the "spectator . . . in the same state that he would be in were the objects real," Laplace wrote, yet the imperfections in rendering and perspective always threaten to destroy the magic.[82] The more you pay attention, however, the more the magic works.

Another reaction to panorama was just the opposite of attention: vertigo. Spectators crowded into specially made round buildings, and some felt themselves swept away, literally dizzy from the sights. Heights induced this sensation as well. Standing on the spire of the Strasbourg Cathedral—home of that marvelous clock in chapter 1—Goethe commented on the "giddiness" and disequilibrium felt when trying to take in the entire scene below. The English physician and physiologist Erasmus Darwin tried to define this curious mental state. When "no steady objects are . . . within the sphere of . . . distinct vision," he argued, vertigo is the consequence. On a mountaintop or in a panorama, all of the orienting objects are far outside the normal field of vision, and so we cannot easily judge where we stand. As a result, we're not sure if we are standing or falling, and the constant effort to steady ourselves produces vertigo. Vertigo, moreover, is a constant threat from the world. Because our bodies are never at rest, what Darwin called the "irritative ideas of objects" are "attended perpetually with irritative ideas of their apparent motions." There is a "continued buzz" that shadows us everywhere, from the "wind in our rooms, the fire, distant conversations, mechanic business." These sounds pulsate through our day, making "a great circle of irritative tribes of motion," ideas producing apparent motion producing more ideas, and so on.[83] Here was a mind at work without those properties of self-organization, unable to attend and thus unable to put itself out of the chaos of things.

Vertigo bore, for Darwin, a family relation to the curious mental state that we met at the beginning of the previous chapter. A reverie, Darwin explained, is also a moment of indistinction:

> We cease to be conscious of our existence, are inattentive to time and place, and do not distinguish this train of sensitive and voluntary ideas from the irritative ones excited by the presence of external objects. . . . At length . . . we return with surprise, or with regret, to the common track of life.

Darwin the clinician had a specific patient in mind, a "very ingenious and elegant young lady . . . about the age of seventeen." Her reveries began shortly after menarche. During them, she conversed with imaginary companions, recited long passages of poetry, walked about the room, drank tea, and so on. After the reverie, "she could never recollect a single idea of what had passed in it."[84] Attention is so dispersed in this dissociative state that memory itself ceases to function. The mind is no longer engaged in reflection; the difference between background buzz and volitional thought has vanished; the very personality of the sufferer has disappeared. The difference between internal and external world falls away.

No wonder that Rousseau, so plagued by the cares of the world, was so enchanted by his reveries on the shores of Lake Bienne. In reverie, Rousseau was "like God," he reported. Alienation between thought and its objects, between nature and mankind, came to end. He exited reverie with regret, but for the rest of us, waking up and allowing our minds to pay attention is the precondition to thought, indeed, to being conscious at all. At the end of the day, said the eighteenth-century cognitive sciences, we must live in a fallen state, distinct from God and nature. Our very minds depend on it, for they grow dynamically out of the very tension between self and the world. They are not given to us beforehand, as vehicles for either salvation or survival. Instead they are made through interactions, often difficult, with things that are unlike us. We are not mere calculating machines, but instead plastic organisms, building ourselves in a dynamic engagement with the things of the world. And if the eighteenth-century move from nothing to something is never free of an element of magic and mystery, this is because this entire project is necessarily incomplete. It resists any full metaphysical account, because the kind of minds that we have cannot be yoked to a determined causal matrix. At least not by us, limited and fallen as we are. Instead the complex orders of human thought are (or must be described as if they are) emergent, subject to spontaneous development, products of time, accident, and the marvelous powers of self-organization.

* 3 *

PROLOGUE

# An Island of Goats

"Never were there more important ninety-nine pages written," it has been said, than a pamphlet published in 1786, and yet it is unlikely that you have encountered any of them. (This assessment being that of the modern editor of the text, however, it may not be 100 percent disinterested.) The author of these very important ninety-nine pages was "a mineralogist, a fossilist, and conchologist," and yet they have nothing to do with geology. Our naturalist—one Joseph Townsend, an impressive six-and-a-half-foot Englishman who was also a physician and a Methodist preacher—had turned his sights on one of the eighteenth century's hot-button social issues: the shape and efficacy of the English poor laws.

Townsend's point was not that the poor laws, a parish-based system for the provision of poor relief that limited the movement of the poor between parishes, were not sufficiently effective. "What is most perplexing," he began his tract, "is, that poverty and wretchedness have increased in exact proportion to the efforts which have been made for the comfortable subsistence of the poor." The very effort on the part of government to intervene on the poor's behalf was the cause of just the problem it was supposed to cure. In order to explain why, Townsend offered a striking natural-historical tableau in two acts.[1]

Taking as his starting point a short comment by a seventeenth-century traveler about a pair of goats left on the South Sea island Juan Fernandez by its eponymous discoverer, our naturalist freely spun a hypothetical scenario of what happened next.[2] The goats happily increased and multiplied until they reached the limit that the island resources could support. "From this unhappy moment they began to suffer hunger. . . . In this situation the weakest first gave way, and plenty was again restored." The goat population now reached a "degree of aequipoise," "nearly balancing at all times their quantity of food." But this balance

> was from time to time destroyed, either by epidemical diseases or by the arrival of some vessel in distress. On such occasions their numbers were considerably reduced; but to compensate for this alarm, and to comfort them for the loss of their companions, the survivors never failed immediately to meet returning plenty. They were no longer in fear of famine: they ceased to regard each other with an evil eye; all had abundance, all were contented, all were happy. Thus, what might have been considered as misfortunes, proved a source of comfort; and, to them at least, partial evil was universal good.

The disruptions of the equilibrium on the island of goats turned out to be only temporary and naturally self-correcting. Reaching back to the early eighteenth century—recall the writers in the prologue to part 1—Townsend concluded act one, perhaps unexpectedly, with a blend of Pope's words in the *Essay on Man* and Mandeville's fabled logic. Like bees, like goats.

Act two, as in any good play, has more action. The Spaniards—according to Townsend's narrative, based on another traveler's account and liberally embellished—became concerned about this well-stocked haven for the pirates of the South Pacific. So they landed a pair of greyhounds on the island, with the intention that the carnivores would bring about "the total extirpation of the goats." But the Spaniards, armed presumably with an early modern rather than a *dix-huitièmiste* imagination, did not anticipate the dynamics of self-organizing systems. So after a period of canine goat-flesh feasts, something happened that they had not foreseen.

> Many of the goats retired to the craggy rocks, where the dogs could never follow them. . . . Few of these, besides the careless and the rash, became a prey; and none but the most watchful, strong, and active of the dogs could get a sufficiency of food. Thus a new kind of balance was established.

What emerged was a new natural equilibrium that now involved *both* populations.

Take a moment to savor the surprise. In this short story Townsend foreshadowed with clear outlines not only Thomas Malthus's insights into what Townsend himself called "the principle of population" but also Charles Darwin's principle of natural selection, respectively twelve and seventy years *avant la lettre*. Malthus himself belatedly recognized this, and paid tribute to Townsend in the preface to the second edition of the *Essay on Population* (1803). Looking forward to these major contributions to nineteenth-century thinking, Townsend's tract is indeed a precocious and unexpected harbinger of their own important variations on the self-organizing theme.

Townsend's interest in this tract, however, lay in humans, not goats. Like Malthus and Darwin later, Townsend too recognized that humans were organisms with natural drives, and drew a direct analogy from the self-organizing dynamics of nature to the order of social and political life. On the island of goats "the weakest of both species were among the first to pay the debt of nature; the most active and vigorous preserved their lives." Just so, "it is the quantity of food which regulates the numbers of the human species." Townsend was especially interested in the circumstances he believed were prevalent then in England, when population growth put pressures on limited food supplies:

> Where things are left to a course of nature, one passion regulates another, and the stronger appetite restrains the weaker. There is an appetite, which is and should be urgent, but which, if left to operate without restraint, would multiply the human species before provision could be made for their support. Some check, some balance is therefore absolutely needful, and hunger is the proper balance; . . . even in such a case, when it is impolitic that all should marry, this should be wholly left to every man's discretion, and to that balance of the appetites which nature has established.[3]

Social relations, economic relations, sexual relations: all should be left to the care of the self-organizing dynamic emerging from individuals' free choice. This dynamic can be relied upon equally well in the human as in the natural world, since in this fundamental respect they are analogous to each other.

Townsend's specific goal, we recall, was a critique of the counterproductive poor laws. The poor laws got in the way of natural order in two ways. First, since, as one learns from the analogy to the island of goats, hunger is the key factor driving the poor to work, "a fixed, a certain, and a constant provision for the poor weakens this spring" and promotes laziness rather than industry. Second, Townsend objected to the provisions in the poor laws that confined the poor to their own parishes and prevented them from seeking employment elsewhere. Consequently, "for want of competition the price of labour to the manufacturer has been much enhanced." Whereas manufacturers "universally agree, that the poor are seldom diligent, except when labour is cheap, and corn is dear," the poor laws produced precisely the opposite effect.[4]

The more fundamental lesson from the island of goats and greyhounds was the folly in attempting an end run around natural order. The poor law, by eliminating the natural force of hunger, "tends to destroy the harmony and beauty, the symmetry and order of that system, which God and nature have established in the world." It is especially significant, moreover, to realize that Townsend's vision of order subsumed all manner of *disorder* as constitutive of its harmonious symmetry. He thus drew on Hesiod's *Works and Days* to differentiate "two kinds of strife and contention among men": violence leading to plunder, which is undesirable, versus the desirable jealous competition in economic and artistic production, which encourages industry and diligence and is generative "of peace, harmony, and plenty." He drew on Mandeville to hail the social benefits of the pursuit of luxury ("a taste for luxury must be productive of industry . . . and promote the welfare of the state"), even as he rather feebly rejected the amoralism associated with Mandeville's "private vices." And in what his modern editor singles out as a seminal innovation, and what seems like a de facto acceptance of Mandevillian sins, Townsend found value to the whole also in the very vices and faults of the poor:

> It seems to be a law of nature, that the poor should be to a certain degree improvident, that there may always be some to fulfil the most servile, the most sordid, and the most ignoble offices in the community. . . . The fleets and armies of a state would soon be in want of soldiers and sailors, if sobriety and diligence universally prevailed.

Thus works the law of nature, and thus it produces the greater stock of aggregate human happiness, even at the expense of the unhappiness of some. Finally, returning to the poor laws, not only were they obstructionist, but they were also bound to fail, since they floundered against laws of nature much more forceful. "The course of nature may be easily disturbed, but man will never be able to reverse its laws." Even if legislators created a disturbance, therefore, "things would soon return into their proper channel."[5]

In the final part of *Invisible Hands* we take our cue from Joseph Townsend. With him, we make the leap from the self-organizing thinking about natural systems of part 2 back to the self-organizing thinking about social affairs. Sometimes the transition is seamless, as Townsend implied with his goats and hounds. Other times we become aware of the differences. It is thus notable that the dominant self-organizing dynamic of the previous two chapters, in the sciences of life and of the mind, is an ever-expanding unbounded one, that of a positive feedback loop. By contrast, Townsend envisions society and the economy as a bounded, long-term self-correcting equilibrium: a very different picture indeed. Chapter 6 examines further both types of self-organizing narratives as they are incorporated into different ways of thinking about human systems. As we have said multiple times, there are many members in the family of self-organizing moves, and while sharing key characteristics they do not at all have to look alike.

More specifically, we are taking our cue from Joseph Townsend by putting his thoughts on the poor laws into their broader historical context. From one perspective Townsend's scenario was indeed an uncanny prefiguration, looking forward to Malthus and Darwin. From another perspective, however, looking back from 1786, his arguments could be seen as not only part of his times but actually rather predictable. When Townsend talked about things returning to their proper channel of their own accord after a disruption; when he summed up his goal as guaranteeing that "our population would be no longer unnatural and forced, but would regulate itself by the demand for labour"; when he insisted that "it is with the human species as with all other articles of trade without a premium; the demand [for labor] will regulate the market": in every such case, and indeed throughout his tract, his readers may well have felt that they had heard it all before. "It is said," observed one contemporary soon thereafter, "and it has been said *till the observation is become trite and hackneyed*, that trade will at all times find its level."[6]

What made such observations so trite and hackneyed was, of course, the impact of Adam Smith's *Wealth of Nations*, published a decade before Townsend's ninety-nine pages. The affinity of Townsend's language to Smith's is easy to spot. Here, for example, is Adam Smith's formulation at the opening of his well-known tenth chapter:

> The whole of the advantages and disadvantages of the different employments of labour and stock must, in the same neighbourhood, be either perfectly equal or continually tending to equality. If in the same neighbourhood, there was any employment evidently either more or less advantageous than the rest, so many people would crowd into it in the one case, and so many would desert it in the other, that its advantages would soon return to the level of other employments. This at least would be the case in a society where things were left to follow their natural course, where there was perfect liberty, and where every man was perfectly free both to chuse what occupation he thought proper, and to change it as often as he thought proper.[7]

Townsend's tract, in Adam Smith's language, was in the same neighborhood. Much more so, it must be said, than in the neighborhood of poor law orthodoxy earlier in the eighteenth century, which had insisted very much on the rôle of legislation and the magistrate—in the words of one of many examples, from 1767—"to restrain men from idleness, intemperance, and disorder, and by some means or other oblige them to be industrious, sober, and peaceable, and to endeavour to the best of their abilities to support themselves and families."[8]

In moving to the vicinity of Adam Smith, moreover, Townsend was part of a much broader trend. A count on the Eighteenth-Century Collection Online database of the relevant uses of the common "hydrological" phrase, when things "find their level" as they self-organize, shows a dramatic tenfold rise in the appearance of this formulation and its variants after the publication of the *Inquiry into the Nature and Causes of the Wealth of Nations* in 1776.[9] That is precisely how an insight becomes trite and hackneyed.

And yet, although this statistic appears to reinforce the history of self-organization as it is commonly told, it is clear by now that we propose to reverse its chronological flow and its causal logic. As Townsend with his reaching back to Mandeville and Pope knew well, and as we do too by now, Adam Smith did not introduce the self-organizing insight, nor for that matter its "hydrological" formulation (recall, for example, Charles Davenant's language in the 1690s as quoted in chapter 2). Nor did Smith by himself make it so popular. Rather, half a century of precedents and a full century of preparation were what had made people so receptive to his formulation, so able and willing to spread it further. Single enunciations rarely make revolutions: without the deeper and wider emergence of notions of self-organization that had prepared the ground for Adam Smith's authoritative encapsulation thereof, it would be hard to explain why it spread so widely so fast.

Chapter 6, therefore, picks up the story of self-organization in the context of human systems, looking back at some of the forms that self-organization took in social and economic thinking during the half century after the 1720s, and then placing Adam Smith's self-organizing thinking itself within this broader framework. Finally, chapter 7 also takes its cue from Joseph Townsend, perhaps against the grain of his declared intentions, by turning attention to the role of the legislator. Townsend appears steadfast in his limited view of political interventions. "If laws alone could make a nation happy, ours would be the happiest nation upon earth." Instead, in truth "the reverse of this we find to be the case," since legislators and magistrates meddle in complex situations that should have been left to their own devices. This does not mean that Townsend imagined a world without political power:

> A wise legislator will endeavour to confirm the natural bonds of society, and give vigour to the first principles on which political union must depend. He will preserve the distinctions which exist in nature independent of his authority, and the various relations which, antecedent to his creation, connected man to man. He will study the natural obligations which arise from these relations, that he may strengthen these connections by the sanction of his laws.[10]

In theory, the legislator can do little more than go with the flow and manage situations that preceded him and unfolded independently of his existence. In practice, however, it turned out that this modest role could grow and, even for Townsend, reintroduce regulation once more through the back door. Townsend's actual plan for elimination of the poor rates called for their replacement with voluntary charity and institutions, central among which were the friendly societies. "The greatest misfortune," however, was the fact that the poor were "altogether left to their own option to join these societies or not," and this weakened their aggregate force. Our anti-interventionist prophet of self-organization therefore suddenly changed his tune. Friendly societies, he now proposed, should be made mandatory, "universal, and subjected to wholesome regulations."[11] Welcome back the *dirigiste* legislator, he who suddenly does not want to leave his operators to their own devices, but wishes rather to set involuntary rules for voluntary contributions.

Was this then simply a flaw of execution in Townsend's plan, a momentary lapse? Perhaps not. In order to see this we would like to return once more to Townsend's remarkable zoological tableau, to draw from it an important lesson for the ideal self-organizing system: its purest form takes place on a self-contained, isolated island. Not for nothing did the island of Juan Fernandez serve Defoe, the master of early self-organization experimentation, as the real-life inspiration for the ur-text of self-sufficiency, *Robinson Crusoe*. Beyond a possible prime mover, a true self-organizing system does not rely and cannot rely on any causation outside itself.

Yet this requirement immediately exposes the imaginary, unattainable nature of a true self-organizing system. An island is part of an archipelago. Even an isolated island is affected by trade routes, by explorers and pirates and Spaniards. Or, say, by sea lions. What Townsend neglected to mention when he retold the story of the dogs who failed to eat all the goats was that his source, the Spaniard Antonio de Ulloa, had another ending to the story. De Ulloa reported that the dogs, having failed to capture goats—for this mid-eighteenth-century author there was no dynamic of a self-producing equilibrium, simply an inadequacy of the dogs to their intended task—survived by feasting on the abundant population of sluggish sea lions. The ecosystem of Juan Fernandez was hardly a closed system of two species.[12]

Similarly, while it may be convenient to imagine nations as closed systems, in truth they too are lodged within an environment of international relations and transnational trade. To achieve full closure, this environment itself must become part of the system, so that nothing remains as an Archimedean point outside it allowing leverage for external forces. So self-organization as a conceptual move must perform a kind of imperial expansion, ballooning out to subsume all of the environments of which the smaller ones are part. Self-organizing logic means not simply that bigger is better but that *biggest is best*. As the marquis de Condorcet was to discover when he searched for a general law in the "seeming chaos" and "astonishing multiplicity" of economic activity, even when starting humbly with a single person one immediately realizes that any such individual is "dependent on all the accidents of nature and [on] every political event," all that can transpire in "this terrible complication of interests, which connects the subsistence and well-being of an obscure individual with the general system of social existence . . . and extends in a manner to the whole globe."[13] No midway stop can be fully satisfactory: before you know it, the system in which the individual is personally enmeshed is full-blown global.

Internally, too, every social and economic arrangement is hemmed in by factors external to itself. This is the reason why the role of legislators and magistrates always remains active, and why a pamphleteer like Townsend needed to insist on correctives to policies that in theory—*his* theory—should not have had the power to deflect the natural course of things. In the real world political will is necessary to set the opening terms, the delimiting conditions, and the basic rules and framework for the operation of any self-organizing system in human affairs, which is thus only *imagined* to be truly self-organizing. Townsend's mandatory regulations for the voluntary friendly societies are one example; chapter 6 will provide some more. As we shall see in chapter 7, this inescapable fact has the surprising effect of liberating political thinking to adopt self-organization for a variety of otherwise wholly incommensurable social and political arrangements.

CHAPTER 6

# The Secret Concatenation of Society

This chapter follows, then, upon Joseph Townsend's example in shifting our focus from self-organization in the conceptualization of life, nature, and cognition to its role in thinking about human agents and human systems.

For some, there was no real difference. Caught in Newtonian overconfidence, there were those who believed—a belief that would have left Newton himself scratching his head—that humans could be analyzed and accounted for with the very same Newtonian principles that applied to natural forces. One of those was a Scottish mathematician named Colin Maclaurin, a protégé of Newton's who until 2008 (!) held the all-time global record for the world's youngest professorial appointment. In 1714, at age sixteen, Maclaurin wrote a paper explaining consumer desire—that newly conspicuous phenomenon much discussed during the early phases of intensive commercialization—through Newtonian mathematics. He wished to show that the forces with which people's minds are attracted to various goods are analogous to those of celestial mechanics. Decades later, however, in his mature account of mainstream Newtonianism, which has long been recognized as the leading eighteenth-century authority on the subject, Maclaurin dwelled instead on the differences:

> The difference between a man and a machine does not consist only in sensation, and intelligence; but in his power of acting also. The balance for want of power cannot move at all, when the weights are equal: but a free agent, . . . when there appear two perfectly alike reasonable ways of acting, has still within itself a power of chusing.[1]

It took Maclaurin many years to realize what for others was no surprise. Humans cannot be subject to straightforward Newtonian logic, because unlike inanimate objects they have *within* themselves the power to act. Unlike natural objects, they are not simply wax in the hands of universal laws. This additional degree of freedom of individuals in a human system—their agency, their swerve—endows them with a level of uncertainty incompatible with a mechanistic model, and thus makes the systems of which they are part more erratic than what self-proclaimed Newtonians could easily handle. Conversely, however, such individuals with an added degree of freedom are precisely the kinds of elements-within-a-larger-system that are handled best through the analytics of self-organization.

The eighteenth century therefore witnessed an increasing sophistication in the descriptions of human-populated and human-driven aggregates with the language of self-organization. It has been often pointed out that foundational concepts of social organization emerged during this period, not least "society" itself, and subsequently "economy"—a narrative often traced back to Bernard Mandeville and Baron Montesquieu, and forward through Enlightenment social thought. Suddenly released from their divine strings and unmoored from a religion-based scheme of organization for the social sphere, these human congregations were now imagined as left to their own devices.[2]

Self-organization was thus just around the corner. As we have seen in the first part of this book, its harbingers began as often tentative forays at the time of Mandeville and Montesquieu. The present chapter looks ahead to the maturing and elaboration over the next two generations of the ways in which eighteenth-century Europeans thought about themselves—in particular, about social and economic configurations involving large numbers of people—in self-organizing terms. This story will finally circle back to Adam Smith, and will place *The Wealth of Nations* within the broader context of Smith's period as well as of his earlier writings.

Before we get there, however, we wish to share with the reader something of a self-organizing experience of our own, one that took place as we geared up to writing what we had found in the second half of the eighteenth century. The process of writing a book like this one, after all, itself moves between the two contradictory poles that organize our argument: design and chaos. On the one hand, what can be more designed and premeditated than the writing of a scholarly book, carefully unfolding an argument or a narrative? On the other hand, the research itself more often than not meanders in accidental and random ways, producing a rather chaotic jumble before one begins to impose order upon it. In this particular case, as we surveyed the chaos, our jumbled piles of evidence suddenly made sense in a way that we did not plan or anticipate: the whole was greater than the sum of its parts.

The parts were the sources that hid in our files from one decade, a decade right in the middle of the half century that separated the outburst of self-organizing experimentation in the 1720s and the aftershocks of the *Wealth of Nations* in the 1770s. The whole was an emergent panorama of the uses of the language of self-organization in the 1750s. Without prior intent, we had in front of us an excellent concatenation of examples demonstrating the spread of self-organizing thinking together with the whole range of questions that accompany this often mysterious conceptualization of order.

Our strategy for showcasing this array of possibilities from this middle period, without being either exhaustive or exhausting, is to offer in the next few pages one example from every year of this decade, each highlighting a different issue or problem. This year-by-year display cannot be described as a methodologically sound sampling, since it was not amassed by design. Furthermore, unlike our argument regarding the 1720s in chapter 3, there is nothing we can see taking place in the 1750s that made this particular decade more susceptible to self-organization than the years before or after. Yet the voices in the following pages demonstrate well, we believe, the *breadth* of self-organization by the middle of the eighteenth century, in two senses: the pervasiveness of this particular language, with its multiple and varied incarnations, across a wide spectrum of cultural and intellectual domains; and the varied range of themes, questions, doubts, and fantasies that accompanied its spread. Because of the history and development of this book project, and also to maintain coherence and convenience, the decade's worth of snippets of self-organization in the next section is limited to Britain. The second part of this chapter, returning to a thematic examination of self-organization in social and economic thought, and chapter 7, which focuses on the political uses of self-organization, expand our purview again to include the Continent and North America.

## A TOUR THROUGH THE FIFTIES

### 1750

The first example for the decade comes from a 1750 publication titled *The Reflector: Representing Human Affairs, as They Are; and May Be Improved.* Like Colin Maclaurin, its author, Peter Shaw, was a scientist and physician who also saw himself qualified to proclaim on the nature of human affairs. Like Maclaurin, too, Shaw was fascinated with the unfettered agency of the eighteenth-century consumer, that freedom of choice no longer tethered to religious imperatives, as we discussed in chapter 2. The individual consumer is a rather odd being:

> Indeed, when we consider the Make of Man, we must acknowledge him an astonishing Masterpiece; but when we reflect upon his Oddities and Caprice, and compare them with the orderly bounded Desires of other Animals, our towering Notions of this noble Creature droop. Man's Life is spent in liking and disliking, in chusing and refusing, the same Things.

Man's choices are capricious, transient, and often silly. Which, however, turns out to be an excellent thing. "If all Men were wise, Society could not well subsist," Shaw insisted; "Difference of Taste makes nothing remain useless in the World." And again:

> There is a great Variety of Tastes in the World; and this Variety keeps Numbers of People employed, who might otherwise remain idle, or hurt Society. We are apt to exclaim against bad Taste, tho' even ourselves, our Friends, and Relations, receive Advantages from it. Nature regulates all Things wisely; and perhaps more suitably to the State of Man, than we superficially imagine.

The self-organizing logic of multiple, often capricious choices in a commercial society appears inexorable. As Shaw put it on another occasion, "it seems to be a Rule, that Men's particular Disadvantages produce general Advantages." Aggregation was key to the regulatory success of nature's invisible hand.[3]

## 1751

Our text for 1751 comes from a different cultural domain: literature. The language of self-organization offered new modes for the literary imagination, new ways to tell stories, which fused with that key new genre of the eighteenth century, the novel. Indeed, in a more literature-based account of the emergence of self-organizing narratives, the novel could have taken center stage. Our example here, Henry Fielding's last novel, *Amelia*, published in 1751, breaks away from Fielding's earlier novels, as contemporaries and critics have often noted. According to the literary critic Jesse Molesworth, the difference lies in the fact that *Amelia* takes place in a "re-conceptualized causal system," "a radical rethinking of the epistemological terrain of the novel" in which "meaning so often severs itself from intended meaning," so that what is left is "the terrible acknowledgment that one's life resembles nothing so much as a disorganized jumble of unconnected events."[4]

*Amelia*'s plot, however, does not give way to such a disorganized jumble of unconnected events without a fight. Fielding poses the problematic in the novel's very first paragraph:

> The various Accidents which befel a very worthy Couple, after their uniting in the State of Matrimony, will be the Subject of the following History. The Distresses which they waded through, were some of them so exquisite, and the Incidents which produced these so extraordinary, that they seemed to require not only the utmost Malice, but the utmost Invention which Superstition hath ever attributed to Fortune: Tho' whether any such Being interfered in the case, or, indeed, whether there be any such Being in the Universe, is a Matter which I by no Means presume to determine in the Affirmative.[5]

The author's declaration of intent, reinforced in didactic asides elsewhere in the novel, is clear. Just as the accidents in the first sentence morph into the incidents in the second, so every random event will be proven to have an observable cause. Fortune is superstition. As it turns out, however, the novel fails to deliver on this didactic message. "Instead of demonstrating that accidents have their roots in 'minute causes'"—this is Molesworth again—*Amelia* in fact "demonstrates that accidents most often occur, as they usually do in life, unprovoked and entirely randomly." *Amelia* absolutely depends on accidents, a word Fielding invokes compulsively. Rather than exposing accidents as incidents, the novel blurs the boundaries between them and never establishes the primacy of the latter over the former. This is never clearer than at the novel's seemingly improbable ending, which for this very reason enraged many a critic. Ultimately, *Amelia*'s plot depends at key moments on random events that defy intention, premeditation, or linear causality, and yet come together to form a coherent whole. A self-organizing plot, we might say.

Nor was *Amelia* the only novel of its kind. Only three years earlier Tobias Smollett's *Roderick Random* had presented the public with its hero living up to his name ("Roderick," writes one modern critic, "inhabits a Lucretian universe of ceaseless change that is at once random and determined.")[6] And before the decade was out readers would be offered the beginnings of the most extreme eighteenth-century example of random, self-organizing plotting, Laurence Sterne's *Tristram Shandy*. For some literary scholars, indeed, this potential for chance or accidental events to disrupt ordered sequences of causes and effects, and yet to aggregate together into a meaningful narrative, had been lurking in the novel form from its very beginning: probability theory and the novel, in their view, developed hand in hand.[7]

## 1752

In the ongoing public discussion of the poor laws, a generation before Joseph Townsend, two pamphleteers debated the best way to construct an effective system of poor relief. Neither was a precocious Smithian: they disagreed principally about how best to mobilize the higher orders to supervise such a system. One appealed to "a true public spirit" that transcends the "absurd jealousies between the landed and trading interests," positing instead "*that those interests are always the same*." The other responded skeptically, insisting that any poor relief plan must rely only on the self-interest of those intended to run it, since "the desire of Gain is in general the strongest in the human Bosom." The latter author then marveled at "how distressed and miserable" this primacy of self-interestedness should have made every member of society, "had not Providence graciously and wisely so connected the Interest of Self with that of others, as to promote both, while one is only seen and pursued."[8] So not quite self-organization, but a strong affirmation of the logic whereby a mechanism outside human influence—here, providence—turns the uncoordinated pulls of individual selfish self-interests into a public good.

## 1753

The beginning of April 1753 brings us one of the sharpest formulations for the self-organizing insight in the context of a human system. It comes from the pen of the grand eighteenth-century master of the witty and the pithy, Dr. Samuel Johnson: "*We are formed for society, not for combination.*" Johnson's essay in the *Adventurer*, one of his first to appear in this journal that had come into being half a year before, explained his meaning. Speculators, Johnson observed, repeatedly imagine bringing a large number of people together to join in a common effort, but "this gigantic phantom of collective power vanishes at once into air and emptiness, at the first attempt to put it into action." The reason is that it runs contrary to human nature: "The different apprehensions, the discordant passions, the jarring interests of men, will scarcely permit that many should unite in one undertaking." We therefore must accept "the impossibility of confining numbers to the constant and uniform prosecution of common interest." Nevertheless, society does not fall apart: Why? Drawing on an analogy from science, Johnson explains that just as the vast bodies of the universe "are regulated in their progress through the ethereal spaces, by the perpetual agency of contrary forces," so with human beings. "The same contrariety of impulse may be perhaps discovered in the motions of men: we are formed for society, not for combination."[9] Human beings pull in all manner of discordant and jarring directions, and thus cannot align themselves together in a combination for a single purpose. And yet this variety of contraries comes together as a different sort of aggregate, one that is not an intentional combination, to form the cohesive and beneficial whole that is society. It would be hard to find a clearer invocation of this vision of self-organization than Dr. Johnson's: for him it entailed nothing less than the very meaning of "society."

In subsequent months, moreover, Dr. Johnson's contributions to the *Adventurer* returned to the same insight again in other forms. In an oft-quoted example, an essay marveling at modern-day London for "its opulence and its multitudes, its extent and variety . . . , the immense stores of every kind of merchandise piled up for sale," continues in a familiar Mandevillian vein. "Not only by these popular and modish trifles, but by a thousand unheeded and evanescent kinds of business, are the multitudes of this city preserved from idleness, and consequently from want." As Johnson surveys this endless variety offered to the capricious choices of consumers while providing universal employment—a capricious variety mirrored, one might note, in the eclectic constitution of the very genre that carried Johnson's observations—he is led to his notable statement about the logic of the system as a whole:

> In the endless variety of tastes and circumstances that diversify mankind, nothing is so superfluous, but that some one desires it; . . . what is thrown away by one is gathered up by another; and the refuse of part of mankind furnishes a subordinate class with the materials necessary to their support.
>
> When I look round upon those who are thus variously exerting their qualifications, I cannot but admire the secret concatenation of society, that links together the great and the mean, the illustrious and the obscure.[10]

*The secret concatenation of society* (literally, a secret chain, echoing Montesquieu's *chaîne secrète*): a hidden logic, undesigned, unintended, and unheeded, that of its own accord ties the endlessly diverse pushes and pulls of London's inhabitants into one wholesome aggregate. In some sense this was the same populous aggregate that John Graunt had discovered already in the seventeenth century, as we recall from chapter 2, but now it was endowed with a much expanded meaning and significance.

## 1754

In 1754 Horace Walpole invented serendipity. But since we have already told that story of the coming of age of a beneficial outlook on sequences of chance events as reflected in a new word, our choice for 1754 is a text that takes us back to that key if not always comfortable element in conceptualizing self-organization: *God.* The writer was James Burgh, a Scottish-born schoolmaster with a formidable progressive reputation still ahead of him. A repeated theme in Burgh's writings was commercial society: how to enjoy its promise without succumbing to its corrupting influences. Mandeville, therefore, was for Burgh the proverbial red rag. How can anyone not realize, he wrote indignantly in *The Dignity of Human Nature* of 1754, that "every deviation"—that is to say, "a deviation from that conduct, which suits a reasonable creature, which is the very definition of moral evil"—is inevitably a "deviation into irregularity [which] would in the end produce universal unhappiness"? Begone Mandeville! "Instead of the sophistical maxim, 'That private vices are publick benefits,' we may establish one more just; 'That the smallest irregularities, unrestrained, and encouraged, tend to produce universal confusion and misery.'" Human affairs are an unstable equilibrium: irregularities are magnified and thus inherently lead to greater evil. The very opposite of a self-organizing vision.

At least, that's what one might think. But this analysis omits the most important presence, that of God. If without God a quick descent into chaos would thus be inevitable, the divine "general and universal perfection" guarantees that this in fact is not the case. Ultimately, then, what kind of order *does* prevail in the world? It is here that Burgh is at his most creative. On the one hand, in our postlapsarian state it is "impossible" to find "universal regularity . . . promot[ing] order, perfection, and happiness": individuals will not all act morally in unison, but many of them will deviate in irregular and seemingly random ways. Yet "to suppose any one part or member to be left out of the [divine] general scheme, left to itself, or to proceed at random, is absurd." Randomness of individual actions does not imply randomness at the level of the system. Just as it is impossible to see humans all acting in moral unison, "it is likewise impossible to conceive a system in which the habitual conformity of reasonable beings to the grand scheme of the universal Governor should not *naturally, and as it were of itself*, produce happiness." Humans can only be expected to strive for an "habitual conformity" to general principles, from which they will often deviate in specific instances. The rest must be left to God, "the Divine wisdom [that] has taken measures for incorporating the irregularities and vices of men into his regular and perfect oeconomy, so that the greatest happiness, that can consist with the freedom of the creature, shall be the result."[11] Steering between the Scylla of the fatalism that might accompany moral uniformity and the Charybdis of moral entropy, Burgh's God has set up and maintains from afar the self-organizing system that ensures the greatest possible amount of human happiness.

## 1755

In 1755 a group of young Scotsmen launched a short-lived periodical, the *Edinburgh Review*, designed to capture Scotland's intellectual excitement by reviewing all Scottish publications every six months, as well as a selection of English ones. The first paragraph of the first article of the first issue, an article reviewing a Scottish work on Peter the Great, began thus:

> The attempt of Peter the Great towards civilizing that vast Empire, of which he was the Sovereign, is perhaps the most singular and interesting object that the history of mankind presents to the view of a Philosopher. Commerce, learning, the art of war and polished manners, have penetrated into most nations by degrees, and have owed their establishment more to *the casual operation of undesigned events, than to the regular execution of any concerted plan*. . . . The Czar of Muscovy is the first man who, unenlightened by science, and uninstructed by example, conceived the vast design of civilizing sixteen millions of savages.[12]

Peter the Great was the exception that proved the rule. And the rule was, historical progress advances through "the casual operation of undesigned events" rather than through concerted effort or deliberate design. The very movement of history was a self-organizing process, in which the effect of aggregation was produced slowly, "by degrees," through the cumulative passage of time.

This vision of human history as self-organizing is significant not only because the new journal placed it so conspicuously at its forefront, but also because of who penned it. The anonymous author of this review was William Robertson, who went on to become one of the Scottish Enlightenment's greatest historians. And when Robertson wrote years later in his important *History of the Reign of the Emperor Charles V* that "upon reviewing the transactions of any active period, in the history of civilized nations, the changes which are accomplished appear wonderfully disproportioned to the efforts which have been exerted," he clearly continued to carry with him this way of thinking.[13] We shall see later that his historical understanding of the balance of powers in Europe was also basically self-organizing.

The fact that Robertson was the cousin of James Burgh, our author for 1754, is probably just a coincidence. (Burgh lived in southern England, and there is no documented evidence of contact between them.) But the fact that Robertson's early expression of a self-organizing vision of history took place in a new intellectual initiative in mid-eighteenth-century Edinburgh was probably not. At least, the critic who dedicated a full pamphlet to reviewing the new *Review* did not think so. Drawing attention to the same passage from page 1 of the *Edinburgh Review* as we did here, this critic pointed out—rightly—that it was "somewhat similar to that passage of Mr Hume's *Essays* quoted in the *Analysis*, p. 31, 'Chance, therefore, or secret unknown causes, must have a great influence on the rise and progress of all the refined arts.'" This juxtaposition was unusually astute, prophetically bringing together the two writers who were to establish themselves as the two great Scottish historians of the eighteenth century. Even as it was written David Hume was busily composing his great historical work, the six-volume *History of England*, in which he instructed his readers "in the great mixture of accident, which commonly concurs with a small ingredient of wisdom and foresight, in erecting the complicated fabric of the most perfect government."[14] We will return in this chapter to Hume, and indeed to his very words that the anonymous critic of 1754 juxtaposed here with Robertson's, when we place Hume at the epicenter of an eighteenth-century Scottish milieu that adopted the self-organizing logic with great enthusiasm. Finally, as we contemplate the interconnectedness of the Scottish literati among whom the self-organizing understanding of history found its most elaborate champions, it is also worth noting another perhaps not altogether coincidental fact: together with Robertson, one of the other collaborators on that first issue of the *Edinburgh Review* was Adam Smith.

## 1756

The 1750s witnessed a peak, prompted in part by lackluster performances in the early stages of the Seven Years' War, in what Thomas Macaulay once described as the English's morbid interest in their own decline. One of the earlier contributions to the genre, with a broader British view, was written by the Irish actor Thomas Sheridan, father of the better-known Richard Brinsley Sheridan. The direct catalyst for Sheridan's writing was the riots in 1754 that had destroyed his Dublin theater and driven him to London, where he published a disquisition of almost six hundred pages entitled *The Source of the Disorders of Great Britain*. Sheridan found the source of these disorders in defective religious education, compounded by the commercial revolution's introduction of luxury, which is "utterly incompatible" with liberty and virtue. The prevention of the imminent collapse thus "seems to be beyond the power of all human means, and can be effected by nothing but divine interposition." Moreover, "whoever examines the constitution of Great Britain, will find that from it's nature it had no principle; and considering the discordant and jarring parts of which it is composed, it must necessarily fall to pieces in a short time, unless they were cemented by religion." An odd solution, perhaps, for the chaos of the British constitution, but one consistent with the tenor of Sheridan's call for increasing religious righteousness—made possible through education—that might then be rewarded by a crisis-solving divine intervention: standard moralists' fare.

But then Sheridan's analysis took a less conventional turn. The danger, it turned out, was not quite so imminent. The one thing that could prevent "the destruction attendant on luxury" was England's system of laws, which came into being "previous to the great opening of commerce" and thus was in place long before luxury became a palpable threat. So, when it is considered "that this system was not framed by any legislator . . . not cultivated and established by the wisdom or design of man, in times of knowledge, but in the days of ignorance, when our forefathers knew not it's particular use and fitness to their country"; when it is considered "that it was not brought in by the hand of power, and supported by authority, but made it's way against the passions, prejudices, interests, and violence of mankind"; when it is considered that any improvements in this system happened "not from any views to publick utility, but from *agents who were doing good in the dark*"; and when it is considered further that at the end of this process Great Britain, "just emerging out of darkness, found herself in possession of a wider and better system of moral laws, than the labour of centuries in the most polished and intelligent nations ever produced": "when all these things are considered, sure there is not any one of the least reflection who must not necessarily see the hand of God throughout."[15]

What a curious reversal of the logic of action and reaction, of the traditional sequence of cause and effect, that one would expect from a moralist's exhortation. The divine interposition necessary for Britain's salvation had already been long in place, in the shape of a system of laws "peculiarly adapted to the particular circumstances of the country" and "calculated in the most exact manner for all the purposes of society." This excellent system emerged of itself, without human design or understanding of its overall operation. It was the accretion of multiple actions of agents—agents prejudiced, interested, and ignorant, with no idea of the effects of their actions—who together and over a long period of time produced order from disorder. (This, we recall, from the pen of a man who had just witnessed a rioting crowd wreck his Dublin theater!) The British constitution, which Sheridan had proclaimed several pages earlier to be ushered to speedy collapse by its "discordant and jarring parts," now turned out to be an excellent bulwark against precisely such an unfortunate demise because of the self-organizing overall harmony of these same jarring parts. And where is God? God's intervention is proven by the very realization that it is not, in fact, necessary.

## 1757

Josiah Tucker, then a vicar in Bristol and later the Dean of Gloucester, agreed. Providence, he insisted in an analysis of British social and economic arrangements that he cleverly packaged as *Instructions for Travellers: 1757*, has no other design in the forming of the commercial system "than this, That private Interest should coincide with public, self with social, and the present with future Happiness." Sounds good; but precisely how this magic would take place was not easy for Tucker to establish. On the one hand he made statements that aligned him closely with the language of self-organization:

> For let the Legislature but take Care not to make *bad Laws*, and then as to *good ones*, they will make themselves: That is, the Self-Love and Self-Interest of each Individual will prompt him to seek such Ways of Gain, Trades, and Occupations of Life, as by serving himself, will promote the public Welfare at the same Time.

As Tucker explained a couple of years before, if a trade, say, becomes "accidentally overstocked," in cases like that "the best and safest Way is to let the Evil alone, and then it will infallibly cure itself."[16] The key, of course, was Tucker's confidence in the infallibility of the self-correction.

But is it really infallible? Despite the self-organizing bravura, deep down Tucker remained unsure to what extent providence needed a more down-to-earth helping hand. If the passion of self-love is left "to proceed without Direction or Controll, it would in a great measure defeat its own Ends," he worried.

> The Passion of Self-Love therefore must be taken hold of by some Method or other; and so *trained* or *guided* in its Operations, that its Activity may never be mischievous, but always productive of the public Welfare. When Things are

> brought to that pass, the Consequence will be, that every Individual (whether he intends it or not) will be promoting the Good of his Country, and of Mankind in general, while he is pursuing his own private Interest.

Tucker here comes across as more of the *dirigiste* that he really was. The role of good government is to give self-interest "such a Direction, that it may promote the public Interest by pursuing its own: And then the very Spirit of Monopoly will operate for the Good of the Whole." Difficult alchemy this. Tucker recognized the claims of the self-organizing, invisible hand logic, and obviously found them appealing; so much so that the doyen of the modern preoccupation with the genealogy of liberal economic thought, Friedrich Hayek, heralded these same passages as a better eighteenth-century formulation of the logic of spontaneous order than that of Adam Smith himself.[17] But in the end Tucker appears hesitant to take the leap and entrust the socioeconomic system to its own devices, without some oversight to ensure that its self-propelling mechanisms would actually push it in the right direction. Tucker was not alone, of course, in pausing in front of the mysteriousness of self-organization, that absence of a clear path to cohesion that can set one's mind at ease that this dynamic of order should actually *work*. When it came to spontaneous order, Josiah Tucker could talk the talk but could not quite bring himself to walk the walk.

## 1758

Our files have nothing to offer for this year. Since this decade-long sequence reflects an emergent summation of materials that we have come across during our research, which we now use as a post-factum device to illustrate the density, spread, and variety of self-organization themes in mid-eighteenth-century British texts, and since we found no source from 1758 among those we had collected prior to settling on this device, we leave this year blank.

## 1759

Without doubt, the 1759 publication most important for the argument of this book is Adam Smith's *Theory of Moral Sentiments*. But since we discuss it shortly in more detail, we'd like instead to bring back Samuel Johnson with his 1759 "tale" *Rasselas*, in which he reflects not directly on the dynamics of human society but rather on the nature of a large system and on the limits of human intervention in it.

The title that Johnson had originally intended for *Rasselas* was *The Choice of Life*, referring to the quest of the prince Rasselas for a free choice leading to happiness. Content is matched by form, which is bent on offering a "diversity of opinions" as a cumulative dialectic method of approaching truth—a point made explicitly in the text. The *limits* of individual agency and choice are thus also a key theme in *Rasselas*. At one point the philosopher Imlac meets an insane astronomer who believes that "I have possessed for five years the regulation of weather, and the distribution of seasons: the sun has listened to my dictates . . . ; the clouds, at my call, have poured their waters." And yet the astronomer, while claiming agency over the weather, also admonishes Imlac, to whom he wants to bequeath this onerous job:

> I have diligently considered the position of the earth and sun, and formed innumerable schemes in which I changed their situation. I have sometimes turned aside the axis of the earth, and sometimes varied the ecliptic of the sun: but I have found it impossible to make a disposition by which the world may be advantaged; what one region gains, another loses by any imaginable alteration.

Even the madman—in the 1780s one physician used this very episode to define "notional insanity"—realizes that while tomorrow's rain, sunshine, clouds, or hail are unpredictable happenstance, the self-regulating balance of the weather system as a whole cannot be disrupted even by his imaginary powers of god-like control.[18] On another occasion this lesson is extended to human systems as well. The prince realizes the foolishness in man's pretension of knowing "the great and unchangeable scheme of universal felicity" and of trying deliberately "to co-operate with the general disposition and tendency of the present system of things."[19] Aggregates exceed the reach of human intervention and understanding.

Johnson's choice of the weather as a self-evident example of a self-organizing system—Kant will make the same move a generation later—was itself a revealing sign of the expansion of self-organizing thinking in the eighteenth century. Looking back, Johnson's choice was itself a new departure. The emergence of the modern science of meteorology, Vladimir Janković tells us, had precisely the same logic and same chronology as our own story here. In its seventeenth-century form meteorology had concerned itself with the interpretation of unusual, unstable, and fleeting events—those "meteors" ranging from storms and earthquakes to waterspouts and even flying dragons—that were taken as privileged philosophical "facts," facts that in their particularity were especially suitable for carrying divine messages. The opposite, in other words, of Johnson's assumptions. The latter, by contrast, came out of the subsequent eighteenth-century transformation of meteorology, in which the early modern view of meteorological events as divinely driven and particularist was gradually replaced with the understanding of weather as an integrated, stable, self-balancing system—the same understanding presupposed by Johnson's tale.

## 1760

We close the decade with a short excerpt, but one that points to a suggestive offshoot of the self-organizing move. An anonymous tract with the self-explanatory title *Some Remarks on the Royal Navy: To Which Are Annexed Some Short but Interesting Reflections on a Future Peace* had this to say about peace:

> As the Ocean when it ceases to be agitated and disturbed by Storms and Tempests which come from afar is sure to find its own Level, so the different Continental Powers of Europe, however jarring at present, will soon find their own proper natural Balance of Power when left to themselves.

Using the familiar hydrological language for the self-organizing dynamic, this tract suggested that when left to themselves international relations—at least in Europe—would reach a natural dynamic equilibrium.

These reflections on international relations were hardly unique. In England, we find the same vision repeated by one "Patriot in Retirement," who combined the same hydrological image with an explicit comparison to the self-organization of the economy: "The ballance of power in Europe, is not to be maintained, by any political equation table: it is as sure to find its level, in time, as any branch of trade." On the Continent, the very same insight was expressed at the very same time by none other than Rousseau. There is no reason, Rousseau wrote, to suppose that anyone had actually done anything of substance in order to support "the boasted balance of power in Europe":

> But whether attended to or not, this balance certainly subsists, and needs no other support than itself, if it were to remain uninterrupted: nay, though it should occasionally be disturbed on one side, it presently recovers itself on the other.

"The present system of Europe," Rousseau explained further, "has attained precisely that degree of solidity, which may keep it in a perpetual agitation, without ever effectually subverting it: . . . because no great revolution can now ever happen." Consequently aspiring princes act with "more ambition than judgment," and yet they act in vain, "like the waves of the ocean"—here it comes again, the hydrological image of dynamic order, four years before Rousseau's reverie on the shore of Lake Bienne—"which incessantly agitate its surface without raising it above the shore."[20]

The trigger for this cluster of pronunciations on international relations seems straightforward. While we were busy completing our tour of the 1750s a major war had erupted between the European powers, and this first global conflict, as it is often called, provoked various ruminations about international order and disorder, war and peace. But this context aside, similar visions of international order as a self-organizing system had been expressed before—the earliest example we have found, you may recall, was from the resonant year 1720—and will be repeated later. Kant, for instance, will make a lot of it, as we shall see. So will the Scotsman Lord Kames, who will explicitly compare the beneficial self-regulation of war and peace to that of the weather, both of which are much better off in the hands of providence than had they been left "to our own direction." His compatriot William Robertson, in his aforementioned *History of the Reign of the Emperor Charles V*, singled out Charles's reign as the period when European powers, "formerly single and disjoined," now came together "to form one great political system . . . wherein it has remained since that time with less variation than could have been expected after the events of two active centuries." Scotland's eminent historian laid out how the dynamics of a self-regulating, self-correcting system held the key to the international history of the previous two hundred years.[21]

What makes this particular application of self-organization especially interesting is that it should not really work. A self-organizing system, as we have seen many times, requires an aggregate of a large number of "nodes," an aggregate obeying, arguably, a logic and a dynamic that are distinct from those of each individual component: the qualitative leap from one fish to the school of fish, or from one individual to a whole society and economy. In this case, however, although these visions of international order have the appearance of self-organizing systems, they envision but a small number of actors, typically a handful of European powers. The key to this exception is that these are not simple actors: they are states, not individuals, and thus have an aggregative quality themselves. When critics of the American declaration of independence pondered what would now prevent Captain Kidd the pirate from declaring himself independent in the same manner, the response—a response essential to establishing the legitimacy of the American Revolution—relied precisely on that mysterious qualitative leap that separated the American colonies as collective actors from Captain Kidd as a particular one. As we consider different permutations within what we have called here the family of self-organizing conceptual moves, therefore, we should consider a place for the international-order variety as one that locates aggregation deeper in the system's components rather than in the system itself.

## SELF-ORGANIZING THEORIES OF ECONOMY AND SOCIETY: THE FRENCH

The previous pages introduced a decade's worth of self-organizing moves, providing some texture for the spectrum of possible forms that the language of self-organization could take in mid-eighteenth-century Britain, and at least a crude sense of its pervasiveness. The remainder of this chapter picks up its main thematic goal within the broader structure of this book, namely, the self-organizing conceptualization of human agents and human systems, primarily in relation to society, economy, and history. In particular, we wish to focus on two clusters of the most sustained thinking about human systems in the eighteenth century: one French, one Scottish. Two different traditions of socioeconomic thought. Two different sets of presuppositions, and with two different political outlooks. And yet both ended up heavily reliant on self-organization. At the same time, the French and the Scottish schools represented two different *kinds* of self-organizing narratives, each with its own vision of the self-organizing dynamic: one primarily an organic model of natural equilibrium, the other primarily an evolutionary model of human history. One reminiscent of Townsend, the other more of Rousseau.

"*Toute la Magie de la Société bien ordonnée est que chacun travaille pour autrui, en croyant travailler pour soi.*" In this pithy 1763 statement—"the whole magic of the well-ordered society is that each man works for others, while believing that he is working for himself"—two leaders of the most original group of political economists in mid-eighteenth-century France, the Physiocrats, encapsulated a basic insight running through their writing.[22] Leaving aside the Physiocrats' well-known contention that all real wealth is derived only from agriculture, their overall worldview drew them to self-organization like moths to the flame. The natural world as they saw it was subject to immutable laws. Economy and society were part of this world, and thus were also subject to objective laws that operated independently of human will. Market economy was a complex system, composed of intricate relationships between multiple variables; so much so that it could be said that everything depended in a kind of circular flow on everything else. Just like the new eighteenth-century understanding of meteorology, so the force of Physiocracy was in its ability to subsume localized contingencies—those perturbations that to the individual eye are often the most visible economic phenomena—within a reliable cyclical pattern. But unlike the observers of the weather, consumed by its daily changes, the individuals in the economic system are not merely spectators with a limited vision: they are themselves also actors in this economic system, those who themselves fuel its disruptions while missing the bigger picture.

The pivotal figure in the Physiocratic group—the "Confucius of the West,"[23] at least in their own eyes—was François Quesnay, an aging physician whose physiological theories, especially with regard to blood circulation and its blockages, are often pointed to as inspiration for his economic thinking. In 1763 Quesnay collaborated with his first convert, the marquis de Mirabeau, on an expository tract titled *La philosophie rurale*, from which we have just quoted the pithy statement with the Mandevillian flavor. Variations of the same insight pepper the group's voluminous writings. "Men take so short-sighted a view of things," Mirabeau wrote on one occasion, "and their greed is so avid, that they would continually go astray if they did not set one another back on the right path through the necessity which faces them all, of tending blindly towards the general good."[24]

So far, rather predictable. But two additional points in Quesnay and Mirabeau's 1763 formulation merit further consideration. The first is the frank invocation of *magic* in the workings of the self-organizing society. If one might have overlooked its significance, the next sentence confirmed that the magical element was right at the heart of the matter:

> This magic, the general character and effects of which unfold through the studies that we are undertaking, shows us that the Supreme Being bestowed upon us as a father the principles of economic harmony, when he condescended to announce and prescribe them to us, as God, as if they were religious laws.

The field of knowledge delineated by the Physiocrats is at bottom an elucidation of the mysterious dynamics of self-organization. As we have seen often, it is precisely in this mysterious *je ne sais quoi* that God is found to inhere. God works not through direct divine intervention but by endowing society and the economy with laws that have the standing of religious laws, while in fact operating—as in Mirabeau's other formulation—through the actions of other men that "set one another back on the right path through the necessity which faces them all."

Consider for another moment this image of the multiple actions of men preventing disorder by setting one another back on the right path. It is the image of a stable if dynamic equilibrium, in which temporary, particular disruptions are corrected by the aggregate effects of the collective. A market-price equilibrium "establishes itself necessarily," wrote another member of the group, Pierre-Paul Le Mercier de la Rivière. "This equilibrium cannot be disrupted *but accidentally*," and whenever this happens it will be corrected "by a necessary effect of [market] competition."[25] Le Mercier de la Rivière's hefty tome—the most complete statement of Physiocratic principles in one work—repeats this assertion often, and on one occasion summarizes the basic insight with a phrase that echoes other Physiocratic formulations: "*Le monde* alors *va de lui-même.*"[26] *The world* thus *runs by itself.*

Mirabeau and Quesnay called it "a phrase of great meaning." As we consider in these chapters different forms that self-organizing stories could take, one of the main points of divergence has to do with the role of *time*. Le Mercier de la Rivière's formulation indicates the basic weight assigned to time in the Physiocratic vision, which is, *grosso modo*, nothing. This model of self-organization requires time to keep the system running, but then largely removes the time variable from the equation. In ideal conditions such an equilibrium would continue ad infinitum, perturbed cyclically but always temporarily. The system is dynamic but self-correcting, and always returns to its initial state. Over time, such a self-organizing system is in fact timeless.

And yet the Physiocrats' world did not quite run all by itself. Here is the second point we want to make about our opening quote from *La philosophie rurale*, and in particular about the expression of its goal as a "well-ordered society" (*société bien ordonnée*). These three words hint at an important departure of Physiocracy from much of classical liberal thinking. Although the Physiocrats advocated free trade, and popularized the phrase *laissez faire, laissez passer*, they never intended to leave the economy completely on its own. Theirs was a *dirigiste* form of economic liberalism: its very first principle, in Quesnay's words, was "that there should be a single sovereign authority, standing above all the individuals in the society and all the unjust undertakings of private interests"—or else "the system of counter-forces" in government would undoubtedly replace harmony with discord. The system required a "legal despot" for a continuous oversight over the conditions that guarantee the perpetuation of its self-organizing equilibrium. Yet this legal despot did not need to legislate, since in a fundamental sense the law was already in place.[27] Conditional self-organization, then, if you will—but we must remember that the context of Physiocratic writing was French absolutism, not British constitutionalism.

The Physiocrats were central to French self-organizing thinking about human affairs at midcentury, though their political influence was limited. At the same time, they encouraged others in France to reflect on these matters, often in direct conversation with them. The Neapolitan abbé Ferdinando Galiani, for instance, spent many years in Parisian intellectual circles opposing the Physiocrats. And yet his own position also drew on the language of self-organization. Thus, writing about monetary policy in 1751, Galiani asserted that a beneficial monetary equilibrium is achieved "not from human prudence and virtue but from the lowest stimuli of all—private gain. . . . Providence has so arranged all things that our base passions are often ordered for the benefit of all." Back in Naples Galiani had been a prize pupil of Giambattista Vico, and thus it is not surprising that he carried through strands of Vico's thinking (as discussed in this book's prologue), just as Mirabeau was the one who had for many years preserved Richard Cantillon's precocious manuscript that we encountered in chapter 3. For Galiani, in fact, the result was more far-reaching than the Physiocratic perspective on human affairs. "Great things," Galiani wrote, "surely have small and invisible beginnings, a slow growth, and impregnable power from the start." Great things for Galiani meant "the remarkable and most useful institutions of civil life," as he went on to explain. "Man can neither perceive an institution at its start, nor arrest its growth; neither can he undo it once it is established." Although "as men see the good ordering of things accomplished, they credit themselves with having wished to institute such things," the truth of the matter is that these institutions grow of themselves, from small and invisible beginnings, beyond human understanding and agency.[28] Sounding much like his teacher in the 1720s, Galiani was inching toward a far-reaching theory of evolutionary social development, indeed a self-organizing narrative of history.

But it was another figure in France, much more influential, sometimes associated with the Physiocrats but not part of the group, who probably pushed self-organizing social thinking further than any other. For us too, his role in our mise-en-scène goes further than most, for it is hard to resist the temptation to see the career of Anne Robert Jacques Turgot as itself something of a parable for the story of this book.

Hear Turgot at age twenty-three. Woe to those nations who fall under "systematizing legislators" like the ancient Lycurgus, whose laws, often wrongheaded, acquire a "fatal immutability":

> More happy are the nations whose laws have not been established by such great geniuses; they are at any rate perfected, although slowly and through a

> thousand detours, without principles, without perspectives, without a fixed plan; chance [*le hasard*] and circumstances have often led to wiser laws than have the researches and efforts of the human mind.

What produces "these slow and successive advances" (*ces progrès lents et successifs*), as the modern editor of these reflections points out, is not the deliberate intent of any individual, nor a divine intervention. These advances occur, it seems, of themselves. Such wise laws—note how Turgot's formulation circumvents the attribution of agency—"know how to steer toward the public good the interests, passions and even vices of private individuals."[29] This is much more than Mandeville. It is self-organization squared: not only do private vices spontaneously aggregate together at a given moment as public benefit, but small localized changes in the organization of society—its laws—spontaneously aggregate together over a long period of time to create the best possible conditions for the Mandevillian alchemy.

Turgot delivered these words in Latin, not French. They were the first of two discourses that he presented as an aspiring young theologian at the Maison de Sorbonne. Third son to an old noble family, Turgot had been destined, groomed, and educated for an ecclesiastical career. His first Sorbonne discourse was appropriately titled "The Advantages Which the Establishment of Christianity Has Procured for the Human Race." Given his topic and audience, therefore, it is not surprising that Turgot compounded this seemingly God-free narrative of gradual political progress with a perhaps not altogether consistent claim about the essential role of Christianity. Only Christianity, he insisted in the same oration, with its vision of all men as brothers, had succeeded in uprooting the injustice that all human societies instilled into their laws in preferring a small number of people over the greater mass of humanity. If society was a long-standing self-organizing mechanism, Christianity was its enabling oil.

By Turgot's second discourse of half a year later, however, the balance had tilted, beginning with its title, by now free of any religious gesture: "Philosophical Review of the Successive Advances of the Human Mind." The argument was that "the whole human race, through alternate periods of rest and unrest, of weal and woe, goes on advancing, although at a slow pace, towards greater perfection."[30] So if in the earlier lecture Turgot had dispensed with legislators à la Lycurgus, representing mythical singular historical moments of law-system making, now his gradualist model of undesigned historical evolution required him to face up to another exceptional figure, the ultimate singularity in human progress: the genius. Newton, say.

Turgot's paragraph on genius begins with a picture painted with brush-strokes that seem too broad for the topic at hand.

> Like the ebb and flow of the tide, power passes from one nation to another, and, within the same nation, from the princes to the multitude and from the multitude to the princes. As the balance shifts, everything gradually gets nearer and nearer to an equilibrium, and in the course of time takes on a more settled and peaceful aspect.

Turgot goes on to explain how this gradual process pushes societies toward greater civilization: less destructive wars, larger social groupings, wider circulation of ideas, and ultimately the more rapid "advancement of arts, sciences, and manners." Finally the stage is set for the arrival of genius:

> Amidst this complex of different events, sometimes favourable, sometimes adverse, which because they act in opposite ways must in the long run nullify one another, genius ceaselessly asserts its influence. Nature, while distributing genius to only a few individuals, has nevertheless spread it out almost equally over the whole mass, and with time its effects become appreciable.[31]

Once again Turgot's self-organizing logic asserts itself in a crescendo of superimposed levels. History's ebbs and flows gradually produce a net human advancement. Outlier events in the long run cancel each other out, leaving as a sum result this slow-moving equilibrium. A flash of genius is precisely such an event: although it appears singular, like the meteor, its real significance is comprehensible only in a statistical aggregate view, like that of the new meteorology. With the realization that flashes of genius, statistically, are equally spread, that their aggregate effects are predictable, and that their cumulative consequence can only be appreciated through the long passage of time, it becomes evident that although it may *appear* that the unique genius pushes society forward, the truth remains, in Turgot's carefully chosen words, that "humanity perfects itself."

And God? What was the divine role in effecting this ongoing perfection of humanity? About halfway through the discourse Turgot emits a passionate cry. "Holy religion, could it be that I should forget you? Could I forget the perfecting of manners, the dissipation at last of the darkness of idolatry, and the enlightenment of men on the subject of the Divinity!" The nongrammatical exclamation mark could hardly mask that this was far from a rhetorical question, and Turgot's audience could perhaps have sensed the true answer. They might not have been overly surprised, therefore, to learn that a few weeks later Turgot decided to quit his theology studies and renounce his ecclesiastical ambitions. In a letter to his father Turgot explained that they were no longer compatible with his philosophical convictions. Instead he decided to pursue a career in service to the crown.[32]

Turgot's championing of disorder grew ever more fundamental. When a few years later he clarified his ideas further in a work titled *On Universal History*, he no longer pointed out only the limitations of the supposed reason and design of the benevolent far-sighted legislator, but rather the limited benefit to society of reason *tout court*. Which is of greater benefit to the progress of human society, Turgot wondered, "the passions, tumultuous and dangerous as they are," or "reason, which is justice itself"? If reason had reigned throughout history, he conjectured, it "would have banished wars and usurpations for ever, and would have left men divided up into a host of nations separated from one another and speaking different languages." Consequently the human race "would have remained for ever in a state of mediocrity. Reason and justice, if they had been more attended to, would have immobilised everything, as has virtually happened in China." Fortunately, however, reason had actually taken a back seat to the passions; and those, disorderly and unpredictable as they may be, "became a mainspring of action and consequently of progress . . . led to a multiplication of ideas, the extension of knowledge, and the perfection of the mind." (The individual mind itself, Turgot added a few pages later, replicates precisely the same process: it begins with "a state of confusion" in which sensations, sounds, colors "assail the mind on every hand, and throw it into a kind of intoxication which is nevertheless the germ of reason.") Turgot's diagnosis concluded in an unmistakable self-organizing vein. Throughout history ambitious people have contributed "to the increase in the happiness of the human race, with which they were not concerned at all. Their passions, even their fits of rage, have led them on their way without their being aware of where they were going."[33]

Turgot was no longer merely a philosopher or pamphleteer, but primarily a public administrator with an increasing interest in the details of economic policy. He continued to suggest that the entire economy is a self-correcting system of interacting variables; as he put it in a letter to David Hume in 1767, if we change one factor, "it is impossible that there should not occur in the entire machine a movement which tends to re-establish the former balance or equilibrium." At the same time, he continued to insist on the limits of what an individual in his position could know or control. Thus an administrator can only "in vain" attempt to calculate economic policy to affect prices, say, when facing the inevitable "multitude of obscure facts, the action of a throng of unknown causes, acting slowly and by degrees, the effect of which cannot be discerned but after it had accumulated through the passing of time." In his oft-cited eulogy for Intendant of Commerce Vincent de Gournay, in truth more a succinct outline of his own economic credo, Turgot defended Gournay against the charge that he had been a "man of system." Surely one could not say that of a man whose

> whole doctrine was founded on the impossibility of directing, through constant rules and continuous observation, a multitude of operations whose very immensity prevents their knowledge, and which, moreover, depend continually on a multitude of ever-changing circumstances that one can neither master nor even predict.[34]

Turgot may well have been talking about himself. Prophetically perhaps, given that fifteen years later, in 1774, he would rise to the position of the French *contrôleur général des finances*, a position that would allow him to begin implementing economic reforms in this putatively "system-less" vein, until the political opposition succeeded in removing him from office two years later.

Turgot, then, is a poster boy for the story of this book. He began as a theologian concerned with the workings of providence. He then left the church, having become uneasy with the implications of a self-organizing narrative of human history for insisting on God's active role in human affairs. He carried the self-organizing logic to increasingly more fundamental levels of social organization. He turned to administration and in particular to economic policy, where he drew the practical conclusions from his self-organizing thinking—in ways that contemporaries like Du Pont de Nemours and Condorcet believed influenced Adam Smith's *Wealth of Nations*—and finally attained the supreme position of power, from which he could try to implement them. Not that carrying through an anti-Lycurgus line when he was actually in charge of policy and reform was an easy task. The tension was evident, for instance, in Turgot's discomfort with the heavy-handed top-down *dirigisme* in managing local affairs that the Physiocrat Du Pont de Nemours advocated, supposedly on Turgot's behalf, in his well-known *Mémoire sur les municipalités* (often attributed, wrongly, to Turgot himself).[35]

If Turgot's career is a parable for the emergence of self-organization, then the final act, in which Turgot's power came to an end, gives this parable one further twist, one that looks forward to the political pliability of self-organization that will preoccupy us again in the next chapter. Several months before Turgot was driven out of office he had his day in court. Literally: a *lit de justice* in early 1776 in which the young Louis XVI registered Turgot's Six Edicts, the pinnacle of the controller-general's reforms, which included the suppression of guilds and the abolition of the corvées (seigneurial work dues). The event, ending weeks of stormy debates between Turgot and the Parlement of Paris, was described by Adam Smith in his marginalia to the published text as the "most valuable monument of a person whom I remember with so much veneration," and more recently by one historian as one of the great set pieces of late-eighteenth-century European politics. What is of interest to us here is not so much Turgot's own case, but rather the defense of the traditional corporate order presented by the other side.

When the remonstrance of the Parlement to the king a few days earlier stated that "these institutions are not of the kind that was formed by chance (*le hasard*) or that can be changed by time," it directly and explicitly rejected Turgot's vision of social and political self-organization, based as it was precisely on the operations of *le hasard* over time. But when the advocate-general Antoine-Louis Séguier presented the impassioned arguments against the abolition of guilds and regulations to the king at the *lit de justice* itself, the sharp-eared listener would have been surprised by a rather different tune:

> Guilds can be considered as so many small republics occupied solely with the general interest of all the members of which they are comprised; and if it is true that the general interest forms itself from the meeting [*réunion*] of the interests of each particular individual, it is equally true that each member, in working for his personal advantage, works necessarily, even without willing it, for the true benefit of the whole community.[36]

In the middle of Séguier's defense of the guilds, they unexpectedly turned out to be the best frameworks for the spontaneous emergence of aggregate benefits, deriving from the self-interest of individuals who have no investment in such an outcome. The final words of the parable thus belong to Turgot's opponents. Even as they rejected his liberal economic policies and brought him into disgrace, their language revealed the underlying success of self-organizing thinking, which had swept right across the political divide.

## SELF-ORGANIZING THEORIES OF ECONOMY AND SOCIETY: THE SCOTS

In his youth, Turgot translated some of the writings of David Hume. His uses of the language of self-organization turn out often to be closely reminiscent of similar passages in Hume. Regardless of whether Turgot came to these ideas independently, or was inspired by his early encounter with Hume's writings, there is little doubt that the closest affinity to Turgot's take on the history of civilization can be found among the Scots of his generation. Looking back from the beginning of the nineteenth century, an eminent Scotsman of a younger generation, the *Edinburgh Review* editor Francis Jeffrey, found the self-organizing insight to have been a defining feature of the Scottish Enlightenment. Jeffrey described "one great object" of those "original authors" as follows:

> to trace back the history of society to its most simple and universal elements,—to resolve almost all that has been ascribed to positive institution into the spontaneous and irresistible development of certain obvious principles,—and to show with how little contrivance or political wisdom the most complicated and apparently artificial schemes of policy might have been erected.[37]

A grand unifying scheme of eighteenth-century Scottish moral philosophy, in other words—Jeffrey was referring here specifically if not exclusively to Lord Kames, Adam Smith, and James Millar—was the demonstration that the history of society and its institutions was the product of a long-term, undesigned, self-organizing evolution.

Scholars have often concurred with Jeffrey's observation. The intellectual historian Duncan Forbes, for one, expresses astonishment at the Scottish Enlightenment writers' recognition of the fact that human progress is "a magnificent result, which is out of all proportion to its cause . . . one that cannot have been directly brought about by the conscious but meagre efforts of isolated individuals." For Forbes this breakthrough of "Adam Smith and his friends," their understanding of the "law of heterogeneity of ends," by which he means the effects of unintended consequences, constituted nothing less than the "deepest insight into the historical process that the rationalist eighteenth century ever attained."[38]

The reader by now is well aware of the broader and longer-term context in which to situate the Scots. At the same time it is also true that as a group the Scottish Enlightenment writers ran further and more consistently with the language of self-organization than anyone else, turning it into a kind of axiomatic component in their account of human history. Since this is the best-known part of our story we can dispense with it here with relative brevity, focusing on three friends at the center of Scottish intellectual life. Before we get to Adam Ferguson and Adam Smith, pride of place must go to the first modern philosopher who analyzed causality—in nature or human affairs—as a thing emergent in time rather than inherent in the world: David Hume.

One of Hume's most famous essays, "Of the Rise and Progress of the Arts and Sciences" (1742), begins with an apparent non sequitur on the difficulties of distinguishing contingent from substantive causes.

> If I were to assign any general rule to help us in applying this distinction, it would be the following, What depends upon a few persons is, in a great measure, to be ascribed to chance, or secret and unknown causes: What arises from a great number, may often be accounted for by determinate and known causes.

Hume goes on to elaborate in some detail the significance of aggregation for the identification of predictable causes, before getting to the point, which turns out to be the incidence and effect of genius. The cultivators of arts and sciences are few and far between, and each one is ruled by chance or secret and unknown causes (a phrase that Hume carefully repeats). But the overall aggregate environment within which they sporadically appear is not. Since it concerns "a whole people," it "may, therefore, be accounted for, in some measure, by general causes and principles."[39] Like the shift from meteors to meteorology, like Turgot's soon-to-follow discussion of the very same point about the incidence of genius, Hume insists on aggregation as the correct statistical environment within which the singularity of flashes of genius can be evaluated, analyzed, and predicted.

But the power of aggregates for Hume is greater than just ensuring regularity. They can also produce the alchemy that turns individual human actions that are not necessarily good into a cumulative beneficial outcome. In the case of certain "natural" social virtues like humanity and benevolence, Hume wrote a few years later in the *Enquiry concerning the Principles of Morals*, the effects of humane or benevolent actions are immediate and direct, and they apply to individual objects. The case is different, however, with other social virtues, "artificial" ones, like justice:

> They are highly useful, or indeed absolutely necessary to the well-being of mankind: but the benefit resulting from them is not the consequence of every

> individual single act; but arises from the whole scheme or system concurred in by the whole, or the greater part of the society. . . . The result of the individual acts is here, in many instances, directly opposite to that of the whole system of actions; and the former may be extremely hurtful, while the latter is, to the highest degree, advantageous.[40]

It is hard to imagine a sharper two-sentence summary of the logic of self-organization.

Furthermore, the way that lack of design produces a beneficial outcome is something to be marveled at, not taken lightly. That the laws of human society have "a direct and evident tendency to public good, and the support of civil society," says Hume, must be acknowledged as "remarkable." If men were actually invested in the public good, they would never have needed such a system of laws in the first place, nor would they have constrained themselves by it. By definition, then, the true origin of the system of laws and justice is not a concern for the public good but rather self-love:

> and as the self-love of one person is naturally contrary to that of another, these several interested passions are oblig'd to adjust themselves after such a manner as to concur in some system of conduct and behaviour. This system, therefore, comprehending the interest of each individual, is of course advantageous to the public; tho' it be not intended for that purpose by the inventors.[41]

Without design, interested passions "adjust themselves" into a harmonious system.

Even further, it is not simply the case that individual interest is likely to be unaligned with public interest, in the manner familiar from Mandeville earlier or Smith later. In the case of justice, the disjuncture of the aggregate and the particular is compounded by the additional discrepancy between the general application of a law and the particularities of any given case. Regardless of the original intent of the law, Hume points out, the public good is often *not* served by the dispensation of justice in any specific circumstance. "A single act of justice is frequently contrary to *public interest*; and were it to stand alone, without being follow'd by other acts, may, in itself, be very prejudicial to society." Yet this too does not matter in the greater scheme of things:

> But however single acts of justice may be contrary, either to public or private interest, 'tis certain, that the whole plan or scheme is highly conducive, or in-

> deed absolutely requisite, both to the support of society, and the well-being of every individual.

Hume's position is a radical and original one. He leaves behind ancient worries about equity and rigor—where the judge plays such a decisive role in the achievement of justice—and instead outsources justice to the collective. He not only folds false convictions, and even false acquittals, into the inexorable advance of justice. He even allows for violations of justice, as when the law correctly applied produces unjust outcomes, to remain part of the same overall progress. Here is the self-organizing alchemy again in full force and twice over, and apparently remarkable enough for Hume to repeat it himself, in full force and twice over, in book 3 of the *Treatise of Human Nature*. In what has been described as "one of the boldest moves in the history of the philosophy of law," Hume mixes together acts of self-love, or at best enlightened self-love with limited benevolence, and acts of particularized circumstantial injustice, in order to produce a general benevolent order in one of the most fundamental and "sacred" elements of social life, namely, justice itself.[42] (By contrast, we may recall, a more orthodox and/or more timid Thomas Sheridan was moved by a Humean realization of the marvelous spontaneous emergence of the English system of justice to a conviction in "the hand of God throughout.")

Our attention is again drawn to the role of *time* in this type of self-organizing story. In Hume's vision of the emergence of justice, time is an essential variable. On another occasion Hume offered an evocative simile for the nature of humanly developed institutions as distinct from divine ones:

> If we survey a ship, what an exalted idea must we form of the ingenuity of the carpenter, who framed so complicated, useful, and beautiful a machine? And what surprise must we entertain, when we find him a stupid mechanic, who imitated others, and copied an art, which, through a long succession of ages, after multiplied trials, mistakes, corrections, deliberations, and controversies, had been gradually improving?[43]

A long succession of ages is the historical condition of possibility for the accumulation of so many messy trials, mistakes, and dead ends into something complex and harmonious, something that is far beyond the comprehension of the men now living to enact it.

The consequences of this realization are not only historical—William Robertson's parallel insight in his *Edinburgh Review* essay on Peter the Great comes to mind—but also directly political. "Sovereigns," Hume insisted,

> must take mankind as they find them, and cannot pretend to introduce any violent change in their principles and ways of thinking. A long course of time, with a variety of accidents and circumstances, are requisite to produce those great revolutions, which so much diversify the face of human affairs.[44]

A wariness of an active push to great revolutions in human affairs is a lesson from history that should be attended to by both rulers and ruled. Hume drove the point home with a well-placed rhetorical flourish at the end of one of the volumes of *The History of England*, the most widely read of all his works in the eighteenth century. We quoted it earlier, but it certainly bears repeating. What civilized people like the English should conclude from surveying the historical evolution of their vaunted constitution is "the great mixture of accident, which commonly concurs with a small ingredient of wisdom and foresight, in erecting the complicated fabric of the most perfect government."[45] A humbling realization (even if Hume himself appeared to bypass it when his *History* reached such benighted figures as Alfred the Great or Edward I). We will have more to say about the political implications of self-organization shortly.

In Hume's view, then, "a long course of time" was necessary for individual acts—some virtuous, some not—to add up to a benign, stable order. But the time-consuming type of self-organization was not the only one found in Hume's thinking. Working to keep God out of his philosophizing, Hume resorted to one variety of self-organizing move after another, at times quite different and yet sharing the characteristics that place them in the same conceptual family. In his economic writings we can find Hume closer to the Physiocrats, imagining a natural equilibrium that requires no human intervention. Such was the case of the balance of trade and the flow of money, which Hume describes in the familiar hydrological language as like "water, [which] wherever it communicates, remains always at a level."[46]

Hume's most original contribution to self-organizing thinking, however, was when he proposed an altogether different dynamic that transformed individual passions and self-love into general moral sentiments, the spontaneously emergent "natural" virtues. This mechanism is *sympathy*. "No quality of human nature is more remarkable," Hume exclaims, "both in itself and in its consequences, than that propensity we have to sympathize with others, and to receive by communication their inclinations and sentiments, however different from, or even contrary to our own." Through sympathy we make the pleasure or pain of another our own, and judge the merit of the behavior that produced this pleasure or pain. Hence we recognize ourselves what is virtuous, even when it is contrary to our individual inclinations. Moreover, "so close and intimate is the correspondence of human souls, that no sooner any person approaches me, than he diffuses on me all his opinions, and draws along my judgment in a greater or lesser degree." Indeed elsewhere Hume described sympathy as spreading like "contagion." Sympathetic communication is spontaneous and immediate, affecting people in a nonlinear fashion, much like the spread of panic among Montesquieu's soldiers or investors. Consequently, unlike Mandeville, for whom virtue had required the suppression of the passions and derived from the artifice of politicians and moralists, Hume pointed out that such artifice was unnecessary in the formation of moral sentiments. Even "men of the greatest judgment and understanding," he wrote,

> find it very difficult to follow their own reason or inclination, in opposition to that of their friends and daily companions. To this principle we ought to ascribe the great uniformity we may observe in the humours and turn of thinking of those of the same nation.

Hume identified a mechanism that without external direction or design harmonizes contrary individual tendencies to produce a stable aggregate uniformity: precisely what is required for social self-organization. This mechanism was soon to be picked up by Adam Smith and others, including Laplace, as we already saw in the previous chapter. And looking back, Mandeville, in failing to appreciate the complex ways in which self-interest led men to unintentional sociability, was simply not self-organizing enough.[47]

When Adam Ferguson's *Essay on the History of Civil Society* appeared in 1767, Hume privately opined that his younger colleague's work was not worth publishing. Perhaps, though in the next chapter we shall see that in terms of the political implications of self-organization, at least, Ferguson was the more sophisticated of the two. To the extent that Ferguson was not the most original thinker in this group, moreover, this may be because of his sometimes close reliance on Hume himself. And yet, perhaps precisely because of this reliance on Hume, Ferguson's writing includes some of the clearest formulations of the work that the language of self-organization could perform.

Thus writes Ferguson: "Every step and every movement of the multitude, even in what are termed enlightened ages, are made with equal blindness to the future; and nations stumble upon establishments, which are indeed the result of human action, but not the execution of any human design." Surely even "the most refined politicians do not always know whither they are leading the states by their projects." And again:

> The establishments of men . . . arose from successive improvements that were made, without any sense of their general effect; and they bring human affairs to a state of complication, which the greatest reach of capacity with which human nature was ever adorned, could not have projected; not even when the whole is carried into execution, can it be comprehended in its full extent.

Reason plays precious little role in social development. The most complex social institutions evolve in a long-term process without any underlying purpose. The cumulative whole exceeds the understanding of individuals, whether involved in the historical process or surveying the history of institutions with the benefit of hindsight.

Much of this is by now familiar. And yet one further element in Ferguson's vision deserves a pause, before we return in chapter 7 to the conclusions he drew for politics. Hume, we recall, compared the long-term self-organization of social institutions to the artifice involved in the design of a ship. Ferguson went even further, finding its analogy in *animal instincts*. Human institutions, "like those of every animal, are suggested by nature, and are the result of instinct, directed by the variety of situations in which mankind are placed." Through "a species of chance" such evolution allows mankind to "arrive at ends which even their imagination could not anticipate, and pass on, like other animals, in the track of their nature, without perceiving its end."[48] Darwin is not too far away. Nor, for that matter, Joseph Townsend.

### ADAM SMITH AND THE INVISIBLE HAND

No image encapsulates the language of self-organization more famously than the invisible hand. Although the image has an earlier history, typically as the hand of providence, it is often taken as Adam Smith's most significant figurative bequest to modern social and economic thought, and it is no coincidence that we adapted it for the title of this book. And yet it is also often noted that the invisible hand actually appears in Smith's voluminous oeuvre only three times, of which one is in an essay on the history of astronomy that appears either irrelevant or opposite in meaning to the other better-known two. The intellectual historian Emma Rothschild has concluded, after a careful review of these three appearances against the backdrop of possible valences of the term, that "the invisible hand was an unimportant constituent of Smith's thought . . . a sort of trinket."[49] It seems to us that this conclusion is based on a rather narrow interpretation. Seen against the backdrop of decades of European reflection on order and organization, the insight that Smith sometimes telescoped through the image of the invisible hand appears central to his thinking, even if less of a breakthrough than is often assumed.

Considering Smith's two greatest works together, the Germans famously coined the term *Das Adam Smith Problem* to denote the seeming difficulty in reconciling Smith's emphasis on intersubjective sympathy in *The Theory of Moral Sentiments* with that on self-interest in *The Wealth of Nations*. Scores of scholars have meanwhile labored to show that the contradiction is illusory. Be that as it may, it seems clear that the language of self-organization does run consistently through both works. Indeed it is already there in their memorable opening passages.

Self-organization is close to the surface in the first paragraphs of *The Theory of Moral Sentiments*. They focus on *sympathy*, which is that mysterious mechanism by which every man, however selfish, experiences some projection of the feelings of another as if they were his own. This, in turn, renders sympathy crucial to Smith's vision of what holds society together. From sympathy we derive approval or disapproval of particular actions in specific cases, and it is only from a multitude of these "particular usages" that the general rules of morality emerge, in and of themselves, undesigned and unintended. Since these rules are so agreeable to our enlightened reason, we are tempted to say that reason is their source. But it is a mistake, Smith insists, "to imagine that to be the wisdom of man, which in reality is the wisdom of God," effected through "natural principles" alone. Sympathetic passions, in fact, diffuse "instantaneously, and antecedent to any knowledge of what excited them" or to "any such general [moral] rule." For Smith, as for Hume, sympathy is that singular mechanism that lets the virtues organize themselves, sets the individual and society into a relationship of mutual constitution, and allows a multiplicity of random interactions between people to generate something complex, new, and wholesome.[50]

Self-organization is also close to the surface in the opening chapter of *The Wealth of Nations*. These memorable passages use the day laborer's woolen coat to indicate the nearly incomprehensible intricacy of production in a civilized country. Civilized society thus turns out to be a mysterious vast organization, largely unconscious of its own existence. Elsewhere, in laying out society's economic dynamic structure, Smith's self-organization language is more explicit. Once economic actors are left to their own devices—here is the Adam Smith of the textbooks—"the obvious and simple system of natural liberty establishes itself of its own accord," and dynamic variables, once disrupted, "soon return to the[ir] level." In one section, for example, Smith discusses the employments in which people choose to invest, based on their expectations of profit. If they happen to overinvest in one sector, profits will go down and the balance will soon be restored.

> Without any intervention of law, therefore, the private interests and passions of men naturally lead them to divide and distribute the stock of every society, among all the different employments carried on in it, as nearly as possible in the proportion which is most agreeable to the interest of the whole society.

Since this self-adjustment tends toward an equilibrium and since this equilibrium is "nearly" the best one possible, Smith further asserts, rulers are "completely discharged from a duty . . . for the proper performance of which no human wisdom or knowledge could ever be sufficient; the duty of superintending the industry of private people, and of directing it towards the employments most suitable to the interest of the society."[51] No Lycurgus need apply.

Toward the end of his life, in revisions to the final edition of *The Theory of Moral Sentiments*, Smith expounded further on why every aspiring Lycurgus gets it wrong. Such a "man of system", "enamored with the supposed beauty of his own ideal plan of government,"

> seems to imagine that he can arrange the different members of a great society with as much ease as the hand arranges the different pieces upon a chess-board. He does not consider that the pieces upon the chess-board have no other principle of motion besides that which the hand impresses upon them; but that, in the great chess-board of human society, every single piece has a principle of motion of its own, altogether different from that which the legislature might chuse to impress upon it.

Emma Rothschild takes Smith's commitment to the principle of self-motion of individuals in this passage as evidence that the invisible hand is ultimately "an un-Smithian idea."[52] In our view, to the contrary, the whole point of a self-organizing "invisible hand" is that it produces aggregate order from the multitude of actions of self-propelling individuals. The players in the social game of human chess have their own principles of motion, their own swerve, and thus do not require—and cannot simply follow—an external hand that directs them. Their harmony can only emerge naturally, of its own accord.

And yet if rulers and legislators only set the conditions for natural liberty to establish itself—a task that gives them a significant role in Smith's economic liberalism—then how do societies progress? This too takes self-organization, albeit, as we have seen in other cases, on a different level and over long stretches of time. Take, for example, the development of European commercial society, as Smith presents it in a chapter titled "How the Commerce of the Towns Contributed to the Improvement of the Country." In countries with no commerce, the great proprietors of land spent their surplus on their dependents and on "rustick hospitality," gaining in return authority over others. Subsequently, foreign commerce and manufactures provided the great proprietors with something else on which to spend their surplus, which was now of benefit only to themselves. Selfish desire took hold, rather than an impulse to share with others, and concomitantly the proprietors' authority over their dependents waned. At this stage their surplus was spent on supporting tradesmen and artificers, who for their part did their best to supply the boundless desires of the rich but were not as beholden to any specific landowner. This is how commercial society was born:

> A revolution of the greatest importance to the publick happiness, was in this manner brought out by two different orders of people, who had not the least intention to serve the publick. To gratify the most childish vanity was the sole motive of the great proprietors. The merchants and artificers, much less ridiculous, acted merely from a view to their own interest, and in pursuit of their own pedlar principle of turning a penny wherever a penny was to be got. Neither of them had either knowledge or foresight of that great revolution which the folly of the one, and the industry of the other, was gradually bringing about.[53]

Precisely what we have come to expect from mature self-organizing thinking.

And the invisible hand? The famous "invisible hand" passage in *The Wealth of Nations* is in fact less about the self-organizing dynamic of individual actions—in the marketplace on any given moment, say—than it is about the underlying disconnect between the motivations of these individuals and the collective benefits of what they do. The wealth of any society, Smith writes in this passage, is the sum of the produce of all its industry. So when an individual works to increase the value of his own produce, he "neither intends to promote the publick interest, nor knows how much he is promoting it." Rather, he is "in this, as in many other cases, led by an invisible hand to promote an end which was no part of his intention."[54]

For the broader self-organizing potential of the image of the invisible hand, on that second level of the long-term historical process, we need to go back to *The Theory of Moral Sentiments*. The luxuries and comforts and pleasures of the great are really but contemptible and trifling, says Smith, but luckily we do not see this. Instead our imaginations associate them with "the regular and harmonious movement of the system" that produced them. This imaginary association is the source of our desires, and those, however misdirected, fuel the industry and progress of mankind. Thus a landowner is induced by desire to produce more food than he can possibly eat. The excess serves many others, including those who provide "all the different baubles and trinkets" that are consumed by the great. The rich, in short, end up sharing the produce of all their improvements with the numerous poor they employ, though their only purpose in all this is "the gratification of their own vain and insatiable desires."

> They are [thus] led by an invisible hand to make nearly the same distribution of the necessaries of life, which would have been made, had the earth been divided into equal portions among all its inhabitants, and thus without intending it, without knowing it, advance the interest of the society.[55]

Here the image of the invisible hand, not especially innovative, signals the full logic of a self-organizing system, transforming multiple trivial individual desires into public good. It also retains a characteristic if unacknowledged mysteriousness: What guarantees that the distribution of necessaries by this invisible hand would be "nearly" that of a fully egalitarian and just world? And what unspoken difficulties might be lurking in the imprecision of this word, which we have seen Smith use before, "nearly"?

Perhaps most interesting of all is actually Adam Smith's least-known use of "invisible hand" in that early essay on astronomy. Ancients and savages, Smith then said, ascribed terrifying irregularities of nature like "comets, eclipses, thunder, lightning, and other meteors" to the intervention "of some invisible and designing power." Yet they failed to realize the presence of "the invisible hand of Jupiter"—a common phrase in Latin literature to designate thunder—in the mundane regular operations of nature, when "fire burns, and water refreshes; [and] heavy bodies descend."[56]

Some scholars have seen this early example as contradicting Smith's later uses of the image in *The Theory of Moral Sentiments* and *The Wealth of Nations*, others as irrelevant.[57] But in our view Smith invokes here the invisible hand precisely in order to affirm the same logic of self-organization. The context is familiar to us from Samuel Johnson's more-or-less-simultaneous *Rasselas*: the eighteenth-century shift from the particular interpretation of singular meteors to the aggregate understanding of meteorology as a self-organizing system. Smith's meaning in this phrase becomes clearer in the rarely quoted next few lines. Having just poked fun at those ancients who refused to seek the invisible hand of Jupiter in mundane natural occurrences, Smith explains why such an understanding was difficult for them. Man had been "the only designing power with which they were acquainted," and man's designs are always intended to alter the course of nature from that "which natural events would take, if left to themselves." Consequently the ancients believed the gods too acted in the same manner, "not to employ themselves in supporting *the ordinary course of things, which went on of its own accord*," but rather to thwart it. But in this the ancients were wrong. The true work of the invisible hand is the smooth running of those natural things that move *of their own accord*. This, in turn, is what makes Jupiter's special providence or interest unnecessary. Precisely the same formulations, of course, which Smith would use soon thereafter to describe a self-organizing economy.

## CODA: A VISIBLE HAND, DISEMBODIED

A final word about the image of the invisible hand. In 1764, five years after Adam Smith's *Theory of Moral Sentiments*, Horace Walpole had a dream. In his dream "a gigantic hand in armour" stood at the top of a great staircase in an ancient castle. The next day Walpole embarked on the feverish writing of a new novel, which he completed in less than two months. *The Castle of Otranto* broke with prevalent conventions of fiction writing while satirizing cultural preoccupations of the day (especially Shakespearean bardolatry). The novel is about how the universe reestablishes order following a politically disruptive usurpation. The unlikely device driving the restoration of order is the preternatural dropping out of the sky of enormous pieces of armor at key moments in the plot—including, to be sure, a gigantic hand in armor at the top of a great staircase, which is doubtless the largest *visible* hand to be found anywhere in the eighteenth century. Walpole's novel thus pokes gentle fun at the idea that order is restored by an invisible hand. The invisible hand in his story actually throws carefully aimed pieces of armor from the sky, including the comically out-of-proportion disembodied mailed hand, as it works to restore the disrupted order. It even appears explicitly by name when the villain is thwarted at a door slammed "with violence by an invisible hand."[58]

And yet Walpole—who, you recall, opened our book with his self-aware invention of "serendipity" a decade earlier—was also celebrating disorder in his experimental novel. As he explained in the preface to the second edition, he wished to counter the excessive realism of modern novels (read Richardson) with imagination and fancy that can create "more interesting situations." Walpole thus rehabilitated the preternatural occurrences and the unlikely coincidences of ancient romances next to the psychological probabilism of the modern novel. In the words of E. J. Clery, Walpole restored a laissez-faire of fictionality to resist the novelistic subordination of human agency in the interests of a moral goal. A disembodied hand is thus not that far from an invisible one: both disrupt the familiar relationship of cause and effect, agency and intent, that are evoked by the human hand.[59]

*The Castle of Otranto* set the stage for a new literary genre that was about to become extremely popular in the late eighteenth century: the Gothic novel. It was a genre, writes Jesse Molesworth, that rehabilitated the power of coincidence while taming it with the guarantees of "predictable unpredictability, assured randomness."[60] Disorder within overall order. At the same time, we might add, Walpole's novel is itself surprisingly orderly, arranged in five equal parts, with armored body parts falling from the sky at regular intervals to form a whole ghost. In form, too, the first Gothic novel sported disorder subsumed within overall clear order.

Small wonder, then, that Walpole described the writing of *The Castle of Otranto* as itself a kind of self-organizing process. "In the evening [after the dream] I sat down and began to write, without knowing in the least what I intended to say or relate. The work grew on my hands."[61] Of its own accord, apparently, without prior intent or design, the first Gothic novel came into being. It was a process of emergence that began with a dream about a gigantic disembodied hand and ended with Walpole's own hands, shorn of their deliberate agency, experiencing the novel's self-propelled growth.

CHAPTER 7

# The Politics of Self-Organization

Richard Payne Knight has been described as the *arbiter elegantiarum* of turn-of-the-century London society, though much of his salon cred was subsequently lost when he spectacularly misjudged the Elgin marbles as unimportant Roman imitations. But in his heyday Knight was the epitome of sometimes arrogant connoisseurship, an erudition he imparted to the British public in two rather ponderous "didactic poems" in the mid-1790s. In *The Landscape* (1794) Knight offered a theory of picturesque landscaping that eschewed the contrived style of the famous eighteenth-century landscape architect Capability Brown for a more fundamental embrace of irregularity. Verdant growth, he insisted, nature's "labyrinth's perplexing maze," should be left to its own devices, "as chance or fate will have it." The resulting "magic combination," bringing about a "congruity of parts combin'd," had been found "in days of yore" when "each free body mov'd, without control, / Spontaneous with the dictates of its soul."[1] If Knight's prescribed landscape has a self-organizing undertone reminiscent of Alexander Pope's "mighty maze! But not without a plan," it is his second didactic poem of two years later, *The Progress of Civil Society*—once more reliant on Pope's *Essay on Man*, this time explicitly—that brings Knight to our attention here. A heroic effort to capture the full development of human society in 3,238 lines of verse, *The Progress of Civil Society* was a luxuriant paean to self-organization, proffered as the fundamental principle of complex social organization, social progress, and social modernity.

Knight begins by dismissing as undeterminable the question of the world's prime mover, whether it is divine or merely "the wild war of elemental strife."

(The poem later acknowledges the influence of Lucretius.) Instead he recommends a more down-to-earth alternative:

> Let us less visionary themes pursue,
> And try to show what mortal eyes may view;
> Trace out the slender social links that bind,
> In order's chain, the chaos of mankind,
> Make all their various turbid passions tend, [turbid = producing confusion]
> Through adverse ways, to one benignant end,
> And partial discord lend its aid, to tie
> The complex knots of general harmony;
> And as the tides of being ebb and flow,
> And endless generations come and go,
> Still farther spread their ever-lengthening chain,
> And bid, 'midst varying parts, the mass unchanged remain.[2]

The self-organizing vision here is again a double one, and indeed, albeit a tad inconsistently, even a triple one. At any given moment, those "slender" social links transform multiple adverse, discordant, and chaotic pulls into overall benign harmony. Over time, the accumulation of many generations spreads further the ever-lengthening chain of harmony's complex knots. And in the long-term aggregate, while the components ebb and flow, the mass remains in unchanging equilibrium.

True to this declaration of intent, the poem redounds with multiple self-organizing moves. It repeatedly tries to illustrate them with metaphors, though few really work, like the confusing chain of harmony's knots in the previous paragraph. Other metaphors range from the skills of the architect and builder through the movements of vessels at sea or of the tide to the innards of intricate machines. With the aid of these metaphors and others the poem explains how millions of individuals all share one driving passion for fame and power, which leads them to different actions that counteract each other's effect.

> So in those crowds which to one object tend,
> All still press towards, and none reach the end:
> While partial discords, to one centre bent,
> Serve but the general union to cement.

The historical evolution of human society began with private property, creating local attachments that "progressively increase[d] of themselves," so that "by degrees, the embryo town began / As wants or habits form'd its artless plan." This dynamic of institutions forming of themselves, without design or directing hand, applied equally to the equilibrium of population, the development of social ranks, the effects of the principle of honor molded by "accident and circumstance," and the contribution to the "social fabric" of the "adverse claims" made by "balanced interests" through which "union still from separation springs." Echoing lines from his recent poem on landscape, Knight describes how in mature society the common good—production, commerce, innovation, prosperity—is served by the passion for "wanton luxury," "selfish avarice," and "mean self-interest":

> Yet still where'er it leads, its windings tend,
> Through ways discordant, to one general end;
> For though each object of pursuit be vain,
> The means employ'd are universal gain.[3]

Knight's account culminates in the origins of law and government; as his prose argument explains, "complicated laws arise from complicated interests, and produce republics better balanced, than if they had been planned by prospective wisdom." And in verse:

> Thus, from self-balanced rights, republics sprung,
> As parts to parts self-constituted hung;
> More nicely poised, than when, with rule and line,
> Vain prescience passion's limits would define;
> Or varying interest's boundless measures span,
> In the small compass of a pedant's plan.

In a particularly elaborate passage about the social system, too long to quote in full, Knight reiterates that in "well-poized and complicated states" individuals with "jarring passions" come together in "complex knots" to form diversified interests, and the more "multiplied" and "dispersed" the parts, the more the "golden tide . . . / Enlarged in energy and substance grows." Complexity, importantly, must not be regulated "too strictly," lest the "well ordered" social system drain away the "native vigour of the roving mind," leading to the "loss of graces [which] art can ne'er restore." This was precisely the same language that Knight used two years earlier to describe the sterilization of landscape by the likes of Capability Brown. Randomness, chance, and contention are essential. In a well-run polity "taste and accident" must always be allowed to mix; laws should only "secure the acquired means of life, / Nor yet suppress each rising germ of strife."[4] And we could go on and on. From self-interest and contrariness through synchronic and diachronic layers of aggregation with ever-expanding complexity to an energized, prosperous self-organized polity: Knight's poem was a didactic primer in the diverse range of panaceas offered by the language of self-organization.

On the one hand, as we remarked about Joseph Townsend writing in 1786, by this point in time authors showing off their self-organization dexterity were often derivative and predictable, even if Knight's insistence and intricate vision (like Townsend's) outdid most. In 1796, the same year as Knight's poem, the learned satirist Thomas James Mathias poked fun at the gist of public debate on this very issue: "Left to themselves all find their level[:] price, / Potatoes, verses, turnips, Greek, and rice."[5] Figures like Prime Minister William Pitt (the specific butt of Mathias's jibe), be they self-professed acolytes of Smith or mere followers of intellectual fashions, indiscriminately applied the self-organizing logic to every issue that crossed their path.

On the other hand, Knight's 1796 was not at all like Townsend's 1786. During the decade in between the two, historical events had radically transformed the tenor of public discussion and the weight attributed to ideas and utterances: the French Revolution, bringing royal execution and bloody Terror in its wake, and the threat of Jacobin revolution in Britain, real or imagined, resulting in unprecedented heights of excitement and fear, political activity and political repression. Although Richard Payne Knight had been a member of Parliament since 1780, he rarely showed as much interest in politics as he did in his intellectual pursuits. Yet he too, like everyone else, was drawn in by the fervor of the 1790s. Both of Knight's didactic poems end with long disquisitions on the French Revolution. Like other Foxite Whigs, he combined sympathy for the antityrannical revolution with horror at its excesses. For Horace Walpole, an old Whig of a more conservative bent, this was enough to conclude from Knight's verses on landscape that their author was one "who Jacobinically would level the purity of gardens, who would as malignantly as Tom Paine or Priestly guillotine Mr [Capability] Brown."[6] In overheated imaginations in the mid-1790s, little separated the blade of the lawn mower from that of the executioner.

Self-organization too was swept up in the maelstrom. Given Knight's avowed sympathy for the revolution, at least in its earlier stages, it was easy to read radical overtones into his insistence on the creative force and energy of "free-born soul[s]" whose movements should not be overly regulated. Even more so, perhaps, when he celebrated the beneficial effects of "the ferment of contention," of "crowds" of individuals with "jarring passions" whose "partial discords . . . serve but the general union." Knight published the poem within months of the so-called Gagging Acts, the culmination of what has been described as Pitt's Reign of Terror, that were designed explicitly to disperse crowds, suppress political contention, and stop the flow of free public speech. In such a freighted context, what could one make of the following lines about what one hopes to see in a judicious government?

> Where laws secure the acquired means of life,
> Nor yet suppress each rising germ of strife;
> But leave some social rights still undefined,
> To stimulate the forces of the mind;
> To rouze its torpor in the keen debate,
> And wake its calm repose with gusts of hate;
> For jealous hate, that emulation breeds,
> The efforts of the mind still upward leads.[7]

The example Knight cites for such a good government is the Greek city-states, contrasted with the despotism of ancient Egypt. But the echoes of contemporary events—in Knight's defense of perfectible social rights, of keen debate, even of the positive effects of gusts of hate, as well as in the opposition to the suppression of every sign of strife—were unmistakable.

George Canning, for one, saw this clearly. A major contributor to the *Anti-Jacobin*, an influential 1790s journal devoted to defending the government and attacking revolutionary ideas, the Pittite wit and future prime minister took on Knight's poem with a parodying imitation titled "The Progress of Man. A Didactic Poem. In Forty Cantos . . . Dedicated to R. P. Knight, Esq." Canning immediately singled out for critique Knight's vision of self-organization. "Then say, how all these things together tend / To one great truth, prime object, and good end?" asks Canning's mock poem, identifying the verses to which it responds by repeating Knight's own rhymes. Contrary to Knight's assertions, Canning responded, order is maintained not by some inexplicable magic but because "to each living thing, whate'er its kind, / Some lot, some part, some station is assign'd." Canning offers multiple examples from the natural world for this kind of order, before concluding:

> But each, contented with his humble sphere,
> Moves unambitious through the circling year;
> Nor e'er forgets the fortune of his race,
> Nor pines to quit, or strives to change, his place.[8]

Forget the unfettered movements of free-moving souls that harbor Jacobinical dangers, forget the beneficial energy of contention, forget the space for random movement or self-organization. The good end is arrived at through organic, natural, stable organization, in which all know their place.

Thomas James Mathias was of the same sentiment. "I have no romantick ideas of virtues without motives, and of actions without regulations." Following his mockery quoted above of the unthinking application of the self-organizing model to everything from rice to Greek, Mathias insisted that he, by contrast, was no naive believer in order emerging of itself. "I feel myself a member of a regulated society, and I would maintain an established order." This order "should be declared, taught, and enforced, by law, by religion, and by education."[9] Mathias's mind's eye conjured up a whole slew of visible, regulating hands. In a mirror image of the logic dominating Knight's text, this opponent of revolution and supporter of the current social and political order could not bring himself—despite the well-known enchantment of the conservative prime minister with Adam Smith—to espouse self-organization.

Almost. A few lines later Mathias also wrote:

> Half a century is insufficient for any *new power or constitution* to find it's level. It is indeed matter of great patience, as well as of deepest concern, to reasonable men, to observe what is *still* carrying on in this country in defiance of every evil which has been felt, and will long continue to be felt, from the introduction of new principles among other nations.

A long-term perspective on political change—one longer than a single generation—allows the conservative critic to eat his self-organizing cake and keep it too. Rice and Greek might not find their level as simply as Smithian believers have it, but slow, gradual political evolution, as distinct from the sudden quakes experienced by other nations, may still find its beneficial level after all.

What happened to Mathias here was neither a coincidence nor a slip of the pen. The language of self-organization was supple enough and capacious enough to be mobilized for incompatible positions even in such a polarized political landscape. If one conservative writer could calm his audience by saying that revolutionary disruption cannot have dire long-term consequences because "when society again settles into form, as it naturally must do, all men will find their proper level" and station, a progressive voice from the other side could find in the same logic equal reassurance that there are no grounds to fear the importation of "French principles" to Britain, since "it is absurd to imagine that any society can embrace doctrines, destructive of all society:

for human fitness will be sure to find its own level."[10] Both of them, from their respective perspectives, were in the right.

Charles Lloyd agreed with them both. A close associate of Samuel Taylor Coleridge and Charles Lamb, Lloyd struggled in the late 1790s with his (not wholly undeserved) reputation as a radical. In 1798 he attacked William Godwin's notions of marriage in an epistolary novel, *Edmund Oliver*, in which Oliver's mentor, Charles Maurice, is the vehicle for Lloyd's own opinions. More than once Maurice voices what sound like direct refutations of the self-organizing logic. "The systems of the present day seem to suppose," Maurice sneers, "that general happiness is the aggregate of individual misery." They are as mistaken as those who "through the whole of the [economic] system . . . recommend industry with a view to personal aggrandizement" but fail to understand that "the whole mass" is thus "tainted by corruption, and let me add, with corruption it ever will be tainted while each man has a selfish and individual aim in society."[11]

On another occasion, Maurice refutes in no uncertain terms the assertion of a young impressionable heroine that "the sacred spark of truth is frequently elicited in the collision of heterogeneous and opposing principles." Not at all so, explains the mentor.

> You will hear[,] Edmund, in the circles of London, . . . that frequent collision is the only mean of eliciting truth. So far am I from admitting this as a fact, that I would exactly reverse the proposition: and insist that no greatness of character, no vastness of conception were ever nursed except in solitude, and seclusion.

Not only is the alchemical reversal of negatives to an ultimate positive mere fantasy, but so is also "this crusading spirit, which modern philosophy encourages, that disposition of giving 'an identity to imaginary aggregates.'" What a modern (or postmodern) formulation, all the way to the mocking inverted commas, centuries before Benedict Anderson! The pseudo-mathematics of aggregation that assume a qualitative leap from the parts to the whole are mere conjuring tricks. They are equivalent, Lloyd says in Maurice's voice, to an effort to "attain an end without using the means." Aggregation, collision, opposition, self-interest: all creatures of the newfangled political theories, self-organizing theories that have by now become a readily recognizable target, and none of which lead to the beneficial effects they are reputed to produce.[12]

A few months later Lloyd addressed *A Letter to the Anti-Jacobin Reviewers*, who apparently were still in need of further reassurance that the author of *Edmund Oliver* was no revolutionary. After all, Lloyd explained, he would be the first to acknowledge that the origins of society were divine rather than by human contract. "I believe that all things find their level," he asserted, somewhat unexpectedly. Society is formed because human beings "enter, not reasoning, into it, from an inevitable impulse, with a safe yet blind instinct. The collision of their minds"—thus the even more unexpected continuation—"will, in process of time, lead to discoveries in the arts, in the sciences, and lastly to philosophy." Attentive readers of *Edmund Oliver* who remembered Maurice's admonitions could be forgiven a gasp of surprise. "In all this process there is nothing of contract," Lloyd continued, and it is only "the immutable laws of the universe" that, "acting at first as instincts in the minds of the agents, impel man to society. Society is, therefore, the offspring of him who governs the world by general laws," and thus "an institution . . . of divine origin." Having thus defended the self-organizing social order that follows only a divine, not a human, plan, it was but one further step to retract and reformulate *Edmund Oliver*'s critique of commercial society. Commerce, Lloyd now stated,

> is as good as any other stimulus (and I suppose better, as far as it extends, or it would not be permitted) . . . to blend self-interest with ingenuity, to interweave the objects of the individual with those of the community; and to mingle the distinct aims of millions (in whom the oppositions of hereditary frailty, and those of acquired mental character, unite) into a homogenous, and mutually protecting mass.[13]

After rejecting in his novel self-organization as radical, newfangled, and misleading, Lloyd now completely reversed his position and sought in self-organization support for his conservative credentials.

Lest he be considered a mere ideological flip-flop striving to please, Lloyd appended an unpublished dialogue, which he claimed to have written the previous spring, i.e., contemporaneously with *Edmund Oliver*, in which the antirevolutionary character says the following:

> Our mad and modern speculatists would reverse the order of nature; would efface, with a sacrilegious impiety, the character which the Maker of this world has so evidently imprest upon it—the character of a *mysterious imperfection*! As for myself, when I consider the infinity of elements, the contrariety of energies, that are at work in this scene, not made for mortal explanation; I am only surprized at their wonderful adaptation and order![14]

It turns out that Lloyd had had a strictly antirevolutionary vision of order all along, one that mysteriously emerges of itself from infinite contrary impulses. After all, he knew something about contrarieties.

What then is the political valence of self-organization? Were Lloyd's inconsistencies simply a sign of his own scrambling to clarify what was in truth an ambiguous political position, or did they—like others we quoted—capture something meaningful in the political potential of these conceptual moves?

The answer seems rather straightforward. Time and again we have seen self-organization lead to what can be called the anti-Lycurgus position, a negative role for which the exemplary legislator of antiquity was frequently invoked by people like Turgot or Ferguson. If laws, government, social institutions, and political constitutions evolve slowly and imperceptibly from a multitude of human actions, without design, forethought, and overall plan directing them toward this outcome—if Lycurgus is by definition a mythical fiction—then self-organization is inherently conservative. Political order that emerges of itself requires only cautious overseers, not proactive visionaries.

It is therefore not surprising that the grandfather of modern conservatism, Edmund Burke, was so fond of the language of self-organization. The only way to advance social and political reform, he insisted in his epoch-shaping *Reflections on the Revolution in France*, is with the aid of long swathes of time, so "that its operation is slow, and in some cases almost imperceptible":

> The work itself requires the aid of more minds than one age can furnish. It is from this view of things that the best legislators have been often satisfied with the establishment of some sure, solid, and ruling principle in government; a power like that which some of the philosophers have called *plastic nature; and having fixed the principle, they have left it afterwards to its own operation.*

Burke's self-awareness of the meaning of self-organization and the baggage it carries with it is remarkable. He reaches back here all the way to the late seventeenth century, to Ralph Cudworth's concept of "plastic nature," which we discussed in chapter 1: that middle principle between providentialism and mechanism that works "magically" as an "Inward Principle," simultaneously creating unity and diversity. Burke's plastic nature in government, once fixed in principle, can be left thereafter "to its own operation." In a period of time greater than any living generation—and thus beyond the forethought and design of any living soul—such an evolutionary process "enable[s] to unite into a consistent whole the various anomalies and contending principles that are found in the minds and affairs of men." Burke's self-organizing vision—we have seen this too before—thus achieves order on two different levels. At any given moment self-organization ensures "that the parts of the system do not clash" even though it encompasses all those multiple contradictory impulses that characterize mankind. And over time it guarantees the emergence of "not an excellence in simplicity, but one far superior, an excellence in composition." That is to say, an excellent complexity: "the happy effect of following nature, which is wisdom without reflection, and above it."[15]

It was not merely the revolutionary aftershocks that prodded Burke, at the end of a lifelong reforming career, into this conservative position. Already in 1782 he said pretty much the same to the assembly of British legislators in the House of Commons. Our constitution, he thundered (one imagines), is not the outcome of a choice or plan by any group of people.

> It is a constitution made by what is ten thousand times better than choice, it is made by the peculiar circumstances, occasions, tempers, dispositions, and moral, civil, and social habitudes of the people, which disclose themselves only in a long space of time. . . . The individual is foolish. The multitude, for the moment, is foolish . . . ; but the species is wise, and when time is given to it, as a species, it almost always acts right.

The process is inexorable—the species as a whole displays deliberation and acts right—but these are aggregate long-term effects that circumvent any purposeful choice. "Interest, habit, and the tacit convention, that arise from a thousand nameless circumstances"—thus again the postrevolutionary Burke in 1795—"produce a *tact* that regulates without difficulty, what laws and magistrates cannot regulate at all."[16]

So far, so predictable: Burke's organic view of society and the self-organizing language blended seamlessly together. Thomas Paine, Burke's most famous interlocutor, recognized this tendency right away. "A government of our own is our natural right," Paine had written already in his *Common Sense* of 1776; "it is infinitely wiser and safer, to form a constitution of our own in a cool deliberate manner, while we have it in our power, *than to trust such an interesting event to time and chance*." And again, in his blockbuster *Rights of Man* of 1791, directly responding to Burke's reflections on the French Revolution:

> Government in a well constituted republic, requires no belief from man beyond what his reason can give. He sees the *rationale* of the whole system, its origin and its operation; and as it is best supported when best understood, the human

> faculties act with boldness, and acquire, under this form of Government, a gigantic manliness.[17]

The gigantic manliness aside, the contrast between Burke's and Paine's views of the bedrock on which stable and beneficial governments rest could not be clearer.

But government was not the be-all and end-all of Paine's political vision, which educated a whole generation in the last decade of the eighteenth century. On the contrary, in the second part of *The Rights of Man*, which was even more popular than the first, Paine maintained that "a great part of what is called government is mere imposition," and therefore the less government a society has the better. But if "governments, so far from being always the cause of means of order, are often the destruction of it," then where does order come from? Paine learned the answer, he told his readers, while observing the American colonists during the revolutionary war with Britain. The American Revolution produced at first a political vacuum, in which "there were no established forms of government" for more than two years. "Yet during this interval, order and harmony were preserved as inviolate as in any country in Europe." How so? "The instant formal government is abolished, society begins to act," Paine explained; "all that part of its organization which it had committed to its government, devolved again upon itself, and acts through its medium." What makes society capable of acting as a gigantic aggregate agent—it is here that Paine's social vision becomes especially interesting for our purpose—is "the diversity of [man's] wants, and the diversity of talents in different men for reciprocally accommodating the wants of each other." This multiplicity generates "the unceasing circulation of interest, which, passing through its million channels, invigorates the whole mass of civilized man." Ultimately "it is to these things, infinitely more than to any thing which even the best instituted government can perform, that the safety and prosperity of the individual and of the whole depends." The bottom line of this Smithian vision is resoundingly self-organizing:

> The more perfect civilization is, the less occasion has it for government, because the more does it regulate its own affairs, and govern itself. . . . If we consider what the principles are that first condense men into society, and what the motives that regulate their mutual intercourse afterwards, we shall find, by the time we arrive at what is called government, that nearly the whole of the business is performed by the natural operation of the parts upon each other.[18]

As is typical of Paine's rhetoric, he states the principles of his thought—here, the self-organizing principle—with unflinching clarity. Surprisingly, then, Burke and Paine found themselves in agreement about how things really work on a fundamental level. Or, to be more precise, on two different fundamental levels. For Burke, the self-organizing historical evolution of the political order was the argument against tampering with it rashly. For Paine, the self-organizing natural resilience of the social order was what ensures its survival through all manner of political experiments. The question of where self-organization operates, in the order of the political framework or in the social formation that underlies it, would remain a key distinction for the future of political thought.

We know by now, however, that the reliance of both Burke and Paine on the language of self-organization in service of their respective positions in their famous debate is in truth not as surprising as it first appears. Signs of this supple, multiple-sided potential of self-organization were there all along. Long before the French Revolution, for example, the Scotsman Adam Ferguson had already foreshadowed precisely these two seemingly incommensurable positions. As it happened, both together. In the previous chapter we heard Ferguson proclaim in the manner of Turgot or Hume that the evolution of human institutions is a slow, gradual process beyond human design or comprehension. Consequently, he too cautioned, stories such as those of Romulus or Lycurgus should be taken with a pound of salt, since in truth "no constitution is formed by concert." It is only with hindsight that "we ascribe to a previous design, what came to be known only by experience, what no human wisdom could foresee, and what, without the concurring humour and disposition of his age, no authority could enable an individual to execute."[19] This is familiar territory.

And yet even as Ferguson's self-organizing view of historical development appeared to produce a dampening effect on political action and reform—how active can one be if social institutions only emerge as if through natural animal instincts (Ferguson's terms, as we have seen)—at the very same time, his self-organizing view of *society* also led him to other statements with diametrically opposite political implications. Animal instincts belong to living things. A society, therefore, which is an aggregate of living active individuals, cannot be expected to have the same kind of order as that of "subjects inanimate and dead." Ferguson explained:

> The good order of stones in a wall, is their being properly fixed in the places for which they are hewn; were they to stir the building must fall: but the good

> order of men in society, is their being placed where they are properly qualified to act.

Consequently "our notion of order in civil society is frequently false," since "we consider commotion and action as contrary to its nature." Instead we must realize that "the rivalship of separate communities, and the agitations of a free people, are the principles of political life." A proper polity according to Ferguson is one in which people agitate and pull in all directions. "A perfect agreement in matters of opinion is not to be obtained in the most select company; and if it were, what would become of society?"

This, then, is the summation of Ferguson's vision of politics in a dynamic society of free agents. "Our very praise of unanimity, therefore, is to be considered as a danger to liberty." "Liberty is maintained by the continued differences and oppositions of numbers." "The public interest is often secure, not because individuals are disposed to regard it as the end of their conduct, but because each, in his place, is determined to preserve his own." Let individuals agitate, disagree, push for their own interest, engage in action and commotion: then and only then will they come together as a society of "free men."[20] Not quite a recipe for political quiescence, then, even if only the invisible hand of historical time can produce institutions serving the public good from the long-term aggregates of such self-interested actions. In short, not only did Ferguson's sociopolitical analysis prefigure Burke as well as Paine, combining different levels of temporality (the momentary, the *longue durée*) and different levels of analysis (political, social); it also demonstrated that self-organization taken seriously, at least in Ferguson's hands, leads to the realization that such different and seemingly contradictory understandings of politics are actually complementary.

Indeed, on the level of theoretical consideration as well as praxis, it is not surprising that politics too became subject to self-organizing thinking in the eighteenth century, belonging together with economy and society to the mainstays of the Enlightenment's analytics of human affairs. "Politics are a science as reducible to certainty as mathematics," insisted the essayist and minor political figure Soame Jenyns in 1757; its study in his view was precisely as a science of statistical aggregates.[21] But it is in the Age of Revolutions that we can observe most readily how the language of self-organization infiltrated calls for political *action* rather than only political theory. The Burke-Paine debate is a telling double example. We follow it now with four more, from other revolutionary settings, representing widely divergent political agendas. Taken together, their range demonstrates the versatile political uses to which the language of self-organization could be put, and was in fact put, by the end of the eighteenth century. Given that the reader by now is an experienced connoisseur of the language of self-organization, the following examples are presented with relative brevity.

Consider for starters one of the foundational texts of the American republic, *Federalist* 10, which James Madison contributed to the 1787 debate on the ratification of the United States constitution. Its topic was factions. The received wisdom of Anglo-American political thought had always been that factions are bad. "When a nation is divided against itself, how great must be the providence that must save it from sinking!" Lord Chesterfield roared half a century earlier in an utterly typical manner. "When the people are broken into parties and factions, worrying and reviling one another, what a fine harvest it yields to the common enemy!" In the previous century King Charles II, in the Declaration of Breda that launched his restoration as the British monarch, had already decreed by royal dictate "that henceforth all Notes of discord, separation and difference of Parties, be utterly abolished among all our Subjects, whom we invite and conjure to a perfect Union among themselves under our Protection."[22] Such pronouncements went back to Machiavelli, Hobbes, and a host of others: factions had long been known to be disruptive, dangerous, malignant formations in the body politic.

In Madison's view such platitudes were futile posturing. "The latent causes of faction," he wrote, are inevitable, "sown in the nature of man," and become all the more pronounced as society progresses.

> A landed interest, a manufacturing interest, a mercantile interest, a monied interest, with many lesser interests, grow up of necessity in civilized nations, and divide them into different classes, actuated by different sentiments and views.

Furthermore, while "the regulation of these various and interfering interests forms the principal task of modern Legislation," it is a task beyond the actual abilities of the interventionist legislator. "It is vain to say, that enlightened statesmen will be able to adjust these clashing interests."[23] Factions and interests will forever struggle with each other, and there is nothing anyone can do about it.

And yet not all hope is lost. In a radical move, parallel with many others we have encountered in the pages of this book and yet no less brilliant for it, Madison's main argument in *Federalist* 10 turned faction into virtue, and indeed into the very foundation upon which the American republic could be stabilized. The key to this alchemical transformation according to Madison was *scale*, in which inheres the advantage of a republic over a smaller democracy. The greater the number of divisions and interests pulling and pushing in different directions, the less likely they are to combine and align in a single line of action to the detriment of others. (Madison's logic brings once again to mind the early-eighteenth-century analysis of the stock market as beneficial only because people's actions in it do not align, as it also does the even earlier celebration by tolerationists from Bayle through Defoe to Locke of the strife between different religious sects that paradoxically results in harmony.) Multiplicity, diversity, variety, complexity: these are the qualities of the political system that ensure a balanced and beneficial aggregate outcome. Furthermore, this overall effect is not arrived at in a linear fashion. The other advantage of a republic over a direct democracy, according to Madison, is that this diversity of interests is not represented directly, in a simple mechanism of public vote, but through indirect representation that translates the cacophony of jarring interests into a unified voice. "It may well happen that the public voice pronounced by the representatives of the people, will be more consonant to the public good, than if pronounced by the people themselves convened for the purpose."

When Madison reformulated the argument again in *Federalist* 51, the self-organizing logic of this model was clearer to him and to us. "This policy of supplying by opposite and rival interests, the defect of better motives, might be traced through the whole system of human affairs, private as well as public." People's well-meaning intentions cannot be relied upon, so the key to the public good must lie in the structural characteristics of the system, establishing order from disorderly and contradictory impulses without the active design of an overseeing legislator. Order emerges in the aggregate—an aggregate that "itself will be broken into so many parts"—and thus "the larger the society . . . the more duly capable it will be of self government."[24]

Madison's vision was one suitable for a nascent republican system. Someone more skeptical about a decentered political structure might have objected to it with words like the following:

> A great society which is divided into a considerable number of others, and these again subdivided into a still larger number, cannot subsist without a central point: unless for a predominating will, there can be no order and no harmony.[25]

These words, insisting, *pace* Madison, that social divisions like those celebrated in *Federalist* 10 must be orchestrated from above for harmony to emerge, were written by the Frenchman Louis-Sébastien Mercier. Mercier was a moderate French revolutionary, a member of the Convention who voted against the execution of King Louis XVI in 1793, precisely the time when these words were published.

And yet Mercier did not have to sacrifice self-organization in order to achieve a vision of politics centralized around a "predominating will." Mercier's 1793 *Fragments of Politics and History* both insisted on the necessity of an autocratic polity and acknowledged the underlying multiplicity of differing individuals and groups.

> The springs which combine so many contrary motions into one, almost resembling order, are not concealed under the throne of the monarch, but really emanate from many individuals. . . . Often in an obscure cottage, an unseen hand prepares the will of the sovereign; for that of kings is usually adopted from their subjects.—The royal edict has been composed long before the public herald proclaims it in the streets: every one has contributed to it, his idea, his wish, nay, his expression; and when announced, it is obeyed and respected, only as it is sanctioned beforehand by the public opinion.[26]

Mercier's resolution of the great number of contrary pulls is unexpected. In this version of political order, agency is shifted from the king who issues a royal decree to the invisible hands of myriad individuals pulling in contrary directions, itself an interesting twist for this familiar image. Somehow all these small, unseen contributions come together to form public opinion, a unified opinion that is expressed in the decrees of the single ruler and becomes the basis of order and harmony.

To be sure, the notion of one collective political will of all subjects had already had a long history in France. The model prevalent before the eighteenth century involved gatherings of the whole nation in primary assemblies who would send written summaries, the *cahiers de doléances*, to the Estates-General, who would then condense them further into single books for the king. The eighteenth century saw various thought experiments in more abstract processes for the formation of a general will from myriad individual ones. Some, like Voltaire or d'Alembert, found in this process a key role for the influence of men of letters and other gifted individuals like themselves. Others, like Malesherbes, believed in the emergence of an educated, well-informed public whose numerous dispassionate readings of printed media would somehow result in a coherent public opinion. Rousseau, in his *Social Contract*, imagined the formation of a general will as a kind of mathematical process whereby "the pluses and minuses" of the particular, individual wills "cancel each other out, [and] the sum of the differences is left, and that is the general will."[27] None of these procedures was conceived as self-organizing. Thus Rousseau insisted on the importance of direct democratic representation of every individual and, contra Madison, on the negative effects of factions and mediations.

What was new in Mercier's account was the mechanism that "convert[s] in sentiment a whole nation into one individual." Mercier readily admitted that this mechanism was mysterious. "The tie which binds several thousands of men to a single individual has always appeared to me inexplicable: as it is drawn tighter, so it relaxes and elongates by a multitude of little unperceived causes." The difficulty in understanding was hardly Mercier's alone: it is a mechanism, he said, which "could not have been discovered by all the sagacity of genius," and which thus defied even the greatest legislator. "O Lycurgus! When thy legislating brain reflected on all the modifications of the human species, didst thou ever obtain a glimpse of such a discordance in political harmony!" O Lycurgus, didst thou ever imagine that there would come an age in which your name would become universal shorthand for the impossibility of your project?

At the same time, Mercier identified a necessary condition for the transmutation of discord into harmony: aggregation. Every particular case of law, of policy or of principle, although "at the first glance [it] may appear to be founded on reflection, has, like every other operation of the human mind, its caprices and its absurdities . . . the offspring of chance." Nevertheless, law and government "grow and multiply in time in an invisible manner, and in majestic silence." Thus we arrive at the familiar comforts of mysterious, nonlinear, invisibly self-organizing accumulation, even if Mercier reassures his readers of the bottom line often enough to betray a certain lingering unease: "Never will chance come at the profound combinations of a good system of laws"; "the author of nature, after diffusing order on all sides, left not to chance the lot of humanity"; and in a Rousseauian vein, "the individual will is often suspicious, but the general will is always good, and can never deceive."[28] The repeated mention of Rousseau here is not gratuitous: in the *Social Contract* Rousseau was preoccupied with many of the same questions—the relationship of the parts and the whole, the creation of a single general will from a multiplicity of private interests, the emphasis on aggregation on a large scale, the difficulty in seeing the whole. ("How can a blind multitude, which often does not know what it wants, because only rarely does it know what is for its own good, undertake, of itself, an enterprise so extensive and so difficult as a system of legislation? Left to themselves, the people always desire the good, but, left to themselves, they do not always see it."[29]) Rousseau did not offer a self-organizing solution, and might have found Mercier's reassurances a bit naive, but he would have immediately recognized Mercier's project.

Recall, however, that Mercier promulgated these reassurances in 1793, in the middle of an unprecedented political roller coaster ride. Mercier therefore reassured his readers that even revolutions are not inherently dangerous. "Even should all the political laws undergo a visible change . . . the state would subsist nevertheless, because human societies are a species of *polypi* which live in all their parts." For his own choice of metaphor Mercier summoned those threshold organisms that proved so exciting during the Enlightenment, as we saw in chapter 4; like polyps, like humans, or rather human social formations. "If between the part which governs, and that which is governed, the law of equilibrium is destroyed, an intestine agitation will ensue, until the equilibrium shall be re-established." And again: "When the [political] equilibrium shall be too violently broken, it will re-establish itself."[30] Despite the caprices of individual rulers, a stable polity self-organizes through long-term evolution. Despite the upheavals of revolutions, a destabilized polity self-organizes through a self-correcting equilibrium. Here the moderate revolutionary Mercier, with his insistent belief in the powers of self-organization—which surface elsewhere in his book to correct other large systems like population or the distribution of wealth or the circulation of assignats or the global system of states—joins at the same time both the organicism of Edmund Burke and the republicanism of Tom Paine.

Further to the right—much further—was Joseph de Maistre. Perhaps the epitome of the antirevolutionary and counter-Enlightenment writer, a man who according to Isaiah Berlin wanted to destroy everything the eighteenth century stood for, de Maistre responded to the French Revolution with his *Considerations on France*, which is the best-known French equivalent to Burke's *Reflections* and indeed often sounds much like it.

Where does political order come from? "One would have to have very little insight into what Bacon calls *interiora rerum* to believe that men could have achieved such institutions by an anterior process of reasoning and that such institutions could be the product of deliberation." Writing in 1796 in direct response to France's new constitution of 1795, de Maistre explained:

> All free constitutions known to the world took form in one of two ways. Sometimes they *germinated*, as it were, in an imperceptible way by the combination of a host of circumstances that we call fortuitous, and sometimes they have a single author who appears like a freak of nature and enforces obedience.

Even in this latter scenario, however, de Maistre did not allow the "freak of nature" to be an omniscient, omnipotent Lycurgus. Rather it is merely the case of a legislator whose "constitutive acts . . . are always only declaratory statements of anterior rights, of which nothing can be said other than that they exist because they exist."[31]

The example, as so often, was the English constitution. The emergence of representative government in England in the thirteenth century

> was not an invention, or the product of deliberation, or the result of the action of the people making use of its ancient rights; but that in reality an ambitious soldier, to satisfy his own designs, created the balance of the three powers after the Battle of Lewes, without knowing what he was doing, as always happens.[32]

As de Maistre reiterated elsewhere, the English constitution was "a work of circumstances, and the number of these circumstances is infinite"—he lists them over several lines—which through "endlessly multiplying combinations, have finally produced after many centuries the most complex unity and the most delicate equilibrium of political forces the world has ever known." For de Maistre this undesigned self-organizing evolution was proof of the involvement of God:

> Since these elements, so cast into space, have fallen into such meaningful order without a single man among the multitudes who have acted on this huge stage knowing what relation his actions had with the whole scheme of things or what the future was to be, it follows that these elements were guided in their fall by an unerring hand, superior to man.[33]

In contrast to this Lucretian yet godly history, the actions of the French revolutionaries were blasphemy.

Yet even for de Maistre the underlying belief in the ordering powers of self-organization had a calming effect when staring into the revolutionary abyss. God, again, was the great guarantor. "In divine works" as contrasted with puny human ones, de Maistre wrote in the opening passages of *Considerations on France*, "the irregularities produced by the work of free agents come to fall into place in the general order." It is precisely in moments of great disruption of the "usual order," moments like the French Revolution, when rather than "a series of effects following the same causes . . . we see usual effects suspended, causes paralyzed and new consequences emerging," that self-organization steps in to save the day. "Never is Providence more palpable" than in those circumstances bereft of any linear causality, in which "divine replaces human action and works alone. This is what could be seen at that moment in France." How so?

> The very villains who appear to guide the Revolution take part in it only as simple instruments; and as soon as they aspire to dominate it, they fall ingloriously. Those who established the Republic did so without wishing it and without realizing what they were creating; they have been led by events: no plan has achieved its intended end.

The providential-yet-distant invisible hand, working through the dynamics of the unintended consequences of self-organizing human actions—since "God, not having judged it proper to employ supernatural means in this field, has limited himself to human means of action"—was what guaranteed the failure of the revolutionaries, the restoration of order, and the salvation of French society.[34]

We have come full circle from earlier Christian providentialism, which attributed directly to God both ordinary order and extraordinary moments of disorder—but with the key difference that here God's way was achieved, in both ordinary and extraordinary circumstances, through self-organization. Reminiscent of Ferguson or of Mercier, a mixture of a Burkean belief in the evolutionary, long-term spontaneous ordering of political structures, together with a Painite belief in the short-term self-correcting equilibrium that protects society from fundamental unraveling, and indeed harnesses moments of acute crisis to its own providential purpose, allowed de Maistre to be as antirevolutionary as Burke without sounding as hysterical. For our purposes, of course, the thing to note is that even de Maistre, the self-declared enemy of the Enlightenment who made it his goal (thus Isaiah Berlin) to destroy everything the eighteenth century stood for, even he could not get rid of one of the century's most tenacious inventions, the logic of self-organization.[35]

Finally, from the opposite corner of the political arena, and sitting with Mercier in the 1793 Convention, came a different sort of revolutionary, a more radical Jacobin and a more radical champion of self-organization: Louis-Antoine Saint-Just. Saint-Just took the anti-Lycurgus line to its logical conclusion, and presented to the Convention in April 1793 a constitutional proposal that in effect negated the Convention's very goal. His view, the first comprehensive theory of what Dan Edelstein has recently described as the Jacobin strand of natural republicanism, was that society was self-governing and thus did not require positive laws and constitutions. Protected from external threats and left to its own devices, society would essentially take care of itself. Saint-Just explained:

> In general, order does not result from movements of force. Nothing is ordered but that which moves of its own accord and obeys its own harmony; force need only keep away what is foreign to this harmony. This principle applies above all to the natural constitution of empires. Laws repel only evil; innocence and virtue are independent on the earth.[36]

Nothing could be clearer. Order cannot be forced or intentionally orchestrated. If for Burke and de Maistre the English constitution was itself a marvelous feat of complex self-organizing aggregation; if for Paine and Mercier, and de Maistre again, the self-organizing and self-correcting tendencies of society limited the possible damage of imperfect constitutional experiments; if for Madison self-organization was a key principle to take into account in the rational shaping of a constitution for the nascent American republic: then for Saint-Just, perhaps most consistently of all, the spontaneous order that emerges through self-organization, at least in wishful thinking, rendered a political constitution altogether unnecessary.

## ALL'S WELL THAT ENDS WELL?

This chapter is in danger of ending on too optimistic a note. Indeed, this is a danger of self-organizing-speak more generally, especially in its eighteenth-century guise. Spontaneous order, as imagined by numerous figures throughout this book, is good order. The invisible hand delivers what is ultimately a beneficial outcome, even if things look less rosy along the way. This was the case for Hume's accumulation of multiple instances of local injustice in court. This was the case for Malthus with his demographic disasters essential for the long-term equilibrium of the species. This was the case for de Maistre, whose view of war was that it is hardly the evil it is usually taken to be, since "humanity can be considered as a tree that an invisible hand is continually pruning, often to its benefit."[37] (Yes, here it is again, the invisible hand.) Self-organization enacts the utilitarian calculus: the greatest good for the greatest number.

As self-organization entered the nineteenth century, however, one also encountered new kinds of doubts. Is spontaneous order always better than its opposite? Is it justified or logical to embrace the fortuitously self-organizing universe as the best of all possible worlds, or even simply as a good world? Many—probably most—continued to repeat these nostrums as self-evident natural and/or divine law. But not all.

Samuel Taylor Coleridge, for one, had his doubts. In his second "Lay Sermon"—an 1817 peroration on political economy that led John Stuart Mill to dismiss the Romantic writer's economic views as those of "an arrant driveler"—Coleridge repeated the standard account of the expansion of English trade and industry since the second half of the eighteenth century, and assured his readers that he was no enemy to commerce. "I am not ignorant that the power and circumstantial prosperity of the Nation has been increasing during the same period, with an accelerated force unprecedented in any country," and that the modern system "called into activity a multitude of enterprizing Individuals and a variety of Talent that would otherwise have lain dormant." Coleridge's grudging appreciation for these achievements then began to slip, however, revealing what was really on his mind:

> We shall perhaps be told too, that the very Evils of this System, even the periodical *crash* itself, are to be regarded but as so much superfluous steam ejected by the Escape Pipes and Safety Valves of a self-regulating Machine: and lastly, that in a free and trading country *all things find their level.*[38]

The logic, the language, the images are all taken for granted. The system, whose evils we are told are necessary ones, is a *self-regulating Machine* in which all things *find their level.*

Coleridge, in fact, believed this description to be true. His goal was not to deny the prevalent wisdom about how the system works, but rather to use it as a preamble for his sermonizing blast, which deserves to be quoted at length:

> But there is surely no inconsistency in yielding all due honor to the spirit of Trade, and yet charging sundry evils, that weaken or reverse its blessings, on the over-balance of that spirit, taken as the paramount principle of action in the nation at large. . . . Thus instead of the position, that all things *find*, it would be less equivocal and far more descriptive of the fact to say, that Things are always *finding*, their level: which might be taken as the paraphrase or ironical definition of a storm, but would be still more appropriate to the Mosaic Chaos, ere its brute tendencies had been enlightened by the WORD. . . . But Persons are

not *Things*—but Man does not find his level. Neither in body nor in soul does the Man find his level![39]

Quite an indictment of self-organization, hook, line, and self-adjusting sinker. Self-organizing order, celebrated as it was by the champions of the system of trade, is analogous to the brute natural order of pre-biblical chaos, untouched by human rationality and divine logos. It is a primitive form. There is no contesting that it *works*, but it is appropriate only for soulless things, not for sentient persons.

Coleridge went on to illustrate this with indignant examples about the human costs, degrading and dehumanizing, of the ways in which the system self-organizes. "Go, ask the overseer, and question the parish doctor," he enjoined in the context of a factory that lay inert for a whole season before things got better again. Ask them "whether the workman's health and temperance with the staid and respectful Manners best taught by the inward dignity of conscious self-support, have found *their* level again!"

In short, self-organization works but can be a bad thing. Coleridge repeated this basic message as often as he could. (Though it does not *have* to be a bad thing: elsewhere, for instance in celebrating the quasi-magical formation of public opinion from multiple individual ones, Coleridge sounds much like the celebratory Frenchmen earlier in this chapter.[40]) Here is one more example, notable for its rhetorical force:

> I have often heard unthinking people exclaim, in observing differences of price in different parts of the country, What has become of Adam Smith's *level*? . . . Water will come to a level without pain or pleasure, and provisions and money will come to a level likewise; but, O God! What scenes of anguish must take place while they are coming to a *level*! But still the sneer against Adam Smith, as to the simple fact, is absurd. The tide in the rivers Trent and Parrot flows in in a *head*. Now if a spectator should exclaim to a writer on fluids, What has become of your *level* now? Would he not answer, *stay* and see![41]

This passage is again suffused with Adam Smith, and like modern writers speaking about Freud, say, it expects its audience to be sufficiently conversant with the language and way of thinking that Smith stands for even if they had not read a single word from *The Wealth of Nations*. Yet Coleridge's argument turns out to be a radical one. The invisible hand is dangerous. People must be protected from the pernicious consequences of spontaneous order.

The self-organizing system is therefore not the end point of a discussion about the public good but a preliminary constraint upon which improving action must be taken. According to Coleridge, two things can save people from its harmful effects. These were two things that the champions of self-organization denied had any importance in the greater scheme of things: *individual agency* and *the state.* As regards the individual, Coleridge ends his "Lay Sermon" with an exhortation to every member of society to become a better person. "Let every man measure his efforts by his power and his sphere of action, and do all he can do! . . . *Let him act personally and in detail* wherever it is practicable." The fetish of the aggregate needed toppling, in the name of individual action and responsibility. The political economists, Coleridge warned on another occasion, "worship a kind of non-entity under the different words, the state, the whole, the society, &c. and to this Idol they make bloodier sacrifices than ever the Mexicans did to Tescalipoca." At the same time, the state was also important in Coleridge's vision: it must step in in order to redress the "imbalance" of the commercial system, and "our manufacturers must consent to [its] regulations."[42] For Coleridge the state was no longer the "idolatrous" imaginary outcome of spontaneous order, but rather a proactive entity invited to protect all citizens from its harmful effects.

It is only to be expected, of course, that the nineteenth century, with its dramatic social transformations and the increasing attention of some to their dire social costs, witnessed more heretics of this kind, those who acknowledged the self-organizing dynamic but not its providential interpretation. As Malthus himself had put it, in the context of regulations of the corn trade: "It is undoubtedly true, that every thing will ultimately find its level, but this level is sometimes effected in a very harsh manner." At the first general meeting of the British and Foreign Philanthropic Society for the Permanent Relief of the Labouring Classes in June 1822, one of those present spoke as follows: "Things will, it is true, sooner or later find their level; but the present generation, the agricultural and manufacturing interests, may first be ruined, by being deprived of their comfort and happiness." This statement was denounced by another participant in the meeting as an unjustified attack on "Turgot, Smith, Malthus, Ricardo &c.," though the editor of the proceedings assured readers that it was nothing of the kind. Twenty years later Charles Bray, a social reformer with Owenite and cooperativist tendencies, insisted on the same point, no longer shying away from the critique of his forebears. "The Economists hold that supply and demand have an equal tendency to find their level with water, it being of no importance that in the operation whole towns are ruined and whole countries half-starved." So even as Bray was "quite willing to concede that the principle laid down by the Political Economists ought to be true,—that they [*sic*] are true in the abstract," this was for him only the beginning point of a discussion of how to act to prevent widespread distress in practice, not the end point reached by conceptual modeling "in the abstract."[43]

In France, Jean Charles Léonard Sismondi, a political economist who had gradually become more interested in general human well-being than simply in material wealth, echoed the same theme. "Let us beware of this dangerous equilibrium theory that re-establishes itself of its own accord!" he wrote in his revisionist *Nouveaux principes d'économie politique* of 1819. "A certain kind of equilibrium, it is true, re-establishes itself in the long run, but it is after a frightful amount of suffering."[44] We have certainly entered a different mental world from that of the eighteenth-century invisible hands. It was a more sobering and sobered-up world that was now willing, perhaps, to contemplate self-organization while facing head-on the terrifying prospect of the absence of God.

# EPILOGUE

Epicurus lives again in the midst of Christendom.

A YOUNG PHILOSOPHER, 1755

In 1755, at virtually the same moment as Horace Walpole's invention of serendipity that opened this book, an ambitious young philosopher tried his hand at a grand speculative history of the cosmos. In it, he pursued a question that should be pretty familiar by now: How do we get something from nothing? How do we get from the state where "nothing had organized itself," from the crude chaos of undifferentiated stuff, to the beautiful and orderly thing we call our universe? His answer will sound familiar too. Through the "organization of chaos," matter fashions itself "by natural development." Over time, "the urge of a planet to form itself out of the complex of elementary matter" turns it on its axis and then produces the moons that surround it. Planets gather together in galaxies: larger systems emerging from the smaller, the universe growing from chaos to order.

Then the philosopher leaps beyond our little corner of the Milky Way and sees

> how the infinite space of divine presence . . . buried in a silent night, is full of matter that will serve as material for worlds yet to be produced, and of the forces to bring them into motion. With a gentle impulse, it begins these movements, with which the immensity of this empty space will someday be animated. Perhaps a succession of millions of years, of centuries, will have passed before the sphere of organized nature in which we find ourselves, blossoms into the perfection which it now has. . . . The sphere of organized nature is incessantly engaged in extending itself. Creation is not the work of a moment.

Here is the language of self-organization in all its messy glory, the "phoenix of nature" emerging out of its own ashes. Few concerns for metaphysical precision, few concerns to determine *how* nature might work in this way, constrain the exuberance of this cosmological story. We witness this alchemical transformation just as our philosopher did, with "profound astonishment" at the fact of something rather than nothing.[1]

The philosopher was Immanuel Kant. Kant, who has been largely absent from our story so far, is a particularly apt coda for this book, as well as a good index of things to come. Kant also helps us to capture something of the ambivalence about self-organization that characterizes its role in our modern era. For the first twenty-five years of his philosophical life—the years when he wrote his galactic ruminations above—Kant played in the same wild fields as Leibniz, Mandeville, Vico, Hume, Rousseau, Smith, Diderot, Turgot, and the battalion of other writers that have filled the pages of this book. The next twenty-five years he dedicated to fencing these fields in, to enclosing the peculiar forms of eighteenth-century thought inside precise and delimited doxa.

We have already seen doubts about the language of self-organization. As we have shown in earlier chapters, this language was often unstable, incomplete, and internally inconsistent. We argued that these ambiguities were in fact strengths, because they allowed this way of thinking to proliferate across an enormous spectrum of the intellectual and cultural terrain, from the pious to the impious, from the liberal to the conservative, from social theorists to doctors and biologists. But these ambiguities also made self-organization particularly vulnerable to critique. The circular arguments, the hidden forces, the mysterious appearances of something from nothing: all of these were easy game for the dubious. And of the dubious, Kant was one of the most searching.

At the same time—and Kant also helps us to mark this—these forms of organization were not so easy to leave behind. Emergence, immanent order, autopoeisis, and the like are still live concepts today, although they may appear different or at times more rigorous than those we have seen in this book. Here in the epilogue, we are not concerned to connect all the dots between past and present. But we would like to suggest that self-organization remains a feature of the landscape of modernity, discovered and rediscovered, perhaps, at just the moments when the limits of deductive science and divine design, when the limits of humanity's control over the orders it inhabits, become too glaring to overlook.

## A KANTIAN CODA

It is an old chestnut in the history of philosophy that David Hume interrupted Kant's "dogmatic slumber" and set him on a path toward the critical philosophy. The year that Kant published his cosmological ruminations (1755), Hume appeared in German translation, and Kant read him relatively soon afterward. By 1763, he was actively wrestling with exactly the questions that we have seen repeatedly put on the table in the postmechanical age. "What I should dearly like to have distinctly explained to me," he mused in a short work, "is how one thing issues from another thing. . . . How am I to understand *the fact that, because something is, something else is*?" Refusing to be "fobbed off" by mere dogmatic assertions about causality and other metaphysical puzzles, as he put it, Kant was determined to come up with a rigorously foundational philosophy.[2] This would be a philosophy free of those circular explanations, those mysterious feedback loops, that incommensurability of experience and theory, those paradoxical moments of spontaneous production, and finally those agitated motions across scientific, religious, philosophical, social, and political domains that were together such characteristic features of the eighteenth century.

It would be, in short, a philosophy free of the profligacy of the language of self-organization, a secure philosophy that Kant called "critical." From 1781, when he published the *Critique of Pure Reason*, Kant would dedicate himself to this massive project. His approach was deceptively simple. Those diaporeses—those knots that so perplexed and intrigued the subjects of our book, that whole architecture of explanation in which causes and effects entangled—are not in the world, he relentlessly insisted. Instead, the knots are in *you*. They are in your faculty of reason itself, and only by critiquing this faculty—only when reason takes on "the most difficult of its tasks, namely, that of self-knowledge"—can philosophy free itself from the "obscurity and contradictions" that have plagued it.[3]

Here Kant both leaned on Hume and moved beyond. His encounter with the Scotsman taught him, already by 1763, that there is nothing in the concept of an effect that entails a necessary cause. "The will of God is something," but the "world which exists is something completely different," so how can these things be connected?[4] They cannot be connected analytically—that is, you cannot deduce the given world from a prior concept of God's will. So they must be connected "synthetically," that is, by the active work of human reason.

For Hume, this synthesis was a contingent thing, the product of custom and repeated experiences. The human commitment to causality did not preexist experience, in other words, but was only ever an *outcome* of experience in time. The result was skepticism about any philosophy that aims to be foundational. Because causality is the unforeseen consequence of iterated activities and contingent conjunctions between things, it serves only as an empirical rule, the upshot of inferences that we learn to apply to the world around ourselves. Philosophy, in a sense, came to an end here, unable to offer secure explanations for even its most foundational concepts.

For Kant, in contrast, philosophy had to *start* here. His critical philosophy was an effort to construct a rigorous metaphysics free of Hume's skepticism. Causality did not emerge over time, in a spontaneous fashion, from our contingent experiences of the world. Causality is a priori, and a necessary law of human experience. "It is only because we subject the sequence of the appearances and thus all alteration to the law of causality that experience itself, i.e., empirical cognition of them, is possible."[5] In Kant's terms, then, causality is a constitutive principle; that is, it preexists every particular experience we have. And it is a universal principle, meaning that there is an *objective* necessity to causality given by the virtue of the fact that we have experiences at all. The orderliness of nature is not threatened, in consequence, but guaranteed by our understanding of it.

Kant would take a similar approach to those causal loops that we saw associated with the phenomenon of life. In 1763, he echoed his contemporaries when he declared that "the structure of plants and animals displays a constitution . . . which cannot be explained by appeal to the universal and necessary laws of nature." Neither preformation nor the "arbitrary inventions" of Buffon and Maupertuis offered a satisfying explanation, for Kant, of how one tree can make another tree. Nor could he see how systems of living organisms could be explained. The "community" of living things "is such that the nature of things are not alien to each other, but are united in a complex harmony." "They spontaneously agree with each other," he marveled.[6]

By 1790, however, Kant broke the circle of self-organization, at least to his own satisfaction. Living beings, he proposed, have what he called "natural purposes"—they have a telos—and so they are, in effect, "*organized* and *self-organizing being[s]*." An organized being "possesses in itself *formative* power of a self-propagating kind," and it forms itself "after one general pattern but yet with suitable deviations, which self-preservation demands according to circumstances." This "concept of a natural purpose . . . leads necessarily to the idea of collective nature as a system in accordance with the rule of purposes."[7]

This sounds like self-organization. But, Kant argues in a typical fashion, this self-organization cannot be rooted in anything definite in the world. The concept of a natural purpose—that a thing generates itself from its own energies—"cannot be given us in experience," nor can it be "comprehended *a priori*" from the nature of matter. Rather, it is a tool of reason, a "regulative concept for the reflective power of judgement, for guiding research into objects of this kind." For this reason, too, Kant rejects the entire apologetic enterprise of eighteenth-century design theory. To believe that you can actually discern God's presence in the world just by observing its wondrous forms, he comments, is to be trapped in a "delusive circle." The world's purposes tell us nothing about God. Rather, they tell us about our own inability to form any concept "of such a world [as this] save by thinking a *designedly working* supreme cause."[8]

Here, as elsewhere, Kant's critical project wanted to subdue that unruly family of self-organizing ideas so familiar in the eighteenth century. Kant not only repurposed the vocabulary of self-organization—analogy, experience, time, harmony, order, contingency, to name a few—but also its founding problematics. The production of something from nothing; the curious relationships between parts and wholes; the contradictions between necessity and freedom: all of these are (in Kant's language) antinomies that derive from the nature of reason itself rather than from the world. What you take for *real* paradoxes and curiosities, in short, are not features of the world. They are constitutive features of *us*, and therefore can be thought about in orderly, rational, and consistent ways.

And yet. Even for the philosopher of Königsberg, the language of spontaneous order did not disappear. Its resources were perhaps too seductive or powerful for that. Three examples. First, back in the domain of biology. In his final work, the incomplete *Opus postumum*, Kant returned to the questions that he supposedly answered in 1790: the nature of living organisms, the use of teleological arguments, the place of forces in nature, and more.[9] His new answers, and here are three, should sound familiar:

1. If our globe . . . were to bring forth, by revolutions of the earth, differently organized creatures, which, in turn, give place to others after their destruction, organic nature could be conceived in terms of a sequence of different world-epochs, reproducing themselves in different forms, and our earth as an organically formed body—not one formed merely mechanically.
2. There must exist a matter which . . . is internally self-moving and serves as the basis of all other movable matter. . . . This material . . . cannot

be regarded as a merely *hypothetical*, but as a given, originally moving, world-material. . . .

3. [The] organizing force [of the earth] has so arranged for one another the totality of the species of plants and animals, that they, together, as members of a chain, form a circle.[10]

These arguments are made not critically but, as Kant would have it, dogmatically. They are less about conditions of knowledge than about how the world is actually organized, or better, self-organized. The world produces itself; matter moves itself; there are purposive forces in the world. We are back in the realm of spontaneous order, with its circles and mysteries.

Our second example comes from the world of human systems. In 1784, the same year as his essay "What Is Enlightenment?," Kant also published the *Idea for a Universal History with a Cosmopolitan Purpose*, which he opened with the following observation. In order to have "History," he wrote, we need to "hope that what strikes us in the actions of individuals as confused and fortuitous may be recognized, in the history of the entire species, as a steadily advancing but slow development of man's original capacities." The free will of individuals, when considered as members of society, vanishes "in accordance with natural laws" behind the certitudes of statistical aggregates (for which Kant invokes, like so many others, the analogy of the weather). These individuals, moreover, do not really need to know—and rarely do know—the big picture: "While they are pursuing their own ends, each in its own way and often in opposition to others, . . . [they] are unconsciously promoting an end which, even if they knew what it was, would scarcely arouse their interest."[11] The history of the human species is self-organizing.

Thus in 1784. Given the chronology we suggest here for Kant's thinking, and for the diffusion of self-organizing thinking more broadly, it is not surprising to find in his early *Observations on the Feeling of the Beautiful and Sublime* (1764) a celebration of the urge of men to "turn everything around [their] *self-interest*" as the best way to "give support and solidity to the whole," when "without intending to do so they serve the common good."[12] But this self-organizing move persisted in his analysis of human affairs, even after he strove to purge its alchemy from the rest of his corpus. So our last example is from 1795, from Kant's vision of international relations in his essay "Perpetual Peace."

> Perpetual peace is *guaranteed* by no less an authority than the great artist *Nature* herself. The mechanical process of nature visibly exhibits the purposive

> plan of producing concord among men, even against their will and indeed by means of their discord. This design, if we regard it as a compelling cause whose laws of operation are unknown to us, is called *fate*. But if we consider its purposive function within the world's development, whereby it appears as the underlying wisdom of a higher cause . . . we call it *providence*.[13]

Here are the main elements all over again: causality and agency, unintended consequences and invisible providence, chaos and order. This is how the affairs of men work and produce order and harmony in complexity, despite all appearances, and we can call it what we like—"fate" or "providence," or even self-organization.

There is one more reason why this is a good place to end our Kant coda together with our discussion of self-organization in the eighteenth century. As readers may recall, this is not the first time they have encountered a vision of international relations as a self-organizing system. In fact, it had several precedents throughout the eighteenth century. As it happens, the earliest such example that we have come across can be found in an anonymous tract published precisely in the year of the international financial troubles that pushed the language of self-organization over the threshold as a meaningful European phenomenon, 1720.[14]

## THE INHERITANCE OF SELF-ORGANIZATION

Kant's "relapse," as it were, is not interesting to us in absolute terms. We have no desire to hold him to some metaphysical standard, even his own. Rather, we take his ambivalence about self-organization as paradigmatic, a stance shared widely in the modern age. It is an ambivalence, furthermore, that transcends the metaphysical puzzles that interested Kant, one generated rather by a much broader situation. Put crudely, self-organization aimed to answer a set of questions long contained by the traditional resources of Christianity. It did so in dialogue with, and dependent upon, the energetic new social, economic, political, and scientific developments that we have described in this book, which together made the questions more exigent than ever before. When new wine is poured into old skins, however, the results can be uncomfortable, and so the answers that self-organization provided strained the very fabric of questions long integral to the common philosophical, religious, and cultural matrix of the West.

Once the hands that organize our world are made invisible, after all, they become disconcerting. Whose hands are these, one wonders? Are they still

God's hands, just more carefully hidden and more mysterious than ever? Or are they something different altogether, a new ordering force whose nature is as yet undetermined? Our inability to provide a satisfying answer to these questions only heightens the uncertainty with which we confront another unresolvable conundrum, namely, do these hands have our best interests in mind? Are the outcomes that they produce *good*? Are the freedoms they seem to offer really freedoms at all? Or is this just a story we tell to reconcile ourselves to a subjection we're unable to understand or relieve?

These questions were no less disconcerting in the nineteenth century than they continue to be today. Kant's idealist contemporaries—Hegel, Fichte, Schelling, and many others—certainly felt their exigency in their philosophical projects.[15] But beyond philosophy, the questions were even more unsettling. Coleridge, as we have seen, already realized that the disparity between the operations of the parts and the order of the whole is pretty hard to embrace when it plays out in an arena like politics, with real winners and losers in the process. He was not alone in this. Political reformers after the catastrophes of the Napoleonic era were disinclined to simply let things run their course. The political and cultural project of "restoration" was, prima facie, something deeply inimical to the messiness of eighteenth-century self-organization.[16] Nor did nineteenth-century utopians like Auguste Comte, Saint-Simon, Robert Owen, or Charles Fourier have much patience for the *longue durée* of self-organization. Their major premise—that human beings could and must design their way to a more prosperous and harmonious future—stood athwart that entire cultural and intellectual complex that we have described in this book.

But even for those who embraced immanent order—English laissez-faire liberals, for example—dilemmas abounded: Could self-organization be trusted to accomplish social and political progress on its own? How to regulate a society whose very virtue lay in its self-regulation? The answers had practical implications for policy makers hoping to design a political system that encouraged self-reliance.[17] They also had theoretical implications for those hoping to make liberty the paramount political value.

In 1869, for instance, when John Stuart Mill published *On Liberty*, he deplored the "despotism of society," grounded in its uncanny ability to make contingent rules appear "self-evident and self-justifying." Precisely because society emerges in a spontaneous fashion, it has the power to "practice a social tyranny more formidable than many kinds of political oppression, since it leaves fewer means of escape." The answer to despotism, in his view, was more liberty. Yet this therapy itself depended on spontaneous order: only if liberty is *freely assumed* can it serve as a "strong barrier of moral conviction" against society. As a consequence, Mill agonized over how to implement freedom as a policy. Diversity of opinion, for example, requires "universal education" but also "diversity of education"—how can we ensure both?[18] Must the state compel education? How can it, in the words of Rousseau, force people to be free?

Standing in apparent contrast to Mill was Karl Marx, with his vision of the inexorable march of history. Long ago Marshall Berman drew attention to the modernist paradox of the *Communist Manifesto*. The bourgeois order is one in which change and chaos are endemic and essential. "Uninterrupted disturbance of all social conditions, everlasting uncertainty and agitation distinguish the bourgeois epoch from all earlier ones." And yet this chaos, rather than subverting society, actually serves to strengthen it. Marx did write his 1841 dissertation on Epicurus, after all, and the language of self-organization was no less powerful in his later, unfinished *Grundrisse* (1858), here speaking about circulation of capital:

> As much as the individual moments of this movement arise from the conscious will and particular purposes of individuals, so much does the totality of the process appear as an objective interrelation, which arises spontaneously from nature; arising, it is true, from the mutual influence of conscious individuals on one another, but neither located in their consciousness, nor subsumed under them as a whole. Their own collisions with one another produce an *alien* social power standing above them, produce their mutual interaction as a process and power independent of them.

As in Mill, however, what self-organization added to political analysis, it subtracted from political prescription. If the historical process that will eventually dismantle bourgeois order is governed not by design but by the self-organizing clash of materially contending interests (the class struggle), as Gareth Stedman Jones has noted, Marx left undecided the most important question of all: How do we know that this historical process will finally deliver a morally and politically desirable result?[19] Marxism and liberalism both, in their own ways, leave this question wide open, but presumably it is the only question that matters to real people in real time.

These examples are cursory, of course, but they at least gesture toward the productive yet uncomfortable inheritance of self-organization in the political domain. Self-organization was, moreover, no less productive and vexing to nineteenth-century efforts to understand the natural order. It is here, with the work of that remarkable Victorian Charles Darwin, that we conclude this book. For Darwin's work and its legacy show just how powerful the clustered ideas of self-organization have become in our modern age, and just how many uncertainties they continue to leave behind.

When the young Darwin returned from his years aboard the HMS *Beagle*, he began a series of notebooks that records his early thoughts about natural systems. It is a wonderful library of Darwin's ruminations about the world's complexity. Here's one entry, written in 1838 in his usual telegraphic style:

> What a magnificent view one can take of the world Astronomical <& unknown> causes, modified by unknown ones. cause changes in geography & changes in climate superadded to change of climate from physical causes.—these superinduce changes of form in the organic world, as adaptation. & these changing affect each other, & their bodies, by certain laws of harmony keep perfect in these themselves.—instincts altar [*sic*], reason is formed, & the world peopled <with Myriads of distinct forms> from a period short of eternity to the present time, to the future—How far grander than idea from cramped imagination that God created . . . the Rhinoceros of Java & Sumatra.[20]

As Darwin's eye gazed over the variety of the world's natural things, he was struck by their dynamic interrelations, the obscurity of the causes that shape them, and the time that it takes for the world we know to come to pass. How much more marvelous, in his view, this emergence of something from nothing than that "cramped" imagination that constricts all theories of divine design.

In his *Origin of Species*, Darwin compassed this majestic landscape of change. Voraciously curious, he collected information from all over the world, and from his own backyard as well. Using these materials he built a powerful model of the natural world, in which order and accident are intimately entwined, in which individuals participate unwittingly in the grand formation of species, in which an aggregate whole of nature exerts titanic yet invisible forces on all creatures great and small. From the "war of nature, from famine and death, the most exalted object which we are capable of conceiving, namely, the production of the higher animals, directly follows," he ended his book, offering tribute both to the complexity of the world's processes and the laws that subtend them: "Whilst this planet has gone cycling on according to the fixed law of gravity, from so simple a beginning endless forms most beautiful and most wonderful have been, and are being, evolved."[21]

Yet for all its visionary ambition, Darwin's work was ever sensitive to our limitations, and to the difficulties in moving between the messy world of experience and the grand harmonies of nature. There are orderly laws of development, he felt sure, but he was altogether less sure of our ability to see their totality. Most mysterious of all, for him, was the very engine that created this whole beautiful world, that thing he called "variation." Variation is not an "inherent . . . contingency," he believed, but its laws are "unknown" and its results thus "infinitely complex." The whole *purpose* of generation, he wrote in his notebooks, is "the passing through whole series of forms to acquire differences: if none are added *object failed*." This quasi-teleology was mysterious even to Darwin. Why does variability happen? He immediately asked the question, but then admitted he did not know.[22] Where, Darwin asked, does change come from in the first place?

This "gap" in the theory—the kind of explanatory gap so typical of self-organizing stories—hardly went unnoticed. Indeed, it was (and is!) the gap in which advocates of design theory deposited their cherished forces of divine oversight. Thus the Irish botanist William Henry Harvey, writing to Darwin in 1860, asked "what are these 'Unknown Laws of Variation'"? Why does an organism "'*sport*' or diversify itself *unexpectedly*"? Why, in short, does an organism change at all? Harvey's answer was simple: "My reason tells me that they are probably fresh revelations of the same Supernatural Power which *originated* our 'primordial form.'"[23] Darwin replied to Harvey at length. But not to these questions. On these, he remained silent.

Darwin's silence could hardly satisfy Harvey, just as the same silence cannot possibly satisfy those pious critics of Darwinism today. God can always reenter stories of immanent order, if in ever-attenuated forms. The very incompletion of explanation, then and now, leaves spaces filled with a "something," an unknown whose name may as well be God. Given this, the continued conflicts between intelligent design and evolution are unlikely to be resolved any time soon, if ever. Proving the negative—that in some distant and obscure place, God is *not* at work in the world—is hardly a burden that Darwin (or anyone else) would seem eager, let alone able, to shoulder.

Nor is this burden so easy to dismiss. Insofar as order continues to impress and perplex us, it demands explanations—whether natural, social, or psychological—which rarely satisfy the cravings for a certainty that once seemed guaranteed in a providential world. The free market, the clash of interests in political society, evolutionary competition, the development of social and moral standards: in all of these, we often feel, there is an order, but its purposes and mechanisms are opaque. We are confident in the organization of things but startled by how things turn out. In the eighteenth century, this sensibility was given a peculiarly modern form, and, for better or worse, it remains a signature element of the modernity that we still inhabit.

# ACKNOWLEDGMENTS

The seed of this book was planted in 2004, and nourished through early conversations with friends at the Center for Eighteenth-Century Studies at Indiana University, especially Michel Chaouli and Fritz Breithaupt. In the decade that followed, it developed with the support of numerous institutions, whose generosity made possible something unusual in the humanities: a genuinely collaborative work. Indiana University, the University of Michigan, the University of California, Berkeley, and the Hebrew University of Jerusalem all offered their material and intellectual resources in the form of time, excellent libraries, and devoted staff. A Collaborative Research Grant from the National Endowment for the Humanities gave crucial support to the early research and writing of the book. We have also been grateful beneficiaries of the generosity of the Herzog August Bibliothek in Wolfenbüttel, the Frederick Burkhardt Fellowship from the American Council of Learned Societies (twice!), the Folger Shakespeare Library, Berkeley's Doreen B. Townsend Center for the Humanities and the Social Science Matrix, the École des Hautes Études en Sciences Sociales in Paris, and the Institute for Advanced Studies at the Hebrew University of Jerusalem.

More significant than institutions, however, are our many friends and colleagues, without whose interest and expertise this book would have been much impoverished. In addition to various forums in our home institutions, we benefited from the attention and interest of audiences at the Johns Hopkins Humanities Center, the University of Oslo, Columbia University, the William Andrews Clark Memorial Library, Stanford University, Rutgers University, Vanderbilt University, the University of Texas–Austin, Texas A&M

University, Rice University, New York University, McGill University, Washington University in Saint Louis, the Eighteenth-Century Seminar at the London Institute of Historical Research, the Israeli Society for the History and Philosophy of Science, Ben Gurion University of the Negev (Israel), and the University of Sydney.

At Indiana University, many colleagues generously commented on numerous apparitions of the invisible hand: Hall Bjørnstad, Fritz Breithaupt, Michel Chaouli, Konstantin Dierks, Jonathan Elmer, Mary Favret, Constance Furey, Mike Grossberg, Oscar Kenshur, Sarah Knott, Deidre Lynch, Andrew Miller, Eyal Peretz, Robert Schneider, Rebecca Spang, Johannes Turk, and Sonia Velázquez. They were joined in Israel by Eitan Bar-Yosef, Menahem Blondheim, Raz Chen-Morris, Shmuel Feiner, Snait Gissis, Ruth HaCohen, Michael Heyd, Nimrod Hurvitz, Yael Shapira, David Shulman, and Moshe Sluhovsky. At Berkeley, David Bates, Carla Hesse, Thomas Laqueur, Mark Peterson, Joanna Picciotto, and Ethan Shagan were unstinting and sharp-eyed readers of the manuscript. Members of the Berkeley Early Modern Sodality and the Inflections group were similarly incisive and patient, among them Oliver Arnold, Albert Ascoli, Beate Fricke, Cori Haydon, Kinch Hoekstra, Victoria Kahn, Jeffrey Knapp, David Marno, Hélène Mialet, Christopher Ocker, Peter Sahlins, Mario Wimmer, and Alexei Yurchak. Alice Goff not only read parts of the manuscript but also assisted with its final assembly. In Jerusalem Ray Schrire helped with final touches.

In the last stages, David Bell, Daniel Rosenberg, and Yair Mintzker were generous readers of the full manuscript, and Charles T. Wolfe added eleventh-hour suggestions and corrections. Finally, although the list can hardly be complete, the book has been shaped by our dialogues with a constellation of interlocutors, including Stefan Andriopolos, David Armitage, Linda Colley, Lorraine Daston, Helen Deutsch, Laura Downs, Simon During, Jim Epstein, Anne Eriksen, Joshua Fesi, Lynn Festa, Ofer Gal, Tony Grafton, Katherine Ibbett, Colin Jager, Colin Jones, Jonathan Kramnick, John Kucich, Jonathan Lamb, David Lieberman, John Lyons, Tomoko Masuzawa, Sarah Maza, Michael McKeon, Éric Méchoulan, Natania Meeker, Martin Mulsow, Jane Newman, Julie Park, Mary Poovey, Jessica Riskin, Nils Frederick Schott, Silvia Sebastiani, Phillip Sloan, Joanna Stalnaker, Pamela Smith, Jörn Steigerwald, Roger Strand, Liana Vardi, Fernando Vidal, Dorothea von Mücke, and Carl Wennerlind.

At the University of Chicago Press, we wish to thank Alan Thomas, our editor, for his patience and good cheer and Joel Score and Randolph Petilos for shepherding the manuscript. Thanks as well to India Cooper for her careful copyediting, and to Marie Deer for her sharp eyes and invaluable help with the index.

Collaboration is unusual, so we want to end our acknowledgments with the unusual gesture of thanking each other. Coauthorship of such a long book by two quite different writers requires extraordinary friendship. We thank one another for keeping this friendship productive and cheerful over a decade of labor together.

# NOTES

## PREFACE

1. This episode is discussed in the opening pages of Robert K. Merton and Elinor Barber, *The Travels and Adventures of Serendipity* (Princeton, NJ: Princeton University Press, 2004), written in the 1950s; 1–2 quoted.

2. Ibid, 5.

3. Joseph de Maistre, *Considerations on France* (1796), in *The Works of Joseph de Maistre*, ed. and trans. Jack Lively (New York: Schocken, 1971), 47.

4. Scott Camazine et al., *Self-Organization in Biological Systems* (Princeton, NJ: Princeton University Press, 2001), 8.

5. Some texts that have been useful for us include the following: on biology, Erich Jantsch, *The Self-Organizing Universe: Scientific and Human Implications of the Emerging Paradigm of Evolution* (Oxford: Pergamon Press, 1980), and Ilya Prigogine, *The End of Certainty: Time, Chaos, and the New Laws of Nature* (New York: Free Press, 1997); on social systems, Niklas Luhmann, *Theories of Distinction: Redescribing the Descriptions of Modernity* (Stanford, CA: Stanford University Press, 2002), *Social Systems*, trans. John Bednarz Jr. with Dirk Baecker (Stanford, CA: Stanford University Press, 1995), and "The Autopoiesis of Social Systems," in *Sociocybernetic Paradoxes: Observation, Control and Evolution of Self-Steering Systems*, ed. Felix Geyer and Johannes van der Zouwen (London: Sage, 1986); on law, Gunther Teubner, *Law as an Autopoietic System*, trans. Anne Bankowska and Ruth Adler (Oxford: Blackwell, 1993); on mind and cognition, Humberto Maturana and Francisco Varela, *Autopoiesis and Cognition: The Realization of the Living* (Dordrecht: D. Reidel, 1980), Jean-Pierre Dupuy, *The Mechanization of the Mind: On the Origins of Cognitive Science*, trans. M. B. DeBevoise (Princeton, NJ: Princeton University Press, 2000), Warren S. McCulloch, *Embodiments of Mind* (Cambridge, MA: MIT Press, 1965), and John von Neumann, "Theory and Organization of Complicated Automata," (1949) in *Theory of Self-Reproducing Automata*, ed. Arthur W. Burks (Urbana: University of Illinois Press, 1966).

6. Ludwig Wittgenstein, *Philosophical Investigations*, trans. G. E. M. Anscombe (New York: Macmillan, 1953), §66.

7. John G. A. Pocock, *The Machiavellian Moment: Florentine Political Thought and the Atlantic Republican Tradition* (Princeton, NJ: Princeton University Press, 1975), 466.

## PART 1 PROLOGUE

1. "Some Reflections on the Persian Letters" (1754), in Charles de Secondat, baron de Montesquieu, *Persian Letters*, trans. C. J. Betts (London: Penguin, 1993), 283, with some alterations to the translation. For a summary of the continuing efforts to unravel the "secret chain" see Theodore Braun, "'La Chaîne secrète': A Decade of Interpretations," *French Studies* 42 (1988): 278–91; and Randolph P. Runyon, *The Art of the Persian Letters: Unlocking Montesquieu's "Secret Chain"* (Newark: University of Delaware Press, 2005).

2. See also Theodore E. D. Braun, "Montesquieu, *Lettres persanes*, and Chaos," in Theodore E. D. Braun and John A. McCarthy, eds., *Disrupted Patterns: On Chaos and Order in the Enlightenment* (Amsterdam: Rodopi, 2000), 79–90.

3. *The New Science of Giambattista Vico: Revised Translation of the Third Edition (1744)*, trans. Thomas G. Bergin and Max H. Fisch (Ithaca, NY: Cornell University Press, 1968), 102–3, paras. 342, 344.

4. Giambattista Vico, *The First New Science*, ed. and trans. Leon Pompa (Cambridge: Cambridge University Press, 2002), bk. 2, chap. 1, p. 39. Vico had toyed before with an aggregative providential "third way": see David L. Marshall, *Vico and the Transformation of Rhetoric in Early Modern Europe* (Cambridge: Cambridge University Press, 2010), 61–64.

5. Bernard Mandeville, *The Fable of the Bees; or, Private Vices, Publick Benefits*, 6th ed. (London, 1732), and *The Fable of the Bees: Part II* (London, 1729), ed. Frederick B. Kaye, 2 vols. (Oxford: Clarendon Press, 1924; rpt. Indianapolis, IN: Liberty Classics, 1988), Remark Y, 1:250; 2:139.

6. *Fable of the Bees*, Remark G, 1:91; 2:139, 188 (and compare 183). *The New Science of Giambattista Vico*, 3rd ed., trans. Bergin and Fisch, 62, para. 133. Note also 24, para. 38, for Vico's possible gesture to Mandeville's well-known formula, when he explained the "public virtue" of the Romans as "nothing but a good use which providence made of such grievous, ugly and cruel private vices."

7. Last emphasis added. Alexander Pope, *An Essay on Man*, ed. Maynard Mack, vol. 3.1 of the Twickenham Edition of the Poems of Alexander Pope (London: Methuen, 1950), epistle 1 and editor's note, 12. Laura Brown, *Alexander Pope* (Oxford: Blackwell, 1985), 90. See also Michael Srigley, *The Mighty Maze: A Study of Pope's "An Essay on Man"* (Uppsala: Uppsala University Press, 1994).

8. Brown, *Alexander Pope*, 80–81. For the ruling passion see especially Pope's *Epistle to Cobham*, lines 174–77, in Alexander Pope, *Epistles to Several Persons*, ed. Frederick W. Bateson, vol. 3.2 of the Twickenham Edition of the Poems of Alexander Pope (London: Methuen, 1951).

## CHAPTER 1

1. Hans Zacher, *Die Hauptschriften zur Dyadik von G. W. Leibniz: Ein Beitrag zur Geschichte des binären Zahlensystems* (Frankfurt a. M.: Klostermann, 1973), 9, 36; for reprints of

the medals, see 236. Gottfried Wilhelm Leibniz, *Wunderbarer Ursprung aller Zahlen aus 1 und 0* (18? May 1696), rpt. in ibid., 234.

2. Aristotle, *Physics* 191a1.29–31, 191b1.13–15, in *The Complete Works of Aristotle*, ed. Jonathan Barnes (Princeton, NJ: Princeton University Press, 1984). All references to Aristotle hereafter are from this edition and will be given in standard form.

3. Aristotle, *On Generation and Corruption* 320a1.2.

4. René Descartes, "Principles of Philosophy" (1644), in *The Philosophical Writings of Descartes*, trans. John Cottingham et al. (Cambridge: Cambridge University Press, 1991), 1:224, 232, 223.

5. René Descartes, letter to Mersenne, 27 August 1639, in *Philosophical Writings* 3:137 (here criticizing Herbert of Cherbury); Descartes, "The World" (1629), in ibid. 1:97, 93, 91.

6. Amos Funkenstein, *Theology and the Scientific Imagination from the Middle Ages to the Seventeenth Century* (Princeton, NJ: Princeton University Press, 1986), 118; Descartes, "Meditations on the First Philosophy" (1641), in *Philosophical Writings* 2:39; Daniel Garber, "Descartes and Occasionalism," in *Causation in Early Modern Philosophy: Cartesianism, Occasionalism, and Preestablished Harmony*, ed. Steven Nadler (University Park: Pennsylvania State University Press, 1993), 14. On voluntarism generally, and the medieval intellectual ferment generated by an insistence on God's absolute power, see Funkenstein, *Theology and the Scientific Imagination*, esp. chap. 3; Francis Oakley, *Omnipotence, Covenant, and Order: An Excursion in the History of Ideas from Abelard to Leibniz* (Ithaca, NY: Cornell University Press, 1984); Oakley, "The Absolute and Ordained Power of God in Sixteenth- and Seventeenth-Century Theology," *Journal of the History of Ideas* 59, no. 3 (1998): 437–61; and most recently, Hester Goodenough Gelber, *It Could Have Been Otherwise: Contingency and Necessity in Dominican Theology at Oxford, 1300–1350* (Leiden: Brill, 2004), esp. chap. 8.

7. "That the mind, which is incorporeal, can set a body in motion is shown to us every day by the most certain and most evident experience;" Descartes, letter to Arnauld, 29 July 1648, quoted in Garber, "Descartes and Occasionalism," 18.

8. Nicolas Malebranche, *The Search after Truth*, trans. Thomas M. Lennon and Paul J. Olscamp (Cambridge: Cambridge University Press, 1997), 450. Malebranche, *Dialogues on Metaphysics and on Religion*, trans. David Scott, ed. Nicholas Jolley (Cambridge: Cambridge University Press, 1997), 119, 115. For other Cartesian occasionalists, see Garber, "Descartes and Occasionalism," and Steven Nadler, "The Occasionalism of Louis de la Forge," in Nadler, *Causation in Early Modern Philosophy*.

9. "There is nothing": Hobbes quoted in Simon Schaffer, "Wallifaction: Thomas Hobbes on School Divinity and Experimental Pneumatics," *Studies in the History and Philosophy of Science* 19, no. 3 (September 1988): 297, spelling modernized. Thomas Hobbes, *Elements of Philosophy* (1655), in *The English Works of Thomas Hobbes*, ed. William Molesworth (London, 1839), 1:69. Hobbes, *The Questions concerning Liberty, Necessity and Chance* (London, 1656), 231, 105. Hobbes, *Leviathan*, ed. Richard Tuck (Cambridge: Cambridge University Press, 1991), 146–47. See also Douglas Jesseph, "Hobbes and the Method of Natural Science," in *The Cambridge Companion to Hobbes*, ed. Tom Sorrell (Cambridge: Cambridge University Press, 1996), 90–91. In general, see Thomas A. Spragens Jr., *The Politics of Motion: The World of Thomas Hobbes* (Lexington: University Press of Kentucky, 1973).

10. Nicolas Malebranche, *Treatise on Nature and Grace*, trans. Patrick Riley (Oxford: Clarendon Press, 1992), 137, emphasis added. On Malebranche's threatening reputation, see Alan Kors, *Atheism in France, 1650–1729* (Princeton, NJ: Princeton University Press, 1990), esp. chap. 11. For his associations with Spinoza, see Jonathan Israel, *Radical Enlightenment: Philosophy and the Making of Modernity, 1650–1750* (Oxford: Oxford University Press, 2001), 485–87, 495–97. Also see Funkenstein, *Theology and the Scientific Imagination*, 290ff.

11. Michael Witmore, *A Culture of Accidents: Unexpected Knowledges in Early Modern England* (Stanford, CA: Stanford University Press, 2001), 61. John Calvin, *Institutes of the Christian Religion*, ed. John T. McNeill, trans. Ford Lewis Battles (Louisville, KY: Westminster John Knox, 1960), 1.16.1, 1.16.9. Michael P. Winship, *Seers of God: Puritan Providentialism in the Restoration and Early Enlightenment* (Baltimore, MD: Johns Hopkins University Press, 1996), 44.

12. Calvin, *Institutes*, 3.22.2, 1.16.3. Thomas Culpeper, *Essays; or, Moral Discourses on Several Subjects* (London, 1671), 25. For a discussion of paradox and negative theology, see Rosalie L. Colie, *Paradoxia Epidemica: The Renaissance Tradition of Paradox* (Princeton, NJ: Princeton University Press, 1966).

13. Barbara Donagan, "Providence, Chance, and Explanation: Some Paradoxical Aspects of Puritan Views of Causation," *Journal of Religious History* 11, no. 3 (June 1981): 387.

14. W. Gearing, *The Eye and Wheel of Providence* (London, 1662), 9, 12, 25, 32, 236.

15. Stephen Charnock, *A Treatise of Divine Providence* (London, 1680), 53. William Strong, *A Treatise Shewing the Subordination of the Will of Man unto the Will of God* (London, 1657), 158. John Wilkins, *A Discourse concerning the Beauty of Providence* (London, 1649), 18.

16. Wilkins, *Beauty of Providence*, 86–87. Alain of Lille, *The Complaint of Nature*, trans. Douglas M. Moffat (New York: Holt, 1908), 25 (our thanks to Oscar Kenshur for this reference). Thomas Browne, *Religio medici* (London, 1642), 133, 135–36. See also Don Parry Norford, "Microcosm and Macrocosm in Seventeenth-Century Literature," *Journal of the History of Ideas* 38, no. 3 (July 1977): 409–28.

17. George Mackenzie, *The Religious Stoic*, 2nd ed. (London, 1693), 18, 17.

18. Bernard Frischer, *The Sculpted Word: Epicureanism and Philosophical Recruitment in Ancient Greece* (Berkeley and Los Angeles: University of California Press, 1982), 61.

19. Pierre Bayle, *Historical and Critical Dictionary*, (London, 1710), s.v. "Lucretius," 2052.

20. Lucretius, *The Nature of Things*, trans. Frank O. Copley (New York: Norton, 1977), 1.1021–28.

21. Thomas Stanley, *The History of Philosophy* (London, 1660), 4:171.

22. Ibid., 174. See also Pierre Gassendi, *Syntagma philosophicum*, in *Opera omnia* (Lyon, 1658), 1:327.

23. Augustine, *Answer to Faustus, a Manichean*, trans. Roland Teske, SJ (Hyde Park, NY: New City Press, 2007), *The Works of Saint Augustine*, part 1, 20:284.

24. Augustine, *De natura boni contra manichaeos*, trans. Richard Stothert, in *Nicene and Post-Nicene Fathers*, ed. Philip Schaff (Peabody, MA: Hendrickson, 1995), 4:359, 352.

25. Bayle, *Historical and Critical Dictionary*, s.v. "Manicheans," 1151. On seventeenth-century Manichaeism, see Milad Doueihi, *Le paradis terrestre: mythes et philosophies* (Paris: Seuil, 2006), 63–93.

26. Augustine, *The City of God against the Pagans*, ed. and trans. R. W. Dyson (Cambridge: Cambridge University Press, 1998), 22.24, p. 1159; 21.12, p. 1070. Ultimately, Bayle thought, no reasons could be advanced sufficient to reconcile the Christian concept of God with the real events of the world. Not reason but revelation alone could do this. Bayle, "Manicheans," 1152. On Bayle and evil, see Susan Neiman, *Evil in Modern Thought: An Alternative History of Philosophy* (Princeton, NJ: Princeton University Press, 2002), 117ff.

27. Edward Stillingfleet, *Origines sacrae* (London, 1660), 378. Hugo Grotius, *Truth of the Christian Religion*, trans. Simon Patrick (London, 1680), 13. On the sea, see Hans Blumenberg, *Shipwreck with Spectator: Paradigm of a Metaphor for Existence* (Cambridge, MA: MIT Press, 1997), and Clarence Glacken, *Traces on the Rhodian Shore: Nature and Culture in Western Thought from Ancient Times to the End of the Eighteenth Century* (Berkeley and Los Angeles: University of California Press, 1967).

28. For the best recent discussion of materialism as an outgrowth of a "complex synergy between theological criticism, Renaissance Aristotelianism, the rise of the new sciences, debates concerning natural religion, [and] the rise of empiricism," see Charles T. Wolfe, "Materialism," in *The Routledge Companion to Eighteenth-Century Philosophy*, ed. Aaron Garrett, (New York: Routledge, 2014), here 92.

29. On Gassendi's "baptism" of Lucretius, see David Norbrook and Reid Barbour, introduction to *The Works of Lucy Hutchinson* (Oxford: Oxford University Press, 2012), 1:xxix–xxx. On Gassendi generally, see Howard Jones, *Pierre Gassendi, 1592–1655: An Intellectual Biography* (Nieuwkoop: De Graaf, 1981); Lynn Sumida Joy, *Gassendi the Atomist: Advocate of History in an Age of Science* (Cambridge: Cambridge University Press, 1987); Margaret Osler, *Divine Will and the Mechanical Philosophy: Gassendi and Descartes on Contingency and Necessity in the Created World* (Cambridge: Cambridge University Press, 1994); Lisa T. Sarasohn, *Gassendi's Ethics: Freedom in a Mechanistic Universe* (Ithaca, NY: Cornell University Press, 1996).

30. Pierre Gassendi, *Exercitationes paradoxicae adversus Aristoteleos*, in *Opera omnia* (Lyon, 1658), 2:102. Parts of this have been translated in *The Selected Works of Pierre Gassendi*, trans. Craig B. Brush (New York: Johnson Reprint, 1972). On Mersenne, see Peter Dear, *Mersenne and the Learning of the Schools* (Ithaca, NY: Cornell University Press, 1988).

31. Jones finds the earliest signs of this interest in a letter to Gassendi's great friend Nicolas-Claude Fabri de Peiresc, dated 25 April 1626; Jones, *Gassendi*, 25.

32. Gassendi, *Syntagma*, in *Opera omnia* 1:367, 334, 343, 336. The Lucretius reference is to *Nature of Things* 1.823–29.

33. Gassendi, *Syntagma*, in *Opera omnia* 1:283ff. (on Aristotle); 1:250–51 (on Plato); 1:334. On Gassendi and the causal powers of atoms, see Antonia Lolordo, *Pierre Gassendi and the Birth of Early Modern Philosophy* (Cambridge: Cambridge University Press, 2007), 140ff.

34. Gassendi, *Syntagma*, in *Opera omnia* 1:336, 337.

35. Gassendi, *Epistola III: De motu impresso a motore translato*, in *Opera omnia* 2:487. Gassendi, *Exercitationes Paradoxicae*, in *Opera omnia* 2:168.

36. Gassendi, *Syntagma*, in *Opera omnia* 1:279. See Gassendi quoted in Osler, *Divine Will*, 53: "There is nothing in the universe that God cannot destroy, nothing that he cannot produce; nothing that he cannot produce, even into its opposite qualities."

37. Richard Kroll, "The Question of Locke's Relation to Gassendi," *Journal of the History of Ideas* 45, no. 3 (July–September 1984): 343. On the intellectual position of Gassendi's Epicureanism vis-à-vis Aristotle, Plato, and Descartes, see Lolordo, *Pierre Gassendi*, esp. chap. 2. For a sustained argument for the relevance of what he calls "neo-Epicureanism" in seventeenth-century England, see Kroll, *The Material Word: Literate Culture in the Restoration and Early Eighteenth Century* (Baltimore, MD: Johns Hopkins University Press, 1991), esp. chaps. 3–5. See also Reid Barbour, *English Epicures and Stoics: Ancient Legacies in Early Stuart Culture* (Amherst: University of Massachusetts Press, 1998), and Catherine Wilson, *Epicureanism at the Origins of Modernity* (Oxford: Oxford University Press, 2008). From the older literature on the Epicurean revival, see Charles Trawick Harrison, "The Ancient Atomists and English Literature of the Seventeenth Century," *Harvard Studies in Classical Philology* 45 (1934): 1–79.

38. See Kroll, "Locke's Relation to Gassendi," 350; Osler, *Divine Will*, 77.

39. In general, and on the issue of dating, see Norbrook and Barbour, introduction to *Works of Lucy Hutchinson*; also see Jonathan Goldberg, "Lucy Hutchinson Writing Matter," *English Literary History* 73 (Spring 2006): 275–301.

40. John Evelyn, *Diary and Correspondence*, ed. William Bray (London, 1850), 1:314; John Evelyn, letter to Jeremy Taylor, 27 April 1646, ibid. 3:73–74. Evelyn knew Gassendi's *Life of Epicurus* and used it as a source in the notes; see, e.g., *An Essay on the First Book of T. Lucretius Carus De Rerum Natura* (London, 1656), 109. Interestingly, Evelyn tended to explain Lucretius's more inflammatory doctrines—that the gods care not for humans, for example—by contextualizing them, arguing that, like Lucretius, *any* reasonable man would have rejected the paganism of the ancient world. See, e.g., ibid., 118. More generally on Evelyn and Epicureanism, see Kroll, *Material Word*, 165ff.

41. Evelyn, *First Book of Lucretius*, 104, 171, 172.

42. Walter Charleton, *Physiologia Epicuro-Gassendo-Charletoniana; or, A Fabrick of Science Natural upon the Hypothesis of Atoms* (London, 1654), 94, 87–88, 103, 124, 126.

43. For a version of this argument, see E. J. Dijksterhuis, *The Mechanization of the World Picture*, trans. C. Dikshoorn (Oxford: Clarendon Press, 1961), 425–26. *Pace* Dijksterhuis, see J. C. D. Clark, "Providence, Predestination, and Progress; or, Did the Enlightenment Fail?" *Albion* 35, no. 4 (Winter 2003): 559–89.

44. John Edwards, *Brief Remarks upon Mr. Whiston's New Theory of the Earth* (London, 1697), 28.

45. On clocks, their inaccuracy, and their complexity, see Carlo Cipolla, *Clocks and Culture, 1300–1700* (New York: Norton, 1978), esp. chap. 1.

46. Robert Boyle, "A Requisite Digression concerning Those That Would Exclude the Deity from Intermedling with Matter," in *Some Considerations Touching the Usefulnesse of Experimental Natural Philosophy* (Oxford, 1663), 71. A virtually identical discussion of self-motion and the limits of mechanism can be found in his *Some Considerations about the Reconcileableness of Reason and Religion* (London, 1675), 38ff., in regard to Hobbes.

47. Robert Boyle, *About the Excellency and Grounds of the Mechanical Philosophy*, published with *Excellency of Theology Compar'd with Natural Philosophy* (London, 1674), 4.

48. See Thomas Aquinas, "Whether God Exists," *Summa theologiae* (New York: Mc-Graw-Hill, 1964), 1a, q.2, art. 3. Samuel Parker, *A Demonstration of the Divine Authority of the Law of Nature, and of the Christian Religion* (London, 1681), xvii–xviii.

49. Richard Bentley, *The Folly and Unreasonableness of Atheism*, 4th ed. (London, 1699), 78, 270, 273, 154.

50. John Ray, *The Wisdom of God Manifested in the Works of the Creation* (London, 1691), 16, 24.

51. Guillaume Du Bartas, *His Divine Weekes and Works*, trans. Joshua Sylvester (London, 1605), 9.

52. John Milton, *Paradise Lost*, ed. Gordon Tesky (New York: Norton, 2005), 2.894–96, 909–16. For Milton on chaos, see A. B. Chambers, "Chaos in Paradise Lost," *Journal of the History of Ideas* 24, no. 1 (January–March 1963): 55–84, and John Rumrich, "Milton's God and the Matter of Chaos," *PMLA* 110, no. 5 (October 1995): 1035–46.

53. Milton, *On Christian Doctrine*, quoted in Rumrich, "Matter of Chaos," 1037.

54. John Ray, *Three Physico-Theological Discourses*, 2nd ed. (London, 1693), 5.

55. Ibid., 35, 37. The lowly mole was a staple in this kind of literature; see, e.g., Henry More, *An Antidote Against Atheisme; or, An Appeal to the Natural Faculties of the Minde of Man, whether there be not a God* (London, 1653), 88.

56. More, *Antidote against Atheism*, 44. Bentley, *Folly of Atheism*, 214 (Bentley discusses the reversal of cause and effect at 142–43). Ray, *Wisdom*, 27–28.

57. Ralph Cudworth, *The True Intellectual System of the Universe* (London, 1678), 29–30, 147.

58. Ibid., 50, 147, 156, 162, 163.

59. Ibid., 147, 149. For Ray on Cudworth, see *Wisdom*, 34. Cudworth here was not engaging with Malebranche, although the claims would apply equally well to him. Rather, he was concerned with Puritans; see Catherine Wilson, *Leibniz's Metaphysics: A Historical and Comparative Study* (Manchester: Manchester University Press, 1989), 164.

60. Cudworth, *True Intellectual System*, 158. Boyle, *Excellency of Theology*, 173. Steven Shapin and Simon Schaffer, *Leviathan and the Air Pump: Hobbes, Boyle, and the Experimental Life* (Princeton, NJ: Princeton University Press, 1985), 184. For the best and most sustained account of this epistemological modesty, see Barbara Shapiro, *Probability and Certainty in Seventeenth-Century England: A Study of the Relationships between Natural Science, Religion, History, Law, and Literature* (Princeton, NJ: Princeton University Press, 1983).

61. Boyle, *About the Excellency and Grounds of the Mechanical Philosophy*, 13, 14.

62. Ray, *Three Physico-Theological Discourses*, 51–52. On microscopes and micro-order, see Catherine Wilson, *The Invisible World: Early Modern Philosophy and the Invention of the Microscope* (Princeton, NJ: Princeton University Press, 1995); also Meinel, "Early Seventeenth-Century Atomism," esp. 189ff.

63. Bernard de Fontenelle, *Conversations on the Plurality of Worlds*, trans. H. A. Hargreaves (Berkeley and Los Angeles: University of California Press, 1990), 11, 44–45. Blaise Pascal, *Pensées* (New York: Modern Library, 1941), §72, pp. 22–23.

64. Glanville quoted in Wilson, *Invisible World*, 64. Richard Westfall, *Force in Newton's Physics: The Science of Dynamics in the Seventeenth Century* (London, 1971), 336.

65. Gassendi, *Syntagma philosophicum*, in Howard B. Adelmann, *Marcello Malpighi and the Evolution of Embryology* (Ithaca, NY: Cornell University Press, 1966), 2:806. Charleton, *Physiologia Epicuro-Gassendo-Charletoniana*, 344. Here see also Keith Hutchinson, "What Happened to Occult Qualities in the Scientific Revolution?" *Isis* 73, no. 2 (June 1982): 233–53.

66. Nehemiah Grew, *Cosmologia Sacra; or, A Discourse of the Universe as It Is the Creature and Kingdom of God* (London, 1701), 12, 11, 33. George Cheyne, *Philosophical Principles of Natural Religion: Natural and Reveal'd*, 2nd ed. (London, 1715), 121, 43, 127–28. On Grew, see also Wilson, *Invisible World*, 182.

67. Isaac Newton, *The Principia: Mathematical Principles of Natural Philosophy* (Berkeley and Los Angeles: University of California Press, 1999), 795–96 (bk. 3, rule 3). The first law reads: "Every body perseveres in its state of being at rest or of moving . . . except insofar as it is compelled to change its state by forces impressed" (416). For a strong expression of this side of Newton, see Margaret Jacob, *The Newtonians and the English Revolution, 1689–1720* (Ithaca, NY: Cornell University Press, 1976), 187.

68. Isaac Newton, Query 31, *Opticks; or, A Treatise of the Reflections, Refractions, Inflections, & Colours of Light* (1730 ed.; New York: Dover, 1952), 398, 400. Bentley, *Folly of Atheism*, 214. Henry Pemberton, *A View of Sir Isaac Newton's Philosophy* (1728), quoted in David Kubrin, "Newton and the Cyclical Cosmos: Providence and the Mechanical Philosophy," *Journal of the History of Ideas* 28, no. 3 (July–September 1967): 329. Newton, *Opticks*, 378.

69. On Newton's providence and God's interventions, see Kubrin, "Newton and the Cyclical Cosmos," and Ayval Ramati, "The Hidden Truth of Creation: Newton's Method of Fluxions," *British Journal of the History of Science* 34 (2001): 430.

70. Newton to Oldenburg, quoted in Kubrin, "Newton and the Cyclical Cosmos," 335. Newton, *Opticks*, 375, 344, 350. For the dating of a draft of Query 31, see Westfall, *Force in Newton's Physics*, 397. On the invisibility of miraculous order, see Peter Harrison, "Newtonian Science, Miracles, and the Laws of Nature," *Journal of the History of Ideas* 56, no. 4 (Oct. 1995): 539.

71. Westfall, *Force in Newton's Physics*, 391.

72. Gottfried Wilhelm Leibniz, *Theodicy: Essays on the Goodness of God, the Freedom of Man, and the Origin of Evil*, trans. E. M. Huggard, ed. Austin Farrar (London: Routledge and Kegan Paul, 1951), 53. The labyrinths had already appeared in 1679, in his essay "On Freedom"; *Philosophical Papers and Letters*, ed. and trans. Leroy E. Loemker (Dordrecht: D. Reidel, 1969), 264.

73. Spinoza, *Ethics*, trans. G. H. R. Parkinson (Oxford: Oxford University Press, 2000), 160, 156. Bayle, *Historical and Critical Dictionary*, s.v. "Buridan," 778.

74. Leibniz, *Theodicy*, 150. 61.

75. Ibid., 310. Leibniz, *Discourse on Metaphysics*, trans. George Montgomery (LaSalle, IL: Open Court, 1988), 20. On Leibniz, causality, and divine wisdom, see Daniel Garber, *Leibniz: Body, Substance, Monad* (Oxford: Oxford University Press, 2009), esp. chap. 6.

76. Leibniz, "On Freedom," 264. Leibniz, letter to Antoine Arnauld, 14 July 1686, in *Correspondence with Arnauld*, trans. George Montgomery (LaSalle, IL: Open Court, 1988), 132.

77. Leibniz, *Theodicy*, 126.

78. Leibniz, "Pacidius to Philalethes: A First Philosophy of Motion," in *The Labyrinth of the Continuum: Writings on the Continuum Problem, 1672–1686*, trans. Richard T. W. Arthur (New Haven, CT: Yale University Press, 2001), 566. Leibniz, *New Essays on Human Understanding*, trans. and ed. Peter Remnant and Jonathan Bennett (Cambridge: Cambridge University Press, 1981), 166.

79. Pierre Bayle, *Various Thoughts on the Occasion of a Comet*, trans. Robert C. Bartlett (Albany, NY: SUNY Press, 2000), 259, 255.

80. Leibniz, letter to Arnauld, 30 April 1687, in *Correspondence*, 193. Leibniz, *Monadology*, trans. George Montgomery (LaSalle, IL: Open Court, 1988), 265, 255, 266. Leibniz, letter to Arnauld, 6 October 1687, in *Correspondence*, 221, 256.

81. Leibniz, letter to Arnauld, 14 July 1686, in *Correspondence*, 127. Leibniz, *Discourse on Metaphysics*, 56.

82. Leibniz, *Theodicy*, 188–89.

83. Spinoza, *Ethics*, 93. John Toland, *Letters to Serena* (London, 1704), 193, 197. Pierre Bayle, *Reponse aux questions d'un provincial* (1704–7), chap. 180, in *Oeuvres diverses de Mr. Pierre Bayle* (La Haye, 1727), 3:882.

## CHAPTER 2

1. Daniel Defoe, *The History of the Union of Great Britain* (Edinburgh, 1709), 1–3.

2. John Whittle, *A Thanksgiving-Sermon . . .* (London, 1706), 2; and compare, among others, Richard Allestree, *The Vanity of the Creature* (London, 1684), 34, and John Spalding, *Synaxis Sacra* (Edinburgh, 1703), 121. *A Collection of Debates in the House of Commons, in the Year 1680: Relating to the Bill of Exclusion . . .* (London, 1725), 125–26. For "deep mystery," see Isaac Sharpe, *Animadversions on Other Passages of Mr. Edmund Calamy . . . Part II* (London, 1704), 10. Statements about the uses of "wheel(s) within wheels" are based on a survey of more than nine hundred English appearances of this phrase and its variants in ECCO, EEBO, and MME.

3. Defoe, *History of the Union of Great Britain*, 5.

4. White Kennett, *The Charity of Schools for Poor Children Recommended in a Sermon . . .* (London, 1706), 16.

5. John Owen, *Exercitations on the Epistle to the Hebrews . . .* (London, 1668), 60.

6. *A Review of the State of the British Nation*, no. 31, 14 June 1709, in *Defoe's Review*, ed. A. Wellesley Secord (New York: Columbia University Press, 1938), 6:122. For Defoe's views on credit, see Sandra Sherman, *Finance and Fictionality in the Early Eighteenth Century: Accounting for Defoe* (Cambridge: Cambridge University Press, 1996).

7. Joseph de la Vega, *Confusion de confusiones* (1688), selected and trans. Hermann Kellenbenz (Boston: Baker Library, 1957), 5, 10–12, 22. Cf. José L. Cardoso, "*Confusion de confusiones*: Ethics and Options on Seventeenth-Century Stock Exchange Markets," *Financial History Review* 9, no. 1 (2002): 109–23.

8. De la Vega, *Confusion de confusiones*, 15 (and cf. 7). For the Spanish, see *Confusion de confusiones*, ed. M. F. J. Smith and Gerardus J. Geers (The Hague: M. Nijhoff, 1939), 55 (inconsistencies of spelling in the original).

9. *Defoe's Review* 8:68 (30 August 1711). Defoe's description of credit was also mocked for its seeming contradiction of philosophical principles. See, for example, Arthur Maynwaring as quoted in Carl Wennerlind, *Casualties of Credit: The English Financial Revolution, 1620–1720* (Cambridge, MA: Harvard University Press, 2011), 185.

10. Daniel Defoe, *An Essay upon Projects* (London, 1697), 1. See, among others, Peter G. M. Dickson, *The Financial Revolution in England* (London: Macmillan, 1967); Larry Neal, *The Rise of Finance Capitalism* (Cambridge: Cambridge University Press, 1990); and Ranald C.

Michie, *The London Stock Exchange* (Oxford: Oxford University Press, 1999), chap. 1, and *The Global Securities Market* (Oxford: Oxford University Press, 2006), chap. 1.

11. Quoted in Geoffrey Clark, *Betting on Lives: The Culture of Life Insurance in England, 1695–1775* (Manchester: Manchester University Press, 1999) 90. For the fluctuations in the market value of the guinea, see the testimony of William Stout of Lancaster, quoted in John K. Horsefield, *British Monetary Experiments, 1650–1710* (Cambridge, MA: Harvard University Press, 1960), 27, and further discussion there.

12. *Angliae Tutamen; or, The Safety of England: Being an Account of the Banks, Lotteries . . .* (London, 1695), 17. Gregorio Leti, *Critique historique, politique, morale, economique, & comique, sur les lotteries . . .* (Amsterdam, 1697), 1:1–2, 168.

13. George Alter and James Riley, "How to Bet on Lives: A Guide to Life Contingency Contracts in Early Modern Europe," *Research in Economic History* 10 (1986): 1–53. Lorraine Daston, *Classical Probability in the Enlightenment* (Princeton, NJ: Princeton University Press, 1988), 3–82. Clark, *Betting on Lives*.

14. Law's French manuscript is reproduced in Antoin E. Murphy, "John Law's Proposal for a Bank of Turin (1712)," *Economies et sociétés, série œconomia* 15 (1991); quoted, 18, 24. For the suggestive connections between alchemy and credit in the seventeenth century, see Carl Wennerlind, "Credit-Money as the Philosopher's Stone: Alchemy and the Coinage Problem in Seventeenth-Century England," in *Oeconomies in the Age of Newton*, ed. Margaret Schabas and Neil de Marchi, supp. to *History of Political Economy* 35 (2003): 234–61 (and on the regent's alchemists, 237).

15. The Dutch merchant is quoted in Violet Barbour, *Capitalism in Amsterdam in the 17th Century* (Ann Arbor, MI: Ann Arbor Paperbacks, 1963), 45. *Considerations on the Present State of the Nation, as to Publick Credit, Stocks . . .* (London, 1720), 17–18. Daniel Defoe, *A Review of the State of the English Nation*, vol. 3, ed. J. McVeagh (London: Pickering and Chatto, 2005), 1:38 (no. 6, 12 January 1706), emphasis added.

16. Defoe, *Review*, vol. 3, ed. McVeagh, 1:41–42 (no. 7, 15 January 1706). Bernard Mandeville, *The Fable of the Bees. or, Private Vices, Publick Benefits* (London, 1714), 103. Thomas Haskell, "Capitalism and the Origins of the Humanitarian Sensibility, Part 2," in *The Antislavery Debate*, ed. Thomas Bender (Berkeley and Los Angeles: University of California Press, 1992).

17. Defoe, *Essay upon Projects*, 8. Richard Gorges to Christopher Hatton, 7 March 1699/1700, quoted in Lee K. Davison, "Public Policy in an Age of Economic Expansion: The Search for Commercial Accountability in England, 1690–1750" (unpublished PhD diss., Harvard University, 1990), 129. Daniel Defoe, *The Villainy of Stock-jobbers Detected . . .* (London, 1701), 22.

18. [Sir Dudley North,] *Discourses upon Trade . . .* (London, 1691), ed. Jacob H. Hollander ([Baltimore, MD: Johns Hopkins University Press,] 1907), postscript, 35; and cf. Samuel Pratt, *The Regulating Silver Coin, Made Practicable and Easie . . .* (London, 1696), 103. Compare also the image of credit in perpetual motion, like the water cycle in nature, in John Cary, *An Essay, on the Coyn and Credit of England* (Bristol, 1696), 27, as discussed in Natasha Glaisyer, "'A due Circulation in the veins of the Publick': Imagining Credit in Late Seventeenth- and Early Eighteenth-Century England," *Eighteenth Century: Theory and Interpretation* 46, no. 3 (2005): 277–97.

19. Defoe, *Review*, vol. 3, ed. McVeagh, 2: 645 (no. 127, 22 October 1706). Compare Defoe's depiction of credit as suffering from "*Falling-sickness*," i.e., epilepsy, which in eighteenth-century understanding implied sudden unpredictable convulsions involving the loss of reason: cf. Sherman, *Finance and Fictionality*, 49, 198n85.

20. Charles Davenant, *Discourses on the Publick Revenues, and on the Trade of England* (London, 1698), 1:31, 38; 2:70, 72, 172; and note the similarity to Defoe's characterization of credit, "it comes with Surprize, it goes without Notice," as above, p. 50. Davenant's language about credit's motion of its own accord was also closely echoed by Charles Povey in his periodical *The General Remark on Trade*, no. 214 (7–9 July 1707).

21. Charles Davenant, *An Essay on the East-India-Trade* (London, 1696), 25 (emphasis added). Davenant was in fact more equivocal about the desirability of legislators' intervention than this oft-repeated quote suggests.

22. Joyce O. Appleby, *Economic Thought and Ideology in Seventeenth-Century England* (Princeton, NJ: Princeton University Press, 1978), 51, 80–81, 93–94, 242, and passim.

23. Thomas Mun, *England's Treasure by Forraign Trade* . . . (1623; London, 1664), 218–19.

24. Davenant, *Discourses on the Publick Revenues*, 1:2, 10, 18, 29 (and compare Josiah Child, *A Discourse about Trade* [London, 1690], 85: "It is the care of *Law-makers* first and principally, to provide for the *People* in gross, not particulars").

25. William Petty, *An Essay concerning the Multiplication of Mankind*, 2nd ed. (London, 1686), 20–21. John Graunt, *Natural and Political Observations, Mentioned in a Following Index, and made upon the Bills of Mortality*, 2nd ed. (London, 1662), unpaginated dedicatory epistle, 17.

26. Edmund Halley, "An Estimate of the Degrees of the Mortality of Mankind," *Philosophical Transactions* 17 (1693): 596–610. On annuities, see Geoffrey Poitras, *The Early History of Financial Economics, 1478–1776: From Commercial Arithmetic to Life Annuities and Joint Stocks* (Cheltenham, UK: Edward Elgar, 2000), 189ff. More generally, see Gerd Gigerenzer et al., *The Empire of Chance: How Probability Changed Science and Everyday Life* (Cambridge: Cambridge University Press, 1989), 20; Ian Hacking, *The Emergence of Probability: A Philosophical Study of Early Ideas about Probability, Induction, and Statistical Inference* (Cambridge: Cambridge University Press, 1975), 102ff.

27. Pierre Bayle, *Various Thoughts on the Occasion of a Comet*, trans. Robert C. Bartlett (Albany, NY: SUNY Press, 2000), 259–60.

28. Graunt, *Natural and Political Observations*, 39, 72. [Antoine Arnauld and Pierre Nicole], *Logic; or, The Art of Thinking* (London, 1685), 245, 243.

29. Samuel Pepys to Isaac Newton, 22 November 1693, in *The Letters of Samuel Pepys, 1656–1703*, ed. G. de la Bédoyère (Woodbridge, UK: Boydell, 2006), 218. Jean Le Clerc, *Reflections upon What the World Commonly Call Good-Luck and Ill-Luck with Regard to Lotteries* (London, 1699), 1–2, 3. More generally, see A. L. Murphy, "Lotteries in the 1690s: Investment or Gamble?" *Financial History Review* 12, no. 2 (2005): 229ff. On lotteries, probability, and the novel, see Jesse Molesworth, *Chance and the Eighteenth-Century Novel* (Cambridge: Cambridge University Press, 2010), esp. chap. 1.

30. Le Clerc, *Reflections upon Good-Luck*, 20, 10, 56.

31. Ibid., 26, 42, 85–86.

32. Thomas Gataker, *A Just Defence of Certaine Passages in a Former Treatise concerning the Nature and Use of Lots* (London, 1623), 70–71. James Balmford, *A Modest Reply to Certain Answeres, which Mr. Gataker B.D. in his Treatise of the Nature, and use of Lotts, giveth to Arguments in a Dialogue concerning the Unlawfulnes of Games consisting in Chance* (London, 1623), 16, 92. On Puritan discussion of chance, see Barbara Donagan, "Providence, Chance, and Explanation: Some Paradoxical Aspects of Puritan Views of Causation," *Journal of Religious History* 11, no. 3 (June 1981): 385–403, and "Godly Choice: Puritan Decision-Making in Seventeenth-Century England," *Harvard Theological Review* 76, no. 3 (July 1983): 307–34.

33. Le Clerc, *Reflections upon Good-Luck*, 194.

34. Blaise Pascal, *Pensées* (New York: Modern Library, 1941), 81 (§233). Hacking, *Emergence of Probability*, 70. On the history of probability theory, see, among others, Hacking, *Emergence of Probability*; Isaac Todhunter, *A History of Mathematical Theory of Probability from the Time of Pascal to That of Laplace* (New York: Chelsea Publishing, 1949); Gigerenzer et al., *Empire of Chance*; Daston, *Classical Probability*; and Anders Hald, *A History of Probability and Statistics and Their Applications before 1750* (New York: Wiley, 1990).

35. On the deep history of probability, see F. N. David, *Games, Gods and Gambling: The Origins and History of Probability and Statistical Ideas from the Earliest Times to the Newtonian Era* (New York: Hafner, 1962). On expectation, see Hacking, *Emergence of Probability*, chap. 11, and Daston, *Classical Probability*, 24.

36. [Daniel Defoe?], *The Gamester: A benefit-ticket for all that are concern'd in the lotteries; or, The best way how to get the 20000 L. prize*, (London, 1719), 12–13 (P. N. Furbank and W. R. Owens reject Defoe's often-stated authorship in their *Defoe De-Attributions* [London: Hambledon, 1994]). [Arnauld and Nicole], *Art of Thinking*, 245. On the effort to produce a new "rationality under uncertainty," see Daston, *Classical Probability*, xii.

37. John Arbuthnot, preface to Christian Huygens, *Laws of Chance* (London, 1692), unpaginated.

38. Lady Mary Wortley Montague, *Letters of the Right Honourable Lady M-y W-y M-e: Written during Her Travels in Europe, Asia, and Africa* (Paris, 1790), 118. [John Arbuthnot], *Mr. Maitland's account of inoculating the small pox vindicated* (London, 1722), 21, and Edward Edlin of Holborn to James Jurin, 15 September 1726, both quoted in A. A. Rusnock, *Vital Accounts: Quantifying Health and Population in Eighteenth-Century England and France* (Cambridge: Cambridge University Press, 2002), 43, 67. On smallpox, see F. Fenner et al., *Smallpox and Its Eradication* (Geneva: World Health Organization, 1988), esp. chap. 5. For the North American inoculation debate, beginning in 1721 along similar lines, see P. Cline Cohen, *A Calculating People: The Spread of Numeracy in Early America* (Chicago: University of Chicago Press, 1982), chap. 3.

39. Jacob Bernoulli, *The Art of Conjecturing*, trans. Edith Dudley Sylla (Baltimore, MD: Johns Hopkins University Press, 2006), 38–39; see also 328. Mary Poovey, *A History of the Modern Fact: Problems of Knowledge in the Sciences of Wealth and Society* (Chicago: University of Chicago Press, 1998), 15. As Hacking shows, to think about induction as a question required new ideas about evidence, specifically a belief that *things* have intrinsic evidentiary value, that they point to something beyond themselves but in a nondeductive manner; see Hacking, *Emergence of Probability*, 31ff.

40. Leibniz, letter to Bernoulli, 26 November 1703, in *Art of Conjecturing*, 39–40.

41. Ibid.

42. Bernoulli, *Art of Conjecturing*, 339. On the inverse problem, see Andrew Dale, *A History of Inverse Probability: From Thomas Bayes to Karl Pearson*, 2nd ed. (New York: Springer, 1999), 7.

43. Bernoulli, *Art of Conjecturing*, 315, 339.

44. Pierre Rémond de Montmort, *Essay d'analyse sur les jeux de hazard*, 2nd ed. (Paris, 1713), xiv. Graunt, *Natural and Political Observations*, 44–45. John Arbuthnot, "An Argument for Divine Providence, taken from the constant Regularity observ'd in the Births of both Sexes," *Philosophical Transactions* 27 (1710–12): 186, 188–89. See also Daniel Headrick, *When Information Came of Age: Technologies of Knowledge in the Age of Reason and Revolution, 1700–1850* (Oxford: Oxford University Press, 2000), 62–64.

45. Hacking, *Emergence of Probability*, 2. Daston, *Classical Probability*, 10, 35, 245. Thomas M. Kavanagh, *Enlightenment and the Shadows of Chance: The Novel and the Culture of Gambling in Eighteenth-Century France* (Baltimore, MD: Johns Hopkins University Press, 1993), 5.

46. Hacking, *Emergence of Probability*, 12; see also Gigerenzer et al., *Empire of Chance*, 13. Abraham de Moivre, *The Doctrine of Chances* (London, 1718), iv v, and *The Doctrine of Chances*, 3rd ed. (London, 1756), 253.

47. Nicholas Bernoulli, letter to Montmort, 23 January 1713, in *Essay d'analyse sur les jeux de hazard*, 388, 392. On Bernoulli's calculations, see E. Shoesmith, "Nicholas Bernoulli and the Argument for Divine Providence," *International Statistical Review* 53, no. 3 (December 1985): 255–59.

48. Kavanagh, *Enlightenment and the Shadows of Chance*, and his *Dice, Cards, Wheels: A Different History of French Culture* (Philadelphia: University of Pennsylvania Press, 2005), esp. 68. Jessica Richard, "Arts of Play: The Gambling Culture of Eighteenth-Century Britain" (unpublished PhD diss., Princeton University, 2002). Also see J. Ashton, *The History of Gambling in England* (London, 1898), chap. 3.

49. Defoe, *Essay upon Projects*, 172, 177.

50. Bernard de Fontenelle, "Eloge de M. Le Marquis de Dangeau," in *Œuvres de Monsieur de Fontenelle*, new ed. (Paris, 1742), 6:124–25, trans. in Kavanagh, *Enlightenment and the Shadows of Chance*, 54. Edward Gibbon, *An Essay on the Study of Literature* (London, 1764), 99.

51. Jeremy Collier, *An Essay upon Gaming, in a Dialogue between Callimachus and Dolomedes* (London, 1713), 10–12.

52. Montmort, *Essay d'analyse sur les jeux de hazard*, xviii. W. Browne, introduction to Christian Huygens, *Libellus de ratiociniis in ludo aleae; or, The Value of All Chances in Games of Fortune* (London, 1714), unpaginated. Richard Mead, *A Discourse concerning the Action of the Sun and the Moon on Animal Bodies* (London, 1708), 6, 11.

53. For Halley on risk pricing, see "Estimate of the Degrees of the Mortality of Mankind," 602. On the general problems with pricing risk, see Daston, *Classical Probability*, 169ff.; Clark, *Betting on Lives*, 102; and Gigerenzer, *Empire of Chance*, 24. On early efforts to price annuities, see Poitras, *Early History of Financial Economics*, 188. Mortuary tontine quoted in Clark, *Betting on Lives*, 76.

54. *Mason v. Keeling* (1700), cited and discussed in J. H. Baker, *An Introduction to English Legal History*, 3rd ed. (London: Butterworths, 1990), 466; and see there for the discussion of *Mitchell v. Allestry*, the turning-point case of 1676. M. J. Prichard, *Scott v. Shepherd (1773) and the Emergence of the Tort of Negligence* (London: Selden Society, 1976), 5, 13, 15–18. This paragraph is indebted to Sandra Macpherson, *Harm's Way: Tragic Responsibility and the Novel Form* (Baltimore, MD: Johns Hopkins University Press, 2010). We are grateful to her for the opportunity to read her work before publication.

55. Macpherson, *Harm's Way*, 57.

56. Defoe, *Essay upon Projects*, 118, 120.

57. Ibid., 119, 122–23. For a brilliant analysis of the friendly societies, see Penelope Ismay, "Between Providence and Risk: Odd-Fellows, Benevolence, and the Social Limits of Actuarial Science, 1820s–1880s," *Past and Present* (forthcoming).

58. C. B. Macpherson, *The Political Theory of Possessive Individualism: Hobbes to Locke* (Oxford: Oxford University Press, 1962), 78. For the history of "agency," spearheaded by developments in seventeenth-century England, see Stephen Darwall, *The British Moralists and the Internal "Ought," 1640–1740* (Cambridge: Cambridge University Press, 1995), 12–18 and passim; Katherine Rowe, *Dead Hands: Fictions of Agency, Renaissance to Modern* (Stanford, CA: Stanford University Press, 1999), 17ff.; and James E. Block, *A Nation of Agents: The American Path to a Modern Self and Society* (Cambridge, MA: Belknap Press of Harvard University Press, 2002), chap. 1.

59. Edward G. Andrew, *Conscience and Its Critics: Protestant Conscience, Enlightenment Reason, and Modern Subjectivity* (Toronto: University of Toronto Press, 2001), 7, 36, 83–84.

60. Ralph Cudworth, *The True Intellectual System of the Universe* (London, 1678), esp. 159–60. Urian Oakes, *The Soveraign Efficacy of Divine Providence . . . as Delivered in a Sermon Preached in Cambridge on Sept. 10, 1677 . . .* (Boston, 1682), 5–7.

61. Sir Thomas Browne, *A Letter to a Friend* (London, 1690; written sometime after 1656 and published posthumously). Browne, also a medical doctor and connoisseur of complexity, used the word to characterize a remarkable, unexpected event: "That *Charles* the Fifth was Crown'd upon the day of his Nativity, it being in his own Power so to order it, makes no singular Animadversion; but that he should also take King *Francis* Prisoner upon that day, was an unexpected Coincidence, which made the same remarkable" (4).

62. Lori Branch, *Rituals of Spontaneity: Sentiment and Secularism from Free Prayer to Wordsworth* (Waco, TX: Baylor University Press, 2006), 7.

63. This paragraph draws on John E. Crowley, "The Sensibility of Comfort," *American Historical Review* 104 (1999): 761–62, and his *The Invention of Comfort* (Baltimore, MD: Johns Hopkins University Press, 2001), 150–51. The *OED* dates this new usage to 1672.

64. *Spectator*, 6th ed. (London, 1723), 3:84 (no. 191, 9 October 1711). Compare the versed evocation of capricious arbitrary choices in the stock exchange in *A Familiar Epistle to Mr. Mitchell . . . by a Money'd Man* (London, 1720), 14–15.

65. John Lambe, *The Liberty of Human Nature* (London, 1684), 4, 9, 14–15. Together with luxury, Lambe also lambasted his audience for their litigiousness.

66. Sir George Mackenzie, *The Moral History of Frugality with Its Opposite Vices . . .* (London, 1691), 40–41.

67. Jan De Vries, "Luxury in the Dutch Golden Age in Theory and Practice," in *Luxury in the Eighteenth Century*, ed. Maxine Berg and Elizabeth Eger (Basingstoke: Palgrave Macmillan, 2003), 51, 53. Maxine Berg, *Luxury and Pleasure in Eighteenth-Century Britain* (Oxford: Oxford University Press, 2005), 24 and passim. See also John Styles, "Manufacturing, Consumption and Design in Eighteenth-Century England," in *Consumption and the World of Goods*, ed. John Brewer and Roy Porter (London: Routledge, 1993), 527–54; and for France, Natacha Coquery, "The Language of Success: Marketing and Distributing Semi-Luxury Goods in Eighteenth-Century Paris," *Journal of Design History* 17, no. 1 (2004): 71–89.

68. Defoe, *Review*, vol. 3, ed. McVeagh, 1:68 (no. 12, 26 January 1706). Across the Channel, note "the charm of novelty; an advantage which, particularly in France, determines the fate of almost all things" in the entry for "Fashion" in Jacques Savary des Bruslons, *Dictionnaire universel de commerce* (1723–30), as quoted in Coquery, "Language of Success," 77.

69. Daniel Defoe, *The Complete English Tradesman* (London, 1726 [1725]), 312–13. Hoh-Cheung Mui and Lorni H. Mui, *Shops and Shopkeeping in Eighteenth-Century England* (London: Routledge, 1989). Claire Walsh, "Shop Design and the Display of Goods in Eighteenth-Century London," *Journal of Design History* 8, no. 3 (1995): 157–76. Cynthia Wall, *The Prose of Things* (Chicago: University of Chicago Press, 2006), chap. 6. For France, cf. Coquery, "Language of Success." Linda L. Peck, *Consuming Splendor: Society and Culture in Seventeenth-Century England* (Cambridge: Cambridge University Press, 2005), suggests, plausibly, that signs of these developments had already been evident earlier in the seventeenth century. Berg, however, and especially Claire Walsh's forthcoming book distinguish between the skilled choices required by earlier shoppers confronted by better and worse consumer options, and the subsequent consumer experience in which this demanding knowledge of hierarchized choices was increasingly replaced with equal-weight choices from large stocks in suitably arranged establishments.

70. Joseph Addison, *The Tatler* (Glasgow, 1754), 188–89 (no. 224, 14 September 1710). Maxine Berg and Helen Clifford, "Selling Consumption in the Eighteenth Century: Advertising and the Trade Card in Britain and France," *Cultural and Social History* 4, no. 2 (2007): 145–70. Nancy Cox and Karin Dannehl, *Perceptions of Retailing in Early Modern England* (Aldershot, UK: Ashgate, 2007), chap. 4. For lottery advertisements, cf. Davison, "Public Policy in an Age of Economic Expansion," 37.

71. Beverly Lemire, *Fashion's Favourite: The Cotton Trade and the Consumer in Britain, 1660–1800* (Oxford: Oxford University Press, 1991), chap. 1 (for the duchesse d'Orleans, p. 9). *The Proposal for a General Fishery*, (1691), 2 (emphasis added), quoted in Appleby, *Economic Thought and Ideology in Seventeenth-Century England*, 211. East India Company, Court of Committees to Ugli, 20 May 1681, quoted in Woodruff D. Smith, *Consumption and the Making of Respectability, 1600–1800* (New York: Routledge, 2002), 52 (and cf. 49–59).

72. *A General History of Trade* (London, 1713), 27.

73. John Cary, *An Essay on the State of England in Relation to Its Trade, Its Poor, and Its Taxes* . . . (Bristol, 1695), 53, and compare 54, 58–9.

74. *The Ladies Library* (London, 1714), 1:81. For the changing and intensifying uses of "fashion" from the 1690s, see Hannah Greig, "Leading the Fashion: The Material Culture of London's *Beau Monde*," in *Gender, Taste, and Material Culture in Britain and North America*

*1700–1830*, ed. John Styles and Amanda Vickery (New Haven, CT: Yale University Press, 2006), esp. 294. George Blewitt, *An Enquiry Whether a General Practice of Virtue Tends to the Wealth or Poverty, Benefit or Disadvantage of a People?* (London, 1725), 37–38. Theresa Braunschneider, *Our Coquettes: Capacious Desire in the Eighteenth Century* (Charlottesville: University of Virginia Press, 2009).

75. Mary Vidal, *Watteau's Painted Conversations* (New Haven, CT: Yale University Press, 1992), chap. 5, and Jay Caplan, *In the King's Wake* (Chicago: University of Chicago Press, 1999), chap. 3. For the public market in art, see John Brewer, *The Pleasures of the Imagination* (New York: Farrar, Straus and Giroux, 1997), and Larry Shiner, *The Invention of Art* (Chicago: University of Chicago Press, 2001). Art auctions, which shifted the onus for choice from experts to consumers, appeared in Europe in the late seventeenth century: see the contributions by Thomas Ketelsen and Brian Cowan in *Art Markets in Europe, 1400–1800*, ed. Michael North and David Ormrod (Aldershot, UK: Ashgate, 1998). Relatedly, on eighteenth-century attempts to theorize agency in visual culture, see Peter de Bolla, *The Education of the Eye* (Stanford, CA: Stanford University Press, 2003).

76. Richard Fiddes, *A General Treatise of Morality* . . . (London, 1724), xl–xlii, xlv, xlix.

77. *Speranders sorgfältiger Negotiant und Wechsler* (Leipzig and Rostock, 1706), quoted in Daniel Rabuzzi, "Eighteenth-Century Commercial Mentalities as Reflected and Projected in Business Handbooks," *Eighteenth-Century Studies* 29, no. 2 (1995–96): 179. And cf. Wall, *Prose of Things*, 88–94. For another example, running for many pages, see J. F., *The Merchant's Warehouse Laid Open* . . . (London, 1696).

78. Montmort, *Essai d'analyse sur les jeux de hazard*, xvi. Digby Cotes, *Fifteen Sermons Preach'd on Several Occasions* (Oxford, 1721), 45–46. [Peter Shaw,] *The Tablet; or, Picture of Real Life* (London, 1762), 256 (orig. in *The Reflector*, 1750).

79. Branch, *Rituals of Spontaneity*, 4 quoted.

80. Thomas W. Laqueur, *Solitary Sex: A Cultural History of Masturbation* (New York: Zone, 2003).

81. Daniel L. Smail, *On Deep History and the Brain* (Berkeley and Los Angeles: University of California Press, 2008), 183–84.

82. William Penn, *A Perswasive to Moderation to Dissenting Christians in Prudence and Conscience* . . . (London, 1685), unpaginated preface. Compare Locke's *First Tract of Government* (1660) in *John Locke: Political Writings*, ed. David Wootton (Indianapolis, IN: Hackett, 2003), 145.

83. Victoria Kahn, "Revising the History of Machiavellism: English Machiavellism and the Doctrine of Things Indifferent," *Renaissance Quarterly* 46 (1993), esp. 540–49. Perez Zagorin, *How the Idea of Religious Toleration Came to the West* (Princeton, NJ: Princeton University Press, 2003), chap. 7. Timothy Stanton, "Locke and the Politics and Theology of Toleration," *Political Studies* 54 (2006): 84–102.

84. John Locke, *"An Essay concerning Toleration" and Other Writings on Law and Politics, 1667–1683*, ed. John R. Milton and Philip Milton (Oxford: Clarendon Press, 2006), 273 (from *An Essay concerning Toleration*, 1667). *John Locke: Political Writings*, 105. Andrew, *Conscience and Its Critics*, 87, 91.

85. Pierre Bayle, *A Philosophical Commentary on These Words of the Gospel . . .* (London, 1708), 1:256; originally *Commentaire philosophique sur ces paroles de Jesus-Christ* [*sic*] (Cantorbery [i.e., Amsterdam], 1686), 2:387–88.

86. Bayle, *Philosophical Commentary* 1:243 and *Commentaire philosophique* 2:363–64.

87. [Sir Charles Wolseley,] *Liberty of Conscience upon Its True and Proper Grounds . . . to Which is Added . . . Liberty of Conscience the Magistrate's Interest* (London, 1668), 52–53; and cf. similarly [John Owen,] *A Peace-Offering in an Apology and Humble Plea for Indulgence* (London, 1667), 33. Penn, *Perswasive to Moderation*, 30–32. For diversity as an argument for toleration during this period, see John Marshall, *John Locke, Toleration and Early Enlightenment Culture* (Cambridge: Cambridge University Press, 2006), 544–60; and for Dutch examples, 354 and chap. 11.

88. William Penn, *One Project for the Good of England . . .* [London, 1679?], 1. [Daniel Defoe,] *The Shortest Way to Peace and Union* (London, 1704), 14. John A. W. Gunn, *Politics and the Public Interest in the Seventeenth Century* (London: Routledge and Kegan Paul, 1969), 152–87; and cf. Albert O. Hirschman, *The Passions and the Interests: Political Arguments for Capitalism before Its Triumph* (Princeton, NJ: Princeton University Press, 1977), 51.

89. Locke, *An Essay concerning Toleration*, 309–10.

## CHAPTER 3

1. Charles de Secondat, baron de Montesquieu, *Reflections on the Causes of the Grandeur and Declension of the Romans . . . Translated from the French* (London, 1734), 90. *Œuvres complètes de Montesquieu*, vol. 2 (Oxford: Voltaire Foundation, 2000), 157. (The 1734 English translation muffled somewhat the aggregative effect of the emergence from multiple discordant sounds of "*l'accord total.*") Compare also the famous passage regarding the pursuit of honor in a monarchy, whereby "everyone contributes to the general welfare while thinking that he works for his own interests," in Montesquieu, *De l'esprit des lois* (1748), bk. 3, chap. 7.

2. The words were Robert Walpole's, as he reflected years later on these events, the aftershocks of which had propelled him to the position of Britain's most powerful prime minister. Robert Walpole, Earl of Orford, *Some Considerations concerning the Publick Funds, the Publick Revenues, and the Annual Supplies . . .* (London, 1735), 31.

3. Colin Jones, *The Great Nation: France from Louis XV to Napoleon* (New York: Columbia University Press, 2002), 61–73, and quoting Richer d'Aube (62).

4. Thomas Gordon, *The Humourist*, 3rd ed. (London, 1724), 19. Daniel Defoe, *Essays upon Several Projects* (London, 1702), 1 (and see above, p. 53).

5. Charles de Secondat, baron de Montesquieu, *Persian Letters*, trans. C. J. Betts (London: Penguin, 1993), letters 135 (translation altered), 142. On Law's schemes, chance, and gambling, see Thomas M. Kavanagh, *Enlightenment and the Shadows of Chance: The Novel and the Culture of Gambling in Eighteenth-Century France* (Baltimore, MD: Johns Hopkins University Press, 1993).

6. Montesquieu, *Persian Letters*, letter 138.

7. Ibid., letter 143, with capitalization adjusted.

8. Thus *Craftsman*, no. 284, 11 December 1731: "If there is any such Thing as a Parallel between two Cases, That of the *South Sea Directors*, in the year 1720, and of the Managers of the *Charitable Corporation*, at present, is certainly such." Vincent Carretta, "Pope's *Epistle to Bathurst* and the South Sea Bubble," *Journal of English and German Philology* 77 (1978): 212–31, 217 quoted. The opposition campaign on this issue ended up being delayed for several months by the better-known issue of the excise. See also Laura Brown, *Alexander Pope* (Oxford: Blackwell, 1985), 109–12.

9. The well-known *Essay on Man* also bears likely traces of the chaotic events of 1720. Humans, Pope says early in the first epistle, are blind to the future, and surprised by the unpredictability and seeming randomness of future events, e.g., when "Atoms or systems into ruin hurl'd, / And now a bubble burst, and now a world." Alexander Pope, *An Essay on Man*, ed. M. Mack, vol. 3.1 of the Twickenham Edition of the Poems of Alexander Pope (London: Methuen, 1982), 1.89–90. Although one can probably understand the systems hurling into ruin without recourse to John Law's System in France, and the bursting of a bubble without the South Sea Bubble, given Pope's interest and personal involvement in those recent occurrences, and given that he published these lines about unpredictable occurrences within weeks of the *Epistle to Bathurst* that demonstrated how much they were on his mind and in the public eye during those very months, it seems reasonable to see in them deliberate allusions to the financial world.

10. "In our Nature, we have a certain Fitness, by which great Multitudes of us co-operating, may be united and form'd into one Body; that . . . shall govern itself, and act on all Emergencies, as if it was animated by one Soul, and actuated by one Will." Elsewhere Mandeville follows his insistence on the need for aggregation that we have already seen ("the Fitness of Man for Society . . . is hardly perceptible in Individuals, before great Numbers of them are joyn'd together"), with an analogy to the making of liquor from a great many grapes; which implies, as he explains, "something that individual Persons are not actually possess'd of, whilst they remain single," but that "is palpably adventitious to Multitudes, when joyn'd together." Bernard Mandeville, *The Fable of the Bees: Part II* (London, 1729), ed. Frederick B. Kaye, 2 vols. (Oxford: Clarendon Press, 1924; rpt. Indianapolis, IN: Liberty Classics, 1988), 2:183, 189 (and cf. 185). Contrast with the 1714 edition of the *Fable*, e.g., 8: "The worst of all the Multitude / Did something for the Common Good." All the quotes from Mandeville above, pp. 5–6, are from the later edition and have no corollary in the first. For the first appearance of the concatenated events, see Mandeville, *The Fable of the Bees; or, Private Vices, Publick Benefits*, 2nd ed., enlarged with many additions (London, 1723), 89.

11. Friedrich Hayek saw in Mandeville's speculations following his *Fable of the Bees* the most significant origin of notions of spontaneous order in the eighteenth century: Friedrich A. Hayek, *The Trend of Economic Thinking: Essays on Political Economists and Economic History*, vol. 3 of *The Collected Works of F. A. Hayek*, ed. William W. Bartley III and Stephen Kresge (Chicago: University of Chicago Press, 1991), chap. 6.

12. *Exchange-alley; or, The Stock-jobber Turn'd Gentleman; with the Humours of Our Modern Projectors* (London, 1720), 4.

13. John Evans, *Past Deliverances and Present Calamities Improved. In a Sermon . . .* (London, 1720), 20–21. Compare, among many others, *The Naked and Undisguis'd Truth, Plainly*

*and Faithfully Told*, 2nd ed. (London, 1721), [1]. *An Essay for Discharging the Debts of the Nation . . . and the South-Sea Scheme Consider'd* (London, 1720), ii.

14. Pope to Atterbury, 23 September 1720, in *The Correspondence of Alexander Pope*, ed. G. Sherburn (Oxford: Clarendon Press, 1956), 2:53. Julian Hoppit, "The Myths of the South Sea Bubble," *Transactions of the Royal Historical Society*, 6th ser., 12 (2002): 141–65.

15. Hoppit ("Myths of the South Sea Bubble") tries to explain this "myth" of the Bubble by placing it in the context of broader political concerns and a contemporary moral panic, as well as by relocating the responsibility for the mythologization to later generations. In thus shifting attention from the specifics of the financial events of 1720 to externals, we believe, Hoppit may have overlooked some of their most salient implications for those who experienced them. (In part, Hoppit bases this relocation on his observation that the very term "South Sea Bubble" is a latecomer, not used before the 1770s. But in fact earlier examples of this usage can be found—including Thomas Bradbury, *Twenty-Eight Sermons . . .* [London, 1723], 171; *The Life of Jonathan Wild . . .* [London, 1725], 63; and the title page of the memoirs of James Houstoun of 1747.)

16. This discussion of the meaning in Britain of the South Sea Bubble and the related financial developments is based on a survey of more than 260 British publications from 1720 and 1721 that refer to these events. Counted together with their multiple editions, this number rises to 365, representing over 13 percent of all titles published in Britain during those two years. (These numbers are based on the inventory of the Eighteenth Century Collection Online in August 2006.)

17. *A Collection of Miscellany Letters, Selected out of Mist's Weekly Journal* (London, 1722–27), 2:165, letter 54.

18. *Spectator*, ed. Donald F. Bond (Oxford: Oxford University Press, 1965), 4:442, no. 543, 22 November 1712. Newton's quote (emphasis added) is often repeated, in slightly different wordings: e.g., in Patrick Brantlinger, *Fictions of State: Culture and Credit in Britain, 1694–1994* (Ithaca, NY: Cornell University Press, 1996), 44, quoting Henry R. Fox Bourne, *The Romance of Trade* (London, 1871), 292. For Newton's involvement in the Bubble, see Hoppit, "Myths of the South Sea Bubble," 149, and the sources cited there.

19. *Journal du Parlement par Deslisle*, interventions of 4 March and 17 June 1718, quoted in Thomas E. Kaiser, "Money, Despotism, and Public Opinion in Early Eighteenth-Century France: John Law and the Debate on Royal Credit," *Journal of Modern History* 63 (1991): 12. Samuel Bradford, *The Honest and the Dishonest Ways of Getting Wealth. A Sermon . . .* (London, 1720), 3, 27, 31. Charles Gildon, *All for the Better; or, The World Turn'd Up-side Down . . .* (London, 1720), 10.

20. Jonathan Swift, "The Run upon the Bankers" (1720), in his *Miscellanies, in Prose and Verse: The Fifth and Sixth Volumes* (London, 1738), 1:208–9.

21. *The South-Sea Scheme Detected; and the Management Thereof Enquir'd Into*, 2nd ed. (London, 1720), 8. Edward Ward, *A Looking-Glass for England: or, The Success of Stock-jobbing Explain'd* (Bristol, 1720). Cf. also John Asgill, *A Brief Answer to a Brief State of the Question, between the Printed and Painted Callicoes, and the Woollen and Silk Manufactures* (London, 1720), 12.

22. Thomas Gordon, *Francis, Lord Bacon; or, The Case of Private and National Corruption and Bribery, Impartially Consider'd*, 5th ed. (London, 1721), 1–2. Erasmus Philips, *An Appeal*

*to Common Sense; or, Some Considerations Offer'd to Restore Publick Credit* (London, 1720), 3. Evans, *Past Deliverances and Present Calamities Improved*, 21 (emphasis added).

23. Elias Bockett, *The Yea and Nay Stock-jobbers; or, The 'Change-Alley Quakers Anatomiz'd* (London, 1720), 6. Voltaire to Nicolas Anne Lefèvre de La Faluère, summer 1716, in *Voltaire's Correspondence*, ed. Theodore Besterman (Geneva: Institut et musée Voltaire, 1953), 1:59–60. Harley's son is quoted in Hoppit, "Myths of the South Sea Bubble," 145. For the Dutch observer see Brantlinger, *Fictions of State*, 57. Sir John Meres, *The Equity of Parliaments, and Publick Faith, Vindicated* . . . , 3rd ed. (London, 1720), 24. Peter G. M. Dickson, *Financial Revolution: A Study in the Development of Public Credit, 1688–1756* (London: Macmillan, 1957), 155.

24. J[ohn] T[heophilus] Desaguliers, *The Newtonian System of the World, the Best Model of Government* (Westminster, 1728), 17. Jonathan Swift, "The Bubble," in *"Gulliver's Travels" and Other Writings*, ed. Clement Hawes (Boston: Houghton Mifflin, 2004), 336, lines 138–45; 338, line 213.

25. *A Collection of Miscellany Letters, Selected out of Mist's Weekly Journal* (London, 1722), 1:267–69, letter 92.

26. *Director, num. I, II, III, IV* (Edinburgh, [1720?]), 3:16–17.

27. George Berkeley, *An Essay towards Preventing the Ruine of Great Britain* (London, 1721), 5. *Mr. Law's Character Vindicated: In the Management of the Stocks in France, with the True Reasons for Their Sinking* (London, 1721), 11.

28. *South-Sea Scheme Detected*, 7, 12–13. John Trenchard and Thomas Gordon, *Cato's Letters*, 5th ed., corrected (London, 1748), no. 6, 10 December 1720, 1:26. Mr. Arundell, *The Directors, a Poem*, 2nd ed. (London, 1720), 17. Sir Theodore Janssen, *A Discourse concerning Banks* ([London?], 1720), preface, v. For a poetic equivalent see *The Last Guinea: A Poem*, 2nd ed. (London, 1720), 13.

29. Montesquieu, *Persian Letters*, letter 143. *The Original Weekly Journal*, published by John Applebee, 1 October 1720, 1863. Sir John Midriff, *Observations on the Spleen and Vapours* . . . (London, 1721), 61.

30. John Aislabie, *Mr. Aislabie's Second Speech on his Defence in the House of Lords* (London, [1721]), 12, 14.

31. Pope to Broome, 14 July [1723], in *The Correspondence of Alexander Pope*, ed. George Sherburn (Oxford: Clarendon Press, 1956), 2:182. Francis Bacon, *The Essays, or Councils, Civil and Moral* (London, 1701), 111. Thomas Fuller, *Gnomologia: Adagies and Proverbs* . . . (London, 1732), 222, proverb 5091. Daniel Defoe, *The Compleat English Tradesman* (London, 1727), 2:118. And compare *Some Considerations Touching the Sugar Colonies, with Political Observations in Respect to Trade* (London, 1732), 11. Deborah Valenze, *The Social Life of Money in the English Past* (Cambridge: Cambridge University Press, 2006), 70. The proverbial use later found its way into Adam Smith's *Wealth of Nations* (bk. 1, chap. 9).

32. Daniel Defoe, *The Chimera: or, The French Way of Paying National Debts Laid Open* (London, 1720), 30, and his *The Case of Mr. Law Truly Stated* (London, 1721), 8, 13, 21.

33. *A Full and Impartial Account of the Company of Mississippi* . . . *Projected and Settled by Mr. Law* (London, 1720), 13. *A Collection of All the Humorous Letters in the London Journal*, 2nd ed. (London, 1721), 33. Berkeley, *Essay towards Preventing the Ruine of Great Britain*, 5. N[icholas] Amhurst, *An Epistle (with a Petition in It) to Sir John Blunt*, 2nd ed. (London, 1720),

10. George Flint, *The Lunatick; or, Great and Astonishing News from Bedlam* (London, [1720?]), 12. Ward, *A Looking-glass for England.* For South Sea Bubble critiques of the blasphemous creation out of nothing, see Silke Stratmann, *Myths of Speculation: The South Sea Bubble and 18th-Century English Literature* (Munich: Fink, 2000), 12.

34. Trenchard and Gordon, *Cato's Letters*, 1126. Roderick Mackenzie, *Now or Never; or, A Familiar Discourse concerning the Two Schemes for Restoring the National Credit* (London, 1721), 29. Voltaire to Nicolas Anne Lefèvre de La Faluère, summer 1716, in *Voltaire's Correspondence*, 1:60. Defoe, *Chimera*, 5–6. Daniel Defoe, *An Essay upon the Public Credit* (1710; London, 1797), 6. For more on the association of the new world of finance and commerce with insubstantiality, see Dror Wahrman, *The Making of the Modern Self: Identity and Culture in Eighteenth-Century England* (New Haven, CT: Yale University Press, 2004), chap. 5, esp. 207–11, and the sources cited there.

35. Trenchard and Gordon, *Cato's Letters*, no. 3, 19 November 1720, 1:12. Thomas Gordon, *An Essay on the Practice of Stock-Jobbing* (London, 1724), 1. Thomas Baston, *Thoughts on Trade, and a Publick Spirit* (London, 1716), 4–5. *An Essay for Establishing a New Parliament Money* (London, 1720), 44. *A Familiar Epistle to Mr. Mitchell . . . by a Money'd Man* (London, 1720), 14. *Observations on the New System of the Finances of France . . . by Mr. Law . . .* (London, 1720), iii. *Craftsman*, no. 47, 2:19 20. *Collection of All the Humorous Letters in the London Journal*, 29. *The Nation Preserved; or, The Plot Discovered: Containing an Impartial Account of the Secret Policy of Some of the South-Sea Directors* (London, 1720), 36.

36. Roger North, *The Gentleman Accomptant: or, An Essay to Unfold the Mystery of Accompts . . .* (London, 1714), sig. b2. *A Letter to a Conscientious Man: Concerning the Use and the Abuse of Riches . . .* (London, 1720), 13.

37. *Original Weekly Journal*, 1 October 1720, 1863.

38. Gildon, *All for the Better*, 13–15, 40 (emphasis added).

39. Ibid., 10–11, 15.

40. Edmund Calamy, *Discontented Complaints of the Present Times Prov'd Unreasonable* (London, 1720), 11, 16, 21, 29–30. Digby Cotes, *Fifteen Sermons Preach'd on Several Occasions* (Oxford, 1721), 45–46, 221–23.

41. George Smyth, *A Sermon Preach'd at the New Meeting-house in Hackney* (London, 1720), 5–7, 10, 20.

42. Calamy, *Discontented Complaints of the Present Times Prov'd Unreasonable*, 32. Smyth, *Sermon Preach'd at the New Meeting-house in Hackney*, 12. Berkeley, *Essay towards Preventing the Ruine of Great Britain*, 4–5. Thomas Greene, *The End and Design of God's Judgments: A Sermon Preach'd before the House of Lords* (London, 1721), 25. Gildon, *All for the Better*, 54.

43. Elisha Smith, *A Religious Consideration of the Pursuits, and Possessions of This World the Best Improvement and Consolation under South-Sea Calamity . . .* (London, 1721), preface (unpaginated) and 9.

44. Evans, *Past Deliverances and Present Calamities Improved*, 21–22. For Evans's losses in the South Sea Bubble, see *DNB*.

45. *The Annals of King George, Year the Fifth* (London, 1720), 130. *A Brief Debate upon the Dissolving the Late Parliament . . .* (London, 1722), 10–15 (this pamphlet is sometimes wrongly attributed to Defoe). *Director, num. I, II, III, IV*, 1:4. *A Memorial of the Contractants with Mr. Aislabie: In a Letter to Licinius Stolo* (London, 1721), 7–8.

46. Gildon, *All for the Better*, 10–11. [Theodore Janssen?], "The Secret History of the South-Sea Scheme," in John Toland, *A Collection of Several Pieces of Mr. John Toland . . . with Some Memoirs of His Life and Writings* (London, 1726), 1:404. *A Collection of the Several Petitions of the Counties, Boroughs, &c. . . . Occasion'd by the Mismanagements of the Late Directors of the South-Sea Company* (London, 1721), 5, 12, and passim.

47. *Advantages Which Have Accrued to the Publick, and to the South-Sea Company, by the Execution of the South-Sea Scheme* ([London], 1726), 18–19. [Thomas Gordon,] *A Short View of the Conspiracy . . . by Cato* (London, 1723), 7. Gordon, *Francis, Lord Bacon*, preface, xiii.

48. Gordon, *Francis, Lord Bacon*, 1–2, 5.

49. *An Argument Proving That the South-Sea Company Are Able to Make a Dividend of 38 Per Cent. for 12 Years* (London, 1720), dedication, 7, and passim.

50. Archibald Hutcheson, *A Collection of Calculations and Remarks Relating to the South Sea Scheme & Stock* (London, 1720), 80, 90, 117, 130. Hutcheson is the hero of Richard Dale, *The First Crash: Lessons from the South Sea Bubble* (Princeton, NJ: Princeton University Press, 2004).

51. *Director* 11, 7 November 1720 (unpaginated). James Milner, *Three Letters, relating to the South-Sea Company and the Bank* (London, 1720), 17 (2nd letter, 15 April 1720). *Considerations on the Present State of the Nation, as to Publick Credit, Stocks, the Landed and Trading Interests* (London, 1720), 2.

52. Sir Humphrey Mackworth, *A Proposal for Payment of the Publick Debts, for Relief of the South Sea Company . . .*, 5th ed. (London, 1720), dedication and ii, v, 37.

53. *The Report of the Committee of the House of Commons, to Whom It Was Referred to Consider the Humble Petitions of Several Creditors and Proprietors of Principal Mony, Annuities and Shares in the Mine-Adventure of England* (London, 1710). Anne L. Murphy, "Lotteries in the 1690s: Investment or Gamble?" *Financial History Review* 12, no. 2 (2005): 243.

54. Daniel Defoe, *A Plan of the English Commerce: Being a Compleat Prospect of the Trade of This Nation . . .* (London, 1728), 265–66.

55. René-Louis de Voyer, marquis d'Argenson, *Essays, Civil, Moral, Literary and Political . . . Translated from His Valuable Manuscripts, and Never Before Made Public* (Dublin, 1789), 159. The *Visa* was part of Law's system for dealing with the problem of debt.

56. René-Louis de Voyer, marquis d'Argenson, *Considérations sur le gouvernement ancien et présent de la France, comparé avec celui des autres états*, 2nd ed., (Amsterdam, 1784), 60–61. Published posthumously, this was the most important work that came out of the Club de l'Entresol in Paris in the 1720s: see Henry C. Clark, *Compass of Society: Commerce and Absolutism in Old-Regime France* (Lexington: University Press of Kentucky, 2007), 225.

57. John Law, "Histoire des finances pendant la Régence," in his *Oeuvres complètes*, ed. Paul Harsin (Paris: Sirey, 1934), 3:299, 415. Kavanagh, *Enlightenment and the Shadows of Chance*, 87. Counter to Harsin and Kavanagh, Antoin E. Murphy (in his *John Law: Economic Theorist and Policy-Maker* [Oxford: Oxford University Press, 1997]) doubts that Law was in fact the author of the "Histoire des finances"; if he is right, then this tract was a retrospective by another informed contemporary, which makes little difference in terms of our argument here. For Law's earlier views, see John Law, *Money and Trade Considered, with a Proposal for Supplying the Nation with Money* (Edinburgh, 1705), e.g., 42.

58. Sir David Dalrymple, *Time Bargains Tryed by the Rules of Equity and Principles of the Civil Law* (London, 1720), 17–19. And compare Sandra Sherman, "Credit, Simulation, and the Ideology of Contract in the Early Eighteenth Century," *Eighteenth-Century Life* 19 (1995): 92; Sherman has already noted an "epistemological confusion" in the financial writings of the early 1720s involving, among other things, doubts about the linearity of causality (86).

59. Dalrymple, *Time Bargains*, 24–25 (emphasis added).

60. *Considerations on the Present State of the Nation*, 58–59.

61. Ibid., 40 (emphasis added). Though see also 62 for a more standard skepticism about the value of "all the mad Actions and Agreements we have made with one another in our *delirious* State."

62. For another example, consider *The Pangs of Credit . . . by an Orphan Annuitant* (London, 1722), 15–16. The "orphan annuitant" claimed that the victims of the South Sea Bubble such as himself were entitled to compensation "since they were not Sufferers *by the Chance, or Accidental Fluctuation of Stocks*, which the [South Sea] Act expressly and positively order'd they should enjoy in common with all other Members of the Company; they suffered by meer Force and Violence"; that is to say, they were victims of pernicious external manipulation, not of those vagaries of chance and accident that were, by contrast, legitimate market forces to "enjoy" or to suffer. The South Sea Act itself, incidentally, had nothing specific to say about chance or accidents—this interpretive move was this author's alone.

63. Tamworth Reresby, *A Miscellany of Ingenious Thoughts and Reflections, in Verse and Prose* (London, 1721), 137–39. For the South Sea affair, see, for example, 308.

64. Flint, *Lunatick*, 13, 15 (emphasis added). Smyth, *Sermon Preach'd at the New Meeting-house in Hackney*, 6.

65. Francis Hutcheson, *An Inquiry concerning Beauty, Order, Harmony, Design* (1725), ed. Peter Kivy (The Hague: Nijhoff, 1973), 47.

66. Richard Cumberland, *A Treatise of the Laws of Nature: Translated . . . by John Maxwell* (1727), ed. Jon Parkin (Indianapolis: Liberty Fund, 2005), 361–62. For the changes in Mandeville's language see above, p. 98.

67. William Wollaston, *The Religion of Nature Delineated* (London, [1724]), 105. [Benjamin Franklin,] *A Dissertation on Liberty and Necessity, Pleasure and Pain* (London, 1725), 11–13. Three years earlier, at age sixteen, Franklin had already demonstrated his interest in probabilistic and aggregate thinking in an essay he wrote under his pseudonym Silence Dogood, recommending an insurance scheme using actuarial calculations (*New-England Courant*, 13 August 1722). That his proposal was lifted unacknowledged from Daniel Defoe (his *Essays upon Several Projects* of 1702) does not detract from its relevance to illuminating Franklin's mindset in the early 1720s. And see Joyce E. Chaplin, *The First Scientific American: Benjamin Franklin and the Pursuit of Genius* (New York: Basic Books, 2006), 29, 45, 80.

68. *The Analysis of the Ballance of Power: Wherein Its Necessity, Origin and History Is Examin'd, and Deduc'd from the Common Principles of Justice* (London, 1720), 38–56.

69. François de Salignac de La Mothe-Fénelon, *Two Essays on the Ballance of Europe: The First Written in French by the Late Archbishop of Cambray, and Translated into English. The Second by the Translator of the First Essay*, [trans. William Grant] (London, 1720), 5–6, 32 (emphasis added). Grant's translation itself subtly added an emphasis on equilibrium to Fénelon's

original text. Where Grant's translation went "unless they unite together to preserve the Ballance," as above, Fénelon had had "*si les autres ne se réunissent pour faire le contrepoids*"—a counterweight that did not already imply a preexisting balance in need of preserving. See the "Supplément" to *Examen de conscience sur les devoirs de la royauté*, in Fénelon, *Oeuvres*, ed. Jacques Le Brun (Paris: Gallimard, 1997), 2:1003.

70. Isaac Gervaise, *The System or Theory of the Trade of the World* (London, 1720; rpt. ed., ed. John M. Letiche, Baltimore, MD: Johns Hopkins University Press, 1954), 7, 15, 18, and passim. See Otto Mayr, *Authority, Liberty, and Automatic Machinery in Early Modern Europe* (Baltimore, MD: Johns Hopkins University Press, 1986), 165–67.

71. Charles Wilson, "International Payments: An Interim Comment," *Economic History Review*, n.s., 15, no. 2 (1962): 368. The rediscovery was Jacob Viner's: see his *Studies in the Theory of International Trade* (1937; New York: Harper and Brothers, 1965), 79. In 1729 Thomas Prior, a Dublin intellectual, published a tract on coin and money with a self-regulating model for the flow of gold and silver that sounded much like Gervaise and may have been informed by him: Thomas Prior, *Observations on Coin in General: With Some Proposals for Regulating the Value of Coin in Ireland* (Dublin, 1729), 13.

72. *A True State of Publick Credit: or, A Short View of the Condition of the Nation, with Respect to Our Present Calamities* (London, 1721), 9 and unpaginated dedication (two emphases added; inconsistent spelling in the original).

73. Mark Blaug, quoted in Antoin E. Murphy, *Richard Cantillon: Entrepreneur and Economist* (Oxford: Clarendon Press, 1986), 1. Gaetan Pirou, quoted in Friedrich A. Hayek, "Richard Cantillon," trans. Michael Ó. Súilleabháin, *Journal of Libertarian Studies* 7, no. 2 (1985): 217–47 (quoted, 225), http://www.econlib.org/LIBRARY/Essays/JlibSt/hykCntl.html. See also Anthony Brewer, *Richard Cantillon: Pioneer of Economic Theory* (London: Routledge, 1992).

74. Richard Cantillon, *Essai sur la nature du commerce en général*, ed. and trans. Henry Higgs (London: Macmillan, 1931), 53, 163–67, 185 (with minor changes to Higgs's translation), and William S. Jevons, "Richard Cantillon and the Nationality of Political Economy" (1881), published with ibid., 348. Hayek is quoted in Murphy, *Richard Cantillon*, 262.

75. Cantillon, *Essai sur la nature du commerce*, 323, 322.

76. Middleton Walker, *A Proposal for Relieving the Present Exigences of the Nation by a Land Credit* (London, 1721), 16. *Richard Steele's "The Theatre," 1720*, ed. John Loftis (Oxford: Clarendon Press, 1962), no. 20, 8 March 1720, 86.

## PART 2 PROLOGUE

1. Rousseau, "Fifth Reverie," in *The Reveries of a Solitary Walker, Botanical Writings, and Letter to Franquières*, trans. Charles E. Butterworth et al., in *Collected Writings of Rousseau*, ed. Christopher Kelly (Hanover, NH: University Press of New England, 2000), 8:42, 45, 46–47. For some nice essays on Rousseau's reveries, see Jean Starobinski and Annette Tomarken, "Rousseau's Happy Days," *New Literary History* 11 (Autumn 1979): 147–66, and Daniel Heller-Roazen, *The Inner Touch: Archaeology of a Sensation* (New York: Zone, 2007), chap. 20.

2. Rousseau, *Discourse on Political Economy*, in *Collected Writings* 3:146. Rousseau, *The Social Contract and Other Later Political Writings*, ed. and trans. Victor Gourevitch (Cambridge: Cambridge University Press, 1997), 41, 69, 70.

3. Rousseau, *The "Discourses" and Other Early Political Writings*, ed. and trans. Victor Gourevitch (Cambridge: Cambridge University Press, 1997), 146, 131. Jean-Jacques Rousseau, *Emile; or, On Education*, trans. Allan Bloom (New York: Basic Books, 1979), 268. Rousseau, letter to Voltaire, 18 August 1756, in *Collected Writings* 3:113.

4. Pierre Bayle, *Réponse aux questions d'un provincial*, in *Oeuvres diverses* (The Hague, 1727), 3:836b (chap. 161). Thomas Burnet, *The Sacred Theory of the Earth* (1691; Carbondale: Southern Illinois University Press, 1965), 44, 71.

5. On the profligacy of Enlightenment natural history see, classically, Ernst Cassirer, *The Philosophy of Enlightenment* (Princeton, NJ: Princeton University Press, 1951), esp. chap. 2. More recently see, among others, Peter Reill, *Vitalizing Nature in the Enlightenment* (Berkeley and Los Angeles: University of California Press, 2005); Jessica Riskin, *Science in the Age of Sensibility: The Sentimental Empiricists of the French Enlightenment* (Chicago: University of Chicago Press, 2002); Paolo Rossi, *The Dark Abyss of Time: The History of the Earth and the History of Nations from Hooke to Vico*, trans. Lydia G. Cochrane (Chicago: University of Chicago Press, 1984); Emma Spary, *Utopia's Garden: French Natural History from Old Regime to Revolution* (Chicago: University of Chicago Press, 2000); Barbara Stafford, *Body Criticism: Imaging the Unseen in Enlightenment Art and Science* (Cambridge, MA: MIT Press, 1991); Mary Ashburn Miller, *A Natural History of Revolution: Violence and Nature in the French Revolutionary Imagination, 1789–1794* (Ithaca, NY: Cornell University Press, 2011).

6. Reill, *Vitalizing Nature*, 9.

7. On description as a project in the Enlightenment, see Joanna Stalnaker, *The Unfinished Enlightenment: Description in the Age of the Encyclopedia* (Ithaca, NY: Cornell University Press, 2010).

## CHAPTER 4

1. Jean-Jacques Rousseau, "Fifth Reverie," in *The Reveries of a Solitary Walker, Botanical Writings, and Letter to Franquières*, trans. Charles E. Butterworth et al., in *Collected Writings of Rousseau*, ed. Christopher Kelly (Hanover, NH: University Press of New England, 2000), 8:42–43.

2. Jack Lesch, "Systematics and the Geometrical Spirit," in *The Quantifying Spirit in the Eighteenth Century*, ed. Tore Frängsmyr, J. L. Heilbron, and Robin E. Rider (Berkeley and Los Angeles: University of California Press, 1990), 75. Joseph Addison, "The Royal Exchange," quoted in Richard Drayton, *Nature's Government: Science, Imperial Britain, and the "Improvement" of the World* (New Haven, CT: Yale University Press, 2000), 269–70. Lisbet Koerner, "Linnaeus' Floral Transplants," *Representations* 47 (Summer 1994): 153. For powerful versions of the imperial argument, see E. C. Spary, *Utopia's Garden: French Natural History from Old Regime to Revolution* (Chicago: University of Chicago Press, 2000); David Philip Miller and Peter Hanns Reill, eds., *Visions of Empire: Voyages, Botany, and Representations of Nature* (Cambridge: Cambridge University Press, 1996); and Londa Schiebinger and Claudia Swan,

eds., *Colonial Botany: Science, Commerce, and Politics in the Early Modern World* (Philadelphia: University of Pennsylvania Press, 2005).

3. On the calamity, see Max Horkheimer and Theodor Adorno, *Dialectic of Enlightenment: Philosophical Fragments* (Stanford, CA: Stanford University Press, 2002), 1. For important revisions, see Elizabeth A. Williams, *A Cultural History of Medical Vitalism in Enlightenment Montpellier* (Aldershot, UK: Ashgate, 2003); Peter Reill, *Vitalizing Nature in the Enlightenment* (Berkeley and Los Angeles: University of California Press, 2005); and more generally, Jessica Riskin, *Science in the Age of Sensibility: The Sentimental Empiricists of the French Enlightenment* (Chicago: University of Chicago Press, 2002). Rousseau, undated letter, ca. April 1774, in *Collected Writings* 8:156. Rousseau, *Reveries*, ibid., 47. On Rousseau and botany, see Bernhard Kuhn, "'A Chain of Marvels': Botany and Autobiography in Rousseau," *European Romantic Review* 17, no. 1 (January 2006): 1–20.

4. Aristotle, *De anima* 415b1.9–11, 415a1.24. Aristotle, *On Plants* 816b1.1, 816a1.37, 816b.1.6–7.

5. See, e.g., Marc J. Ratcliff, *The Quest for the Invisible: Microscopy in the Enlightenment* (Farnham, UK: Ashgate, 2009), 232ff.

6. See, e.g., Reill, *Vitalizing Nature*, and Williams, *Medical Vitalism*. See also, among others, Domenico Bertoloni Meli, *Mechanism, Experiment, Disease: Marcello Malpighi and Seventeenth-Century Anatomy* (Baltimore, MD: Johns Hopkins University Press, 2011); Jacques Roger, *The Life Sciences in Eighteenth-Century French Thought*, ed. Keith R. Benson, trans. Robert Ellrich (Stanford, CA: Stanford University Press, 1997); François Duchesneau, "Vitalism in Late Eighteenth-Century Physiology: The Cases of Barthez, Blumenbach and John Hunter," in *William Hunter and the Eighteenth-Century Medical World*, ed. W. F. Bynum and Roy Porter (Cambridge: Cambridge University Press, 1985).

7. Aristotle, *Metaphysics* 1013a1.29–34.

8. Michael Frede, "The Original Notion of Cause," in *Essays in Ancient Philosophy* (Minneapolis: University of Minnesota Press, 1987), 126. Aristotle, *Parts of Animals* 639b1.13. See also, among others, Alan Code, "The Priority of Final Causes over Efficient Causes in Aristotle's *PA*," in *Aristotelische Biologie: Intentionen, Methoden, Ergebnisse*, ed. Wolfgang Kullmann and Sabine Föllinger (Stuttgart: Steiner, 1997), 127–43, and Robert Bolton, "The Material Cause: Matter and Explanation in Aristotle's Natural Science," in ibid., 102.

9. On the timing of Leibniz's engagement with final causes, and its relation to his meetings with Spinoza, see Daniel Garber, *Leibniz: Body, Substance, Monad* (Oxford: Oxford University Press, 2009), esp. chap. 6. In general on Leibniz and Aristotle, see Christia Mercer, *Leibniz's Metaphysics: Its Origins and Development* (Cambridge: Cambridge University Press, 2001).

10. Rousseau, *Reveries*, in *Collected Writings* 8:64. Rousseau, *Fragments on Botany*, in ibid., 252, 250. "The soul is an actuality [ἐντελέχεια] of the first kind of a natural body having life potentially within it"; Aristotle, *De Anima* 412a1.27.

11. Aristotle, *Parts of Animals* 640b1.22–23 640a1.33. Code, "Priority of Final Causes," 136.

12. On Wolff, see Peter McLaughlin, *What Functions Explain: Functional Explanations and Self-Reproducing Systems* (Cambridge: Cambridge University Press, 2000), 16, and Hein van den Berg, "The Wolffian Roots of Kant's Teleology," *Studies in History and Philosophy of Biological and Biomedical Sciences* 44 (2013): 726–28. John H. Zammito, "Teleology Then

and Now: The Question of Kant's Relevance for Contemporary Controversies over Function in Biology," *Studies in History and Philosophy of Biological and Biomedical Sciences* 37 (2006): 757. On organisms and immanent purposes, see McLaughlin, *What Functions Explain*, 176–78. We are using the term "heuristic" in a much looser way than historians and philosophers of biology interested in Kant's later strict distinction between regulative and constitutive concepts of purposiveness tend to do—in our view, the latter distinction tried to clarify a constitutional ambiguity in earlier biological speculations that gave purposes real causal force without precisely specifying their origins or nature, or overly specified these with reference to divine intention.

13. Rousseau, *Emile*, 276. John Hunter, *A Treatise on the Blood, Inflammation, and Gun-Shot Wounds* (London, 1794), 90. On Hunter, see Duchesneau, "Vitalism in Late Eighteenth-Century Physiology."

14. Hunter, *Treatise on the Blood*, 77, 78.

15. Ibid., 84.

16. Ibid., 78. Boerhaave quoted in Roger, *Life Sciences*, 168.

17. Hunter, *Treatise on the Blood*, 88–89, 84.

18. Ibid., 92, 90. On the complex issue of function in organisms, see Timo Kaitaro, "Can Matter Mark the Hours? Eighteenth-Century Vitalist Materialism and Functional Properties," *Science in Context* 21, no. 4 (2008): 581–92.

19. Harvey quoted in Walter Pagel, *William Harvey's Biological Ideas: Selected Aspects and Historical Background* (Basel: S. Karger, 1967), 255. See also Geoffrey Gorham, "Mind-Body Dualism and the Harvey-Descartes Controversy," *Journal of the History of Ideas* 55, no. 2 (April 1994): 227; C. Webster, "Harvey's 'De Generatione': Its Origins and Relevance to the Theory of Circulation," *British Journal for the History of Science* 3, no. 3 (June 1967): 262–74; Benjamin Goldberg, "A Dark Business, Full of Shadows: Analogy and Theology in William Harvey," *Studies in History and Philosophy of Biological and Biomedical Sciences* 44 (2013): 419–33; and, more generally, his "William Harvey, Soul Searcher: Teleology and Philosophical Anatomy" (PhD diss., University of Pittsburgh, 2012).

20. Stahl quoted in Sarah Carvallo, "Leibniz vs. Stahl: A Controversy Well Beyond Medicine and Chemistry," in *The Practice of Reason: Leibniz and His Controversies*, ed. Marcelo Dascal (Philadelphia: John Benjamins, 2010), 108. The first systematic effort to conceptualize a "tradition" of vitalism was, to our knowledge, Hans Driesch, *Der Vitalismus als Geschichte und als Lehre* (Leipzig: Barth, 1905), which begins its early modern section with Harvey and Stahl. More recent writers have, however, taken issue with both Driesch's genealogy of the tradition and the primacy of Stahl, seeing the latter as principally an animist. See, e.g, Guido Cimino, "Introduction: La problématique du vitalisme," in Guido Cimino and François Duchesneau, *Vitalisms from Haller to the Cell Theory* (Florence: Olschki, 1997), 8, and Roselyne Rey, *Naissance et développement du vitalisme en France de la deuxième moitié du 18e siècle à la fin du Premier Empire* (Oxford: Voltaire Foundation, 2000), 3.

21. Ralph Cudworth, *The True Intellectual System of the Universe* (London, 1678), 162, 160.

22. On these machines, see Domenico Bertoloni Meli, *Thinking with Objects: The Transformation of Mechanics in the Seventeenth Century* (Baltimore, MD: Johns Hopkins University Press, 2006).

23. We use the word "biology" with caution, since the term was not in common use until the early nineteenth century (see Peter McLaughlin, "Naming Biology," *Journal of the History of Biology* 35, no. 1 [Spring 2002]: 1–4). Nor, for that matter, was "the life sciences." We hope readers will forgive the anachronisms.

24. Erna Lesky, *Die Zeugungs- und Vererbungslehren der Antike und ihr Nachwirken* (Wiesbaden: Steiner, 1951), 1296 and more generally on the ancient theory of the seed. See also Anthony Preus, "Galen's Criticism of Aristotle's Conception Theory," *Journal of the History of Biology* 10, no. 1 (Spring 1977): 66, 71.

25. Elizabeth B. Gasking, *Investigations into Generation, 1651–1828* (Baltimore, MD: Johns Hopkins University Press, 1967), 42, 38–39 (on early preformationism). More generally, on the history of the microscope, see Ratcliff, *Quest for the Invisible*; on generation and the microscope in the Netherlands, see Edward G. Ruestow, *The Microscope in the Dutch Republic: The Shaping of Discovery* (Cambridge: Cambridge University Press, 1996), esp. chap. 9. On the various distinctions among the preformationists, see Joseph Needham, *A History of Embryology* (Cambridge: Cambridge University Press, 1934), esp. 170–200.

26. Leibniz, *Monadology*, trans. George Montgomery (LaSalle, IL: Open Court, 1988), 266 (§64), 256 (§22). Nicolas Malebranche, *The Search after Truth*, trans. Thomas M. Lennon and Paul J. Olscamp (Cambridge: Cambridge University Press, 1997), 27 (1.6.1). On Leibniz, see François Duchesneau, "Leibniz versus Stahl on the Way Machines of Nature Operate," in *Machines of Nature and Corporeal Substances in Leibniz*, ed. Justin E. H. Smith and Ohad Nachtomy (Dordrecht: Springer, 2011), 11–28, and Pauline Phemister, "Monads and Machines," in ibid., 39–60. On Malpighi, see Bertoloni Meli, *Mechanism, Experiment, Disease*, esp. chap. 8.

27. On piety, see Reill, *Vitalizing Nature*, 58–58, and Roger, *Life Sciences*, 274. Malebranche, *Search after Truth*, 27 (1.6.1). On the issue of forms, see Roger, *Life Sciences*, 260ff.

28. Aristole, *On the Soul* 412a1.28. Gaskell, *Investigations into Generation*, 44, 45.

29. Maria Teresa Monti, preface to *Commentarius de formatione cordis in ovo incubato*, by Albrecht von Haller (Basel: Schwabe, 2000), ciii. On Haller, see also Reill, *Vitalizing Nature*, chap. 4.

30. Monti, preface to *Commentarius de formatione cordis*, cxxiv, cxxx, cxliii (on Caspar Friedrich Wolff's objection to Haller's idea of "weak visibility"). For more on Wolff, see below, p. 165. For discussions of the metaphysics of the unseen, see Gasking, *Investigations into Generation*, 102, and Ruestow, *Microscope in the Dutch Republic*, 228, 234.

31. Diderot, *Pensées sur l'interprétation de la nature*, in *Oeuvres philosophiques*, ed. Michel Delon (Paris Gallimard, 2010), 327 [LVI].

32. Bertoloni Meli, *Mechanism, Experiment, Disease*, 229. Marcello Malpighi, *De formatione pulli in ovo*, in Howard B. Adelmann, *Marcello Malpighi and the Evolution of Embryology* (Ithaca, NY: Cornell University Press, 1966), 2:935. Charles Bonnet, *Considérations sur les corps organisés* (Amsterdam, 1762), 169. On Bonnet, see also François Duchesneau, "Charles Bonnet's Neo-Leibnizian Theory of Organic Bodies," in *The Problem of Animal Generation in Early Modern Philosophy*, ed. Justin E. H. Smith (Cambridge: Cambridge University Press, 2006), 285–314.

33. Réaumur quoted in Roger, *Life Sciences*, 304. On the many connections between preformationism and mechanism, see Georges Canguilhem, "The Social Model," in *A Vital*

*Rationalist: Selected Writings from Georges Canguilhem*, trans. Arthur Goldhammer, ed. François Delaporte (New York: Zone, 1994), 296. Also see Jessica Riskin, "The Divine Optician," *American Historical Review* 116, no. 2 (April 2011): 365.

34. Letter from Bonaventura Cavalieri to Torricelli, December 1641, quoted in Paolo Mancosu and Ezio Vailati, "Torricelli's Infinitely Long Solid and Its Philosophical Reception in the Seventeenth Century," *Isis* 82, no. 1 (March 1991): 51.

35. Walter Pagel, *William Harvey's Biological Ideas: Selected Aspects and Historical Background* (Basel: Karger, 1967), 233; quoting Harvey, 234, 253, 255. William Harvey, *Anatomical Exercitations concerning the Generation of Living Creatures* (London, 1653), 223. See also Walter Pagel, *New Light on William Harvey* (Basel: Karger, 1976), 25, and more generally, Ann Thomson, *Bodies of Thought: Science, Religion, and the Soul in the Early Enlightenment* (Oxford: Oxford University Press, 2008), esp. chap. 3.

36. On spontaneism, see Ratcliff, *Quest for the Invisible*, esp. chap. 6, which also includes (125ff.) a discussion of Catholic neo-Aristotelianism. On the anti-Harvey camp, which included Francesco Redi and Malpighi, see Bertoloni Meli, *Mechanism, Experiment, Disease*, 183ff.

37. Ratcliff, *Quest for the Invisible*, esp. chap. 3. On Needham, see Shirley A. Roe, "John Turberville Needham and the Generation of Living Organisms," *Isis* 74, no. 2 (June 1983): 159–84; Renato G. Mazzolini and Shirley A. Roe, introduction to *Science against the Unbelievers: The Correspondence of Bonnet and Needham, 1760–1780*, ed. Mazzolini and Roe (Oxford: Voltaire Foundation, 1986); Roger, *Life Sciences*, 399–420; and Ratcliff, *Quest for the Invisible*, chap. 6.

38. John Turberville Needham, *Observations upon the Generation, Composition, and Decomposition of Animal and Vegetable Substances* (London, 1749), 31. On Needham and Voltaire, see Mazzolini and Roe, introduction, *Science against the Unbelievers*, 77–95.

39. Needham, *Observations*, 35, 8. Seed theorists did, in fact, try to account for monsters, supposing, for example, the accidental crushing of a seed in development; see Roger, *Life Sciences*, 318ff.

40. Bonnet, *Considérations*, 15. Carolus Linnaeus, *Systema naturae in quo naturae regna tria, secundum classes, ordines, genera, species, systematice proponuntur* (Stockholm, 1740), 67.

41. For the complexity of the Linnaeus-Buffon disagreement, see Phillip Sloan, "The Buffon-Linnaeus Controversy," *Isis* 67, no. 3 (September 1976): 356–75. For Foucault on Linnaeus, see *The Order of Things* (New York: Vintage, 1970), 140–41. On the charge of arbitrary classification, see also Harriet Ritvo, *The Platypus and the Mermaid and Other Figments of the Classifying Imagination* (Cambridge, MA: Harvard University Press, 1997) 23. "Mule species": Linnaeus in a letter to Albrecht von Haller quoted in James L. Larson, "The Species Concept of Linnaeus," *Isis* 59, no. 3 (Autumn 1968): 294. Linnaeus, *Systema naturae* (Stockholm, 1767), 2:9. Buffon quoted in Phillip R. Sloan, "Buffon, German Biology, and the Historical Interpretation of Biological Species," *British Journal for the History of Science* 12, no. 2 (July 1979): 119, and in Jacques Roger, *Buffon: A Life in Natural History*, trans. Sarah Lucille Bonnefoi (Ithaca, NY: Cornell University Press, 1997), 318.

42. Linnaeus, *Systema naturae* (Stockholm, 1767), 1.2:1326–27 (on this, see Ratcliff, *Quest for the Invisible*, 191–92).

43. On Trembley and especially the circulation of his polyps among natural scientists, see Marc J. Ratcliff, "Abraham Trembley's Strategy of Generosity and the Scope of Celebrity in the Mid-Eighteenth Century," *Isis* 95, no. 4 (December 2004): 555–75; Ratcliff, *Quest for the Invisible*, chap. 5; and Aram Vartanian, "Trembley's Polyp, La Mettrie, and Eighteenth-Century French Materialism," *Journal of the History of Ideas* 11, no. 3 (June 1950): 259–86. *Histoire de l'Académie Royale des Sciences, Année MDCCXLI* (Paris, 1744), 33–34. Johann Christian Reil, *Rhapsodieen über die Anwendung der psychischen Curmethode auf Geisteszerrüttungen* (Halle, 1803), 484.

44. Haller quoted in Joseph Needham, *A History of Embryology* (Cambridge: Cambridge University Press, 1934), 179.

45. Pierre-Louis Moreau de Maupertuis, *Vénus physique*, in *Oeuvres des Maupertuis* (Lyon, 1768), 2:66. Maupertuis, *Système de la Nature*, in ibid., 160–61. On Maupertuis generally, see Mary Terrall, *The Man Who Flattened the Earth: Maupertuis and the Sciences in the Enlightenment* (Chicago: University of Chicago Press, 2002). On Maupertuis and self-organization, see the excellent article by Charles T. Wolfe, "Endowed Molecules and Emergent Organization: The Maupertuis-Diderot Debate," *Early Science and Medicine* 15 (2010): 38–65.

46. Maupertuis, "Essai de cosmologie," quoted in Roger, *Life Sciences*, 381.

47. Maupertuis, *Vénus physique*, in *Oeuvres* 2:147; on the question of heredity, 2:68ff. Wolfe, "Endowed Molecules," esp. 56–57.

48. Roger, *Life Sciences*, 330. For the debate among preformationists about monsters, see Maria Teresa Monti, "Epigenesis of the Monstrous Form and Preformistic 'Genetics' (Lémery-Winslow-Haller)," *Early Science and Medicine* 5, no. 1 (2000): 3–32. On Lémery, see also Michael Hagner, "Enlightened Monsters," in *The Sciences in Enlightened Europe*, ed. William Clark et al. (Chicago: University of Chicago Press, 1999), 188. Monsters—and generally morbid states in medicine—were central to the seventeenth-century debate about epigenesis and preformation as well; see Bertoloni Meli, *Mechanism, Experiment, Disease*, 143–49.

49. Shirley A. Roe, *Matter, Life, and Generation: Eighteenth-Century Embryology and the Haller-Wolff Debate* (Cambridge: Cambridge University Press, 1981), 112. See also Elke Witt, "Form—A Matter of Generation: The Relation of Generation, Form, and Function in the Epigenetic Theory of Caspar F. Wolff," *Science in Context* 21, no. 4 (208): 649–64. On Wolff and monsters, see Hagner, "Enlightened Monsters," 183 (on Peter I), 193–96 (on Wolff in St. Petersburg). More generally, see Ritvo, *Platypus and the Mermaid*, and Lorraine Daston and Katharine Park, *Wonders and the Order of Nature, 1150–1750* (New York: Zone, 1998).

50. [M. la Grange], *Lucrece, Traduction nouvélle* (Paris, 1768), 1.407–8. On La Grange, see Natania Meeker, *Voluptuous Philosophy: Literary Materialism in the French Enlightenment* (New York: Fordham University Press, 2006), 48ff.

51. "*Latens Lucretianum systema, quo animalia non structa esse, sed ex attractione aliqua confluxisse dicuntur* . . ." Monti, preface to *Commentarius de formatione cordis*, cvii.

52. George Cheyne, *Philosophical Principles of Religion: Natural and Reveal'd*, 2nd ed. (London, 1715), 40, 41, 121.

53. Johann Gottfried Herder, *Ideen zur Philosophie der Geschichte der Menschheit*, ed. Martin Bollacher (Frankfurt a. M.: Deutscher Klassiker, 1989), 636.

54. See Ann Thomson, *Materialism and Society in the Mid-Eighteenth Century: La Mettrie's*

*"Discours préliminaire"* (Geneva: Droz, 1981), chap. 1, and Jonathan Israel, *Enlightenment Contested: Philosophy, Modernity, and the Emancipation of Man, 1670–1752* (Oxford: Oxford University Press, 2006), 794ff. For the argument that La Mettrie has to be seen inside the French Enlightenment medical tradition, see Kathleen Wellman, *La Mettrie: Medicine, Philosophy, and Enlightenment* (Durham, NC: Duke University Press 1992).

55. Julien Offray de la Mettrie, "Treatise on the Soul," in *"Machine Man" and Other Writings* (Cambridge: Cambridge University Press, 1996), 43 ("passive principle"), 49 ("motive force"), 51 ("matter is mobile"). La Mettrie, *Machine Man*, in ibid., 7 ("perpetual motion"), 33 ("folly"), 34 ("clock"). On Epicurus, see Charles T. Wolfe, "A Happiness Fit for Organic Bodies: La Mettrie's Medical Epicureanism," in *Epicurus in the Enlightenment*, ed. Neven Leddy and Avi S. Lifschitz (Oxford: Voltaire Foundation, 2009), 69–83. Meeker, *Voluptuous Philosophy*, esp. chap. 3. See also Jessica Riskin, "Mr. Machine and the Imperial Me," in *The Super-Enlightenment: Daring to Know Too Much*, ed. Dan Edelstein (Oxford, 2010), 75–94, although we would not agree that La Mettrie is a "proto-Romantic" (76).

56. Yves Citton, *L'envers de la liberté: L'invention d'un imaginaire spinoziste dans la France de la Lumières* (Paris: Éditions Amsterdam, 2006), 64. Geoffroy quoted in Mi Gyung Kim, *Affinity, That Elusive Dream: A Genealogy of the Chemical Revolution* (Cambridge, MA: MIT Press, 2003), 134. On the complex relation between epigenesis and "Spinozism," see Charles T. Wolfe, "Epigenesis as Spinozism in Diderot's Biological Project," in *The Life Sciences in Early Modern Philosophy*, ed. Ohad Nachtomy and Justin E. H. Smith (Oxford: Oxford University Press, 2014). On the development of atomic theory and the structure of the molecules, see Alan J. Rocke, *Image and Reality: Kekulé, Kopp, and the Scientific Imagination* (Chicago: University of Chicago Press, 2010). In general—although we disagree in particulars—see Robert E. Schofield, *Mechanism and Materialism: British Natural Philosophy in an Age of Reason* (Princeton, NJ: Princeton University Press, 1970).

57. Kim, *Affinity*, 264. Cullen quoted in A. L. Donovan, *Philosophical Chemistry in the Scottish Enlightenment: The Doctrines and Discoveries of William Cullen and Joseph Black* (Edinburgh: Edinburgh University Press, 1975), 101. On Bergman and chemistry, see Reill, *Vitalizing Nature*, 83ff.

58. The list of "paradoxical" synonyms is from Giulio Barsanti, "Les phénomèmes 'étranges' et 'paradoxaux' aux origins de la première revolution biologique (1740–1810)," in Cimino and Duchesneau, *Vitalisms*, 67–68. Williams, *Medical Vitalism*, 175. *Encyclopédie, ou Dictionnaire raisonée des sciences, des arts et des métiers* (Paris, 1751), s.v. "Chymie" and "Mort (*Hist. nat. de l'homme*)." J.-C. Delamétherie quoted in Duchesneau, "Territoires et frontiers du vitalisme (1750–1850)," in Cimino and Duchesneau, *Vitalisms*, 307.

59. Georges-Louis Leclerc de Buffon, *Histoire générale des animaux*, in *Oeuvres complètes*, ed. Stéphane Schmitt (Paris: Champion, 2008), 2:99, 356, 367 (see Aristotle, *On Generation and Corruption* 318a). Reill, *Vitalizing*, 47. François Jacob, *The Logic of Life: A History of Heredity* (New York, 1973), 80. On cause, effect, and reciprocity, see also Duchesneau, "Vitalism in Late Eighteenth-Century Physiology," 283. For a possible experimental basis of this idea of the living molecule, see Phillip Sloan, "Organic Molecules Revisited," in *Buffon 88*, ed. Jean-Claude Beaune et al. (Paris and Lyon: Vrin, 1992), 415–38.

60. Herder, *Ideen* 6:101, 104. Martin S. Staum, *Cabanis: Enlightenment and Medical Philosophy in the French Revolution* (Princeton, NJ: Princeton University Press, 1980), 87. Barthez, *Nouveaux Elements* (1778), quoted in Williams, *Medical Vitalism*, 267. Baader quoted in Reill, *Vitalizing Nature*, 85. For Herder on causation, see H. B. Nisbet, *Herder and the Philosophy and History of Science* (Cambridge: Modern Humanities Research Association, 1970). On Montpellier, see also Charles T. Wolfe and Motoichi Terada, "The Animal Economy as Object and Program in Montpellier Vitalism," *Science in Context* 21, no. 4 (2008): 537–79.

61. Johann Friedrich Blumenbach, *An Essay on Generation* (London, 1792), 20; on the Koehlreuter experiments, 56. Blumenbach, *Handbuch der Naturgeschichte*, 3rd ed. (Göttingen, 1788), 12. Blumenbach, *Beyträge zur Naturgeschichte*, 2nd ed. (Göttingen, 1806), 24, 43, 27, 14. On Blumenbach, see Reill, *Vitalizing Nature*, and Phillip R. Sloan, "Preforming the Categories: Eighteenth-Century Generation Theory and the Biological Roots of Kant's A Priori," *Journal of the History of Philosophy* 40, no. 2 (2002): 229–53.

62. Blumenbach quoted in Timothy Lenoir, "Kant, Blumenbach, and Vital Materialism in German Biology," *Isis* 71, no. 1 (March 1980): 83. Blumenbach, *Beyträge*, 124. For a sharp criticism of Lenoir, especially in relation to Kant, see Robert J. Richards, "Kant and Blumenbach on the *Bildungstrieb*: A Historical Misunderstanding," *Studies in History and Philosophy of Biological and Biomedical Sciences* 31, no. 1 (2000): 11–32. More generally, on the ubiquity of these occulted causes in the life sciences, see Charles T. Wolfe, "On the Role of Newtonian Analogies in Eighteenth-Century Life Science: Vitalism and Provisionally Inexplicable Explicative Devices," in *Newton and Empiricism*, ed. Zvi Biener and Eric Schliesser (Oxford: Oxford University Press, 2014), 223–61.

63. [Georg Christoph Tobler], "Die Natur," in Johann Wolfgang Goethe, *Sämtliche Werke*, ed. Hendrik Birus et al. (Frankfurt a. M.: Deutscher Klassiker, 1989), 25:11–13. Herder, *Ideen* 648.

64. Aristotle, *Physics* 220b15–16. On the unresolved perplexities of time here, see Gareth B. Matthews, *Socratic Perplexity and the Nature of Philosophy* (Oxford: Oxford University Press, 1999), 113ff.

65. Blumenbach, *Essay on Generation*, 14. Reill, *Vitalizing Nature*, 170. Roger, *Life Sciences*, 393. On the temporalization of the life sciences in the later eighteenth century, see also Wolf Lepenies, *Das Ende der Naturgeschichte* (Munich: Hanser, 1976). For the link between preformationism and progressive notions of history, see Francesca Rigotti, "Biology and Society in the Age of Enlightenment," *Journal of the History of Ideas* 47, no. 2 (April–June 1986): 215–33.

66. Diderot, *Le rêve de d'Alembert*, in *Oeuvres philosophiques*, 364, 365, 369. Diderot, *Pensées sur l'interprétation de la nature* (LVII), in ibid., 329.

67. Diderot, *Le rêve de d'Alembert*, in *Oeuvres philosophiques*, 371. Diderot, *Lettre sur les aveugles*, in ibid., 162–63.

68. Maupertuis, *Systême de la nature*, in *Oeuvres* 2:170, 164.

69. Carl Friedrich Kielmeyer, *Über die Verhältniße der organischen Kräfte unter einander in der Reihe der verschiedenen Organisationen, die Geseze und Folgen dieser Verhältniße* (1793; rpt. Marburg an der Lahn: Basilisken, 1993), 3, 37 (emphasis added).

70. Herder, *Ideen* 6:280.

71. Ibid., 31, 191.

72. Ibid., 31, 32. Immanuel Kant, *Allgemeine Naturgeschichte und Theorie des Himmels*, ed. Fritz Krafft (Munich: Kindler, 1971), 124. On Bichat, see Elizabeth Haigh, *Xavier Bichat and the Medical Theory of the Eighteenth Century* (London: Wellcome Institute, 1984), esp. chap. 6. Bichat's key work on life and death, *Recherches physiologique sur la vie et la mort*, was published in 1800.

73. Buffon quoted in Sloan, "Buffon, German Biology," 117 ("constant destruction"), 119 ("sole beings"). Buffon, *Histoire générale des animaux* 2:135 ("constant form").

74. Herder, *Ideen* 6:166, 168, 87–88.

75. Diderot, *Pensées sur l'interprétation de la nature* (L), in *Oeuvres philosophiques*, 321. Maupertuis, *Système de la nature*, in *Oeuvres* 2:170–71. Diderot acknowledges the debt, of course, since Bordeu plays the part of the doctor ministering to the dreaming d'Alembert in the book. See Diderot, *Le rêve de d'Alembert*, in *Oeuvres philosophiques*, 361. On Diderot and Bordeu more generally, see Anne Vila, *Enlightenment and Pathology: Sensibility in the Literature and Medicine of Eighteenth-Century France* (Baltimore, MD: Johns Hopkins University Press, 1998), chap. 5. For the original bee metaphor, see Reill, *Vitalizing Nature*, 133; Williams, *Medical Vitalism* (chap. 4 on Bordeu); and Wolfe and Terada, "Animal Economy," 550–51.

76. Aristotle, *Metaphysics* 982b12–13. Hunter, "Observations on Physiology," in *Essays and Observations on Natural History, Anatomy, Physiology, Psychology, and Geology* (London, 1861), 1:203–4.

77. Foucault, *Order of Things*, 273, 277, 275.

78. Aristotle, *Physics* 217b30. Aristotle, *Metaphysics* 996a17, 995a30, 995a28. On *diaporesai*, see Arthur Madigan, introduction to Aristotle's *Metaphysics Book B and Book K 1–2* (Oxford: Oxford University Press, 1999), xxi–xxii. Thanks to Éric Méchoulan for this formulation.

## CHAPTER 5

1. Georg Lichtenberg, *Vermischte Schriften* (Göttingen, 1801), 2: 21–22. Lichtenberg, *Sudelbücher*, in *Schriften und Briefe* (Munich: Carl Hanser, 1968), 1:625 (F1159). On Lichtenberg and dreams, see J. P. Stern, *Lichtenberg: A Doctrine of Scattered Occasions* (Bloomington: Indiana University Press, 1959), 230ff.

2. Johann Christian Reil, *Rhapsodieen über die Anwendung der psychischen Curmethode auf Geisteszerrüttungen* (Halle, 1803), 92–93.

3. Jean-Jacques Rousseau, "Fifth Reverie," in *The Reveries of a Solitary Walker, Botanical Writings, and Letter to Franquières*, trans. Charles E. Butterworth et al., in *Collected Writings of Rousseau*, ed. Christopher Kelly (Hanover, NH: University Press of New England, 2000), 8:45.

4. Our treatment here shares much with one sketched out by David Bates, "Super-epistemology," in *The Super-Enlightenment: Daring to Know Too Much*, ed. Dan Edelstein (Oxford: Oxford University Press, 2010), 53–74. On psychology, see Fernando Vidal, *The Sciences of the Soul: The Early Modern Origins of Psychology*, trans. Saskia Brown (Chicago: University of Chicago Press, 2011); and more generally, see Ernst Cassirer, *The Philosophy of the Enlightenment*, trans. Fritz C. A. Koelln and James P. Pettegrove (Princeton, NJ: Princeton University

Press, 1951), esp. chaps. 1, 3. For sensitive treatments of empiricism as a philosophical and scientific project, see Anne Vila, *Enlightenment and Pathology: Sensibility in the Literature and Medicine of Eighteenth-Century France* (Baltimore, MD: Johns Hopkins University Press, 1998), and Jessica Riskin, *Science in the Age of Sensibility: The Sentimental Empiricists of the French Enlightenment* (Chicago: University of Chicago Press, 2002). On description in the eighteenth century, see Joanna Stalnaker, *The Unfinished Enlightenment: Description in the Age of the Encyclopedia* (Ithaca, NY: Cornell University Press, 2010), and the essays in John Bender and Michael Marrinan, ed., *Regimes of Description: In the Archive of the Eighteenth Century* (Stanford, CA: Stanford University Press, 2005).

5. René Descartes, "First Meditation," in *Meditations on First Philosophy*, in *The Philosophical Writings of Descartes*, trans. John Cottingham et al. (Cambridge: Cambridge University Press, 1991), 2:13, 15. Descartes, "Third Meditation," in ibid., 24. See also Vidal, *Sciences of the Soul*, 75–76.

6. John Locke, *An Essay concerning Human Understanding*, ed. Peter H. Nidditch (Oxford: Oxford University Press, 1975), 111–14.

7. Ibid., 107, 226–27.

8. Ibid., 149, 226.

9. Gottfried Wilhelm Leibniz, *New Essays on Human Understanding*, trans. and ed. Peter Remnant and Jonathan Bennett (Cambridge: Cambridge University Press, 1981), 53, 54. On Leibniz's theory of mind and "small perceptions," see Alison Simmons, "Changing the Cartesian Mind: Leibniz on Sensation, Representation, and Consciousness," *Philosophical Review* 110, no. 1 (January 2001): 61ff., and Daniel Heller-Roazen, *The Inner Touch: Archaeology of a Sensation* (New York: Zone, 2007), 184–85.

10. Vidal, *Sciences of the Soul*, 76. Descartes, "Sixth Meditation," in *Philosophical Writings* 2:56. Nima Bassiri, "Material Translations in the Cartesian Brain," *Studies in History and Philosophy of Biological and Biomedical Sciences* 43, no. 1 (March 2012): 249. Descartes, *Passions of the Soul*, in *Philosophical Writings* 1:341, 330.

11. Vidal, *Sciences of the Soul*, 76. Locke, *Essay*, 137.

12. John D. Spillane, *The Doctrine of the Nerves: Chapters in the History of Neurology* (Oxford: Oxford University Press, 1981), 20; Vesalius quoted at 43.

13. For an effort to work though the mind-matter issue in more traditionally philosophical terms, see John W. Yolton, *Thinking Matter: Materialism in Eighteenth-Century Britain* (Minneapolis: University of Minnesota Press, 1983).

14. Thomas Willis, "The Anatomy of the Brain," in *The Practice of Physick* (London, 1684), 75, 76, 104. The volume has separate pagination for each work. Generally on Willis, see Georges Canguilhem, *La formation du concept de réflexe aux xviie et xviiie siècles* (Paris: Presses Universitaires de France, 1955), chap. 3, and Robert G. Frank Jr., "Thomas Willis and His Circle: Brain and Mind in Seventeenth-Century Medicine," in *The Languages of Psyche: Mind and Body in Enlightenment Thought*, ed. G. S. Rousseau (Berkeley and Los Angeles: University of California Press, 1990), 107–46. On Willis and "solid-nerve histology," see Wes Wallace, "The Vibrating Nerve Impulse in Newton, Willis and Gassendi: First Steps in a Mechanical Theory of Communication," *Brain and Cognition* 51, no. 1 (2003): 76.

15. Willis, "Anatomy," 103. Willis, "Two Discourses concerning the Souls of Brutes," in *Practice of Physick*, 5, 6. On the sensitive soul, see Rina Knoeff, "Reins of the Soul: The Centrality of the Intercostal Nerves to the Neurology of Thomas Willis and to Samuel Parker's Theology," *Journal of the History of Medicine and Allied Sciences* 59, no. 3 (2004): 424; on the relationship here between this double soul and the by-now-familiar Gassendi, see John Wright, "Locke, Willis, and the Seventeenth-Century Epicurean Soul," in *Atoms, Pneuma, and Tranquility: Epicurean and Stoic Themes in European Thought*, ed. Margaret J. Osler (Cambridge: Cambridge University Press, 1991).

16. Willis, "Treatise of Fevers," in *Practice of Physick*, 79. Luc Brisson and Jean-François Pradeau, "Plotinus," in *A Companion to Ancient Philosophy* (Oxford: Oxford University Press, 2006), 585, 592. Thomas Aquinas, *Summa theologiae* 3.2.5. On hypostasis in an older Christian context, see Gerhard Friedrich, ed., *Theologisches Wörterbuch zum Neuen Testament* (Stuttgart: Kohlhammer, 1969), s.v. "Hypostasis."

17. Willis, "Of Musculary Motion," in *Practice of Physick*, 36. Willis, "Anatomy," 74, 106.

18. Willis, "Souls of Brutes," 30.

19. Vidal, *Sciences of the Soul*, 92; see also Robert J. Richards, "Christian Wolff's Prolegomena to *Empirical and Rational Psychology*: Translation and Commentary," *Proceedings of the American Philosophical Society* 124, no. 3 (June 1980): 227–39.

20. George Rousseau, "'Brainomania': Brain, Mind and Soul in the Long Eighteenth Century," *British Journal for Eighteenth-Century Studies* 30, no. 2 (2007): 161–91. In general, on eighteenth-century blends of body and mind, see Roy Porter, "*Barely Touching*: A Social Perspective on Mind and Body," in Rousseau, *Languages of Psyche*, 45–80; for a different take on the mind-matter issue, see Yolton, *Thinking Matter*, esp. chaps. 2, 5, and 8.

21. Robert Whytt, *An Essay on the Vital and Other Involuntary Motions of Animals*, in *Works* (Edinburgh, 1768), 65, 134, 142. Generally, see R. K. French, *Robert Whytt, the Soul, and Medicine* (London: Wellcome Institute, 1969), and, more recently, Nima Bassiri, "The Brain and the Unconscious Soul in Eighteenth-Century Nervous Physiology: Robert Whytt's *Sensorium Commune*," *Journal of the History of Ideas* 74, no. 3 (July 2013): 425–48.

22. Whytt, "Essay," 97, 134.

23. Gary Hatfield, "Remaking the Science of Mind: Psychology as Natural Science," in *Inventing Human Science: Eighteenth-Century Domains*, ed. Christopher Fox et al. (Berkeley and Los Angeles: University of California Press, 1995), 187. Whytt, *Observations on the Sensibility and Irritability of the Parts of Men and Other Animals*, in *Works*, 292, 290. For the Whytt-Haller dispute, see French, *Robert Whytt*, 68ff.

24. Whytt, "Essay," 145, 122 (emphasis added). Vidal, *Sciences of the Soul*, 97, 329–30.

25. Emanuel Swedenborg, *The Brain, Considered Anatomically, Physiologically and Philosophically*, trans. R. L. Tafel (London, 1882), 1:60–61.

26. Ibid., 76, 77, 79–80, 98.

27. Ibid., 101.

28. Julien Offray de la Mettrie, *Machine Man*, in *"Machine Man" and Other Writings* (Cambridge: Cambridge University Press, 1996), 31. La Mettrie, *Treatise on the Soul*, in ibid., 51.

29. La Mettrie, *Treatise on the Soul*, 57, 59, 51 (an approximate quotation of Leibniz), 62, 63.

30. Ibid., 68, 61. The French doctor Antoine Le Camus made similar arguments respectable in the middle of the next decade, as Anne Vila shows (*Enlightenment and Pathology*, 81ff.).

31. Canguilhem, *La formation du concept de réflexe*, 62, 65. Willis, "Souls of Brutes," 36. La Mettrie, *Treatise on the Soul*, 51.

32. Rush in a letter to Thomas Jefferson, quoted in Richard C. Allen, *David Hartley on Human Nature* (Albany, NY: SUNY Press, 1999), 15. David Hartley, *Observations on Man, His Frame, His Duty, and His Expectations* (London, 1749), 1:500, 503, 84, 83. For a similar theologically loaded version of psychology, see Vidal, *Sciences of the Soul*, 333ff. (on Charles Bonnet).

33. Hartley, *Observations on Man*, 21–22, 33–34.

34. Ibid., 58, 61.

35. Ibid., 70, 71, 77.

36. Cassirer, *Philosophy of the Enlightenment*, 5, 13.

37. David Hume, *Treatise of Human Nature* [1739–40], ed. L. A. Selby-Bigge (Oxford: Clarendon Press, 1978), 87.

38. Ibid., 102 (emphasis added).

39. Ibid., 93, 103, 104, 106.

40. Richard Scholar, *The* Je-Ne-Sais-Quoi *in Early Modern Europe: Encounters with a Certain Something* (Oxford: Oxford University Press, 2005), 3. *A Review of the State of the British Nation* 31 (14 June 1709), in *Defoe's Review*, ed. A. Wellesley Secord (New York: Columbia University Press, 1938), 6:122. See chapter 2 for further discussion of Defoe.

41. David Hume, *An Enquiry concerning Human Understanding*, ed. Tom L. Beauchamp (Oxford: Oxford University Press, 2000), 25, 26, 27, 51. For figures on a beach, see Clarence Glacken, *Traces on the Rhodian Shore: Nature and Culture in Western Thought from Ancient Times to the End of the Eighteenth Century* (Berkeley and Los Angeles: University of California Press, 1967).

42. Hume, *Enquiry*, 11, 30, 54. Hume, *Treatise*, 60. See here Annette Baier, "The Energy in the Cause," in *Death and Character: Further Reflections on Hume* (Cambridge, MA: Harvard University Press, 2008), 234.

43. Annette Baier, "The Life and Mortality of the Mind," in *Death and Character*, 149. Hume, *Enquiry*, 17, 79.

44. On analogy as a method in Enlightenment cognitive theory, see Bates, "Super-epistemology"; on the idea of correlation applied to the *Encyclopédie* and the project of cross-referencing, see John Bender and Michael Marrinan, *The Culture of Diagram* (Stanford, CA: Stanford University Press, 2010), esp. chap. 1. More generally on analogy and cognition, see Barbara Stafford, *Visual Analogy: Consciousness as the Art of Connecting* (Cambridge, MA: MIT Press, 1999). On analogy in the life sciences, see Charles T. Wolfe, "On the Role of Newtonian Analogies in Eighteenth-Century Life Science: Vitalism and Provisionally Inexplicable Explicative Devices," in *Newton and Empiricism*, ed. Zvi Biener and Eric Schliesser (Oxford: Oxford University Press, 2014), 223–61.

45. Leibniz, *New Essays*, 473. Étienne Bonnot de Condillac, *Essay on the Origins of Human Knowledge*, trans. Hans Aarsleff (Cambridge: Cambridge University Press, 2001), 33, 25. For Condillac as a reader of Leibniz, see Laurence L. Bongie's edition of, and introduction to, Condillac's *Les Monades* (Oxford: Voltaire Foundation, 1980). On Condillac and mind, see Bates,

"Super-epistemology," 65–66. More generally, for a refreshing view of Condillac, see Aarsleff's introduction to the volume above.

46. Condillac, *Essay*, 49, 103. For an interpretation of Condillac—and sensationalist thought more generally—as unqualifiedly reductive, see John C. O'Neal, *The Authority of Experience: Sensationist Theory in the French Enlightenment* (University Park, PA: Penn State University Press, 1996).

47. Leibniz, *New Essays*, 473. Maupertuis, *Systême de la Nature*, in *Oeuvres des Maupertuis* (Lyon, 1768), 2:172.

48. Condillac, *Essay*, 88, 89.

49. For the language of the prize-question, and generally on the competition, see Cordula Neis, *Anthropologie im Sprachdenken des 18. Jahrhunderts: Die Berliner Preisfrage nach dem Ursprung der Sprache (1771)* (Berlin: De Gruyter, 2003), 95; more generally, see Avi Lifschitz, *Language and Enlightenment: The Berlin Debates of the Eighteenth Century* (Oxford: Oxford University Press, 2013).

50. Condillac, *Essay*, 114–16.

51. Jean-Jacques Rousseau, "Discourse on the Origin and Foundation of Inequality Among Men," in *The "Discourses" and Other Early Political Writings*, ed. and trans. Victor Gourevitch (Cambridge: Cambridge University Press, 1997), 146, 159.

52. Johann Gottfried Herder, *Abhandlung über den Ursprung der Sprache*, in *Werke* (Frankfurt a.M.: Deutscher Klassiker, 1985), 1:710, 697, 733.

53. For the latter, see *The Archeology of the Frivolous: Reading Condillac*, trans. John P. Leavey Jr. (Pittsburgh, PA: Duquesne University Press, 1980).

54. On apperception, see Heller-Roazen, *Inner Touch*, 201ff. Christian Gottfried Schütz, "Betrachtungen über die verschiedenen Methoden der Psychologie," in *Herrn Karl Bonnets . . . Analytischer Versuch über die Seelenkräfte* (Bremen and Leipzig, 1771), 2:199. Trembley quoted in Vidal, *Sciences of the Soul*, 148. Trembley was the nephew of our polyp researcher, Abraham.

55. Richard Price, introductory letter to Thomas Bayes, "An Essay towards Solving a Problem in the Doctrine of Chances," *Philosophical Transactions* 53 (1764): 372, 373.

56. Condorcet quoted in Keith Michael Baker, *Condorcet: From Natural Philosophy to Social Mathematics* (Chicago: University of Chicago Press, 1975), 187. Condorcet, "A General View of the Science of Social Mathematics" (1793), in *Selected Writings*, ed. Keith Michael Baker (Indianapolis, IN: Bobbs-Merrill, 1976), 194.

57. Condorcet, "De la natur du calcul des probabilités," in *Arithmétique politique: Textes rares ou inédits (1767–1789)* (Paris: Institut national d'études démographiques–PUF, 1994), 289–91. Condorcet, "General View of the Science of Social Mathematics," 201.

58. Hume, *Treatise*, 126. Lorraine Daston, *Classical Probability in the Enlightenment* (Princeton, NJ: Princeton University Press, 1988), 88 quoted, and for a longer treatment of the conflict about inoculation; on Hume and probability, see 199–200.

59. Price, appendix to Bayes, "Doctrine of Chances," 407.

60. Ibid., 409, 411. Daston, *Classical Probability*, 256.

61. Richard Price, *Four Dissertations*, 4th ed. (London, 1777), 393, 395–97.

62. Hartley, *Observations on Man* 1:339, 343.

63. Pierre-Simon Laplace, *Philosophical Essay on Probabilities*, trans. Andrew I. Dale (New York: Springer-Verlag, 1995), 2. On Laplace, see Stephen M. Stigler, *The History of Statistics: The Measurement of Uncertainty before 1900* (Cambridge, MA: Harvard University Press, 1986), esp. chap. 3. For Stigler's translation of the 1774 paper that solved the inverse problem, see Pierre Simon Laplace, "Memoir on the Probability of the Causes of Events," *Statistical Sciences* 1, no. 3 (August 1986): 364–78. On the priority discussion, see also Daston, *Classical Probability*, 268; on Laplace and error, see ibid., 271.

64. Laplace, *Philosophical Essay*, 100, 98, 100 (emphasis added).

65. Ibid., 114, 115, 101, 102.

66. Ibid., 107.

67. Laurence Sterne, *The Life and Opinions of Tristram Shandy, Gentleman* ([York], 1760 [1759]), 1:1–3.

68. Christian Wolff, *Vernünfftige Gedancken von Gott, der Welt, und der Seele des Menschen, auch allen Dingen überhaupt* (1720) in *Gesammelte Werke* (Hildesheim: Ohms, 1983), 1.2:150. On Wolff, see Vidal, *Sciences of the Soul*, 130. More generally, on attention, see Gary Hatfield, "Attention in Early Scientific Psychology," in *Visual Attention*, ed. Richard D. Wright (Oxford: Oxford University Press, 1998), 3. Also see David Braunschweiger, *Die Lehre von der Aufmerksamkeit in der Psychologie des 18. Jahrhunderts* (Leipzig, 1899).

69. Condillac, *Essay*, 20–21.

70. Gottfried Wilhelm Leibniz, *Discourse on Metaphysics*, trans. George Montgomery (La-Salle, IL: Open Court, 1988), 57.

71. Condillac, *Essay*, 24, 25, 41. [Denis Diderot], *Encyclopédie, ou Dictionnaire raisonné des sciences, des arts et des métiers* (Paris, 1751), 1:840, s.v. "Attention." On Diderot and attention in eighteenth-century art, see Michael Fried, *Absorption and Theatricality: Painting and Beholder in the Age of Diderot* (Berkeley and Los Angeles: University of California Press, 1980), esp. chap. 1.

72. Herder, *Abhandlung* 1:722. Herder, *Ideen* 6:151.

73. Bonnet quoted in Vidal, *Sciences of the Soul*, 145–46. Charles Bonnet, *Essai analytique sur les facultés de l'ame*, 2nd ed. (Copenhagen and Geneva, 1769), 124, 127.

74. For Bonnet on mechanism, see *Essai analytique*, 130. Alexander Crichton, *An Inquiry into the Nature and Origin of Mental Derangement* (London, 1798), 1:234, 250–51, 252, 255.

75. Reil, *Rhapsodieen*, 47, 98, 99, 101, 107–8. On Reil more generally, see Thomas Broman, *The Transformation of German Academic Medicine, 1750–1820* (Cambridge: Cambridge University Press, 1996), 86ff.; Robert J. Richards, "Rhapsodies on a Cat-Piano, or Johann Christian Reil and the Foundations of Romantic Psychiatry," *Critical Inquiry* 24, no. 3 (Spring 1998): 700–736; and Richards, *The Romantic Conception of Life: Science and Philosophy in the Age of Goethe* (Chicago: University of Chicago Press, 2002), esp. chap. 7.

76. On this, see Dora Weiner, "The Madman in the Light of Reason: Enlightenment Psychiatry," in *History of Psychiatry and Medical Psychology*, ed. Edwin R. Wallace IV and John Gach (New York: Springer, 2008); and more generally, Roy Porter, *Mind-Forged Manacles: A History of Madness in England from the Restoration to the Regency* (Cambridge, MA: Harvard University Press, 1987).

77. Reil, "Untersuchungen über den Bau des grossen Gehirns im Menschen," *Archiv für Physiologie* 9 (1809): 138. Reil, *Rhapsodieen*, 97.

78. Reil, *Rhapsodieen*, 123, 124, 125, 132.

79. Ibid., 34. Baader quoted in Doris Kaufmann, *Aufklärung, bürgerliche Selbsterfahrung und die "Erfindung" der Psychiatrie in Deutschland, 1770–1850* (Göttingen: Vandenhoeck & Ruprecht, 1995), 90. Herder, *Ideen* 6:181, 182 (emphasis added).

80. Reil, *Rhapsodieen*, 146, 157, 150, 164, 187. On the neurosis, see Porter, "*Barely Touching*," 60.

81. Reil, *Rhapsodieen*, 92. Leibniz, *Philosophical Papers and Letters*, ed. and trans. Leroy E. Loemker (Dordrecht: D. Reidel, 1969), 114–15. Reil, *Rhapsodieen*, 88, 81.

82. Laplace, *Philosophical Essay on Probabilities*, 108. Saussure quoted in Stephan Oettermann, *The Panorama: History of a Mass Medium* (New York: Zone, 1997), 35.

83. Goethe quoted in Oettermann, *Panorama*, 13 (see here too for the giddiness associated with the panorama). Erasmus Darwin, *Zoonomia; or, The Laws of Organic Life* (London, 1796), 233, 238, 239.

84. Darwin, *Zoonomia*, 224, 225–27.

## PART 3 PROLOGUE

1. Joseph Townsend, *A Dissertation on the Poor Laws by a Well-Wisher to Mankind (1786)*, ed. Ashley Montagu (Berkeley and Los Angeles: University of California Press, 1971), 2, 4, 19–20.

2. Ibid., 36–38.

3. Ibid., 37–38, 43–44; on hunger see also 26.

4. Ibid., 29, 36.

5. Ibid., 33–36, 39–41.

6. Ibid., 46, 65. *The Millennium: A Poem in Three Cantos* (London, 1800–1801), 2:86n. (emphasis added).

7. Adam Smith, *An Inquiry into the Nature and Causes of the Wealth of Nations*, ed. Roy H. Campbell and Andrew S. Skinner (Indianapolis, IN: Liberty Fund, 1981), 1:116.

8. *An Inquiry into the Management of the Poor* (London, 1767), 38.

9. The search for "find n3 level" (i.e., "find" and "level" within three words of each other), performed in January 2005, produced 456 hits. Identifying only the relevant ones with a self-organizing meaning, and enumerating them by text rather than by hit, resulted in 10 relevant uses pre-1776 and 98 between 1776 and 1800.

10. Townsend, *Dissertation on the Poor Laws*, 20, 26.

11. Ibid, 20, 64.

12. Antonio de Ulloa, *A Voyage to South America: Describing at large, the Spanish Cities, Towns, Provinces, &c* (London, 1760), 2:218, 222.

13. Condorcet, *Outlines of an Historical View of the Progress of the Human Mind* (London, 1795), 236; and compare his *Esquisse d'un tableau historique des progrès de l'esprit humain*, 4th ed. (Paris, 1798), 148.

## CHAPTER 6

1. Colin Maclaurin, *An Account of Sir Isaac Newton's Philosophical Discoveries* (1748), ed. Larry L. Laudan (New York: Johnson Reprint, 1968), 83. For Maclaurin's earlier career, see *DNB*.

2. Keith Baker, "Enlightenment and the Institution of Society: Notes for a Conceptual History," in *Main Trends in Cultural History*, ed. Willem Melching and Wyger Velema (Amsterdam: Rodopi, 1992), 95–120. David A. Bell, *The Cult of the Nation in France: Inventing Nationalism, 1680–1800* (Cambridge, MA: Harvard University Press, 2001), chap. 1. Edward J. Hundert, *The Enlightenment's Fable: Bernard Mandeville and the Discovery of Society* (Cambridge: Cambridge University Press, 1994). For "economy," see Margaret Schabas, *The Natural Origins of Economics* (Chicago: University of Chicago Press, 2005), chap. 1.

3. [Peter Shaw,] *The Reflector: Representing Human Affairs, as They Are; and May Be Improved* (London, 1750), 181, 256, 317. And see further iterations on 108–9, 161–65, but also the proclaimed repudiation of Mandeville on 22–23.

4. Jesse Molesworth, *Chance and the Eighteenth-Century Novel: Realism, Probability, Magic* (Cambridge: Cambridge University Press, 2010), 170–71, 180–82.

5. Henry Fielding, *Amelia* (London, [1751]), 1:1.

6. Tobias Smollett, *The Adventures of Roderick Random* (London, 1748), 1:2. Leopold Damrosch Jr., *God's Plot and Man's Stories* (Chicago: University of Chicago Press, 1985), 286.

7. With Molesworth see also Sandra Macpherson, *Harm's Way: Tragic Responsibility and the Novel* (Baltimore, MD: Johns Hopkins University Press, 2009), 3 and passim. In the French context see, for example, Thomas M. Kavanagh, *Enlightenment and the Shadows of Chance: The Novel and the Culture of Gambling in Eighteenth-Century France* (Baltimore, MD: Johns Hopkins University Press, 1993), e.g., 108, 114, 122, 235–36, 248. Cf. also Rüdiger Campe, *The Game of Probability: Literature and Calculation from Pascal to Kleist*, trans. Ellwood Wiggins (Stanford, CA: Stanford University Press, 2013).

8. [Charles Gray,] *Considerations on Several Proposals, Lately Made, for the Better Maintenance of the Poor*, 2nd ed. (London, 1752), iii. *A Letter to the Author of Considerations on Several Proposals for the Better Maintenance of the Poor* (London, 1752), 11–12.

9. *Adventurer* 45, 10 April 1753, in Samuel Johnson, *"The Idler" and "The Adventurer,"* Yale Edition of the Works of Samuel Johnson, vol. 2, ed. Walter J. Bate (New Haven, CT: Yale University Press, 1963), 358–59.

10. *Adventurer* 67, 26 June 1753, in ibid., 384–86. Cf. also 128, 26 January 1754, 478–78.

11. James Burgh, *The Dignity of Human Nature* (London, 1754), 191–92, 207, 213 (emphasis added). This line of thinking is reminiscent of Benjamin Franklin's youthful proof of the existence of God (pp. 25–26 above) it is thus worth noting that Franklin had published essays of Burgh's in America in the 1740s and subsequently became his friend. Cf. also [James Burgh,] *Crito; or, Essays on Various Subjects* (London, 1766), 2:156–58.

12. *Edinburgh Review* 1 (1755), 1–2.

13. William Robertson, *The History of the Reign of the Emperor Charles V* (London, 1769), 3:430.

14. *An Examination of the Edinburgh Review, Numb. I, Especially of Art. VI . . .* (n.p., n.d.), [1]. David Hume, *The History of England, from the Invasion of Julius Cæsar to the Accession of Henry VII . . .* (London, 1762), 2:446 (the words quoted here are the final words of the whole work, which would then be subsumed in the longer history of England).

15. Thomas Sheridan, *British Education; or, The Source of the Disorders of Great Britain . . .* (London, 1756), 72–78 (emphasis added).

16. Josiah Tucker, *Instructions for Travellers: 1757* [London?, 1757], 31–32, 48. Tucker, *The Elements of Commerce, and Theory of Taxes* [Bristol?, 1755], 87.

17. Tucker, *Elements of Commerce*, 6–7, [9]. Friedrich A. Hayek, "The Result of Human Action but Not of Human Design," in his *Studies in Philosophy, Politics and Economics* (Chicago: University of Chicago Press, 1967), 100–101; and compare Jacob Viner, *Studies in the Theory of International Trade* (1937; New York: Harper, 1965), 31–32.

18. Samuel Johnson, *The History of Rasselas, Prince of Abyssinia* (1759), Yale Edition of the Works of Samuel Johnson, vol. 16, ed. Gwin J. Kolb (New Haven, CT: Yale University Press, 1990), 144–48. Thomas Arnold, *Observations on the Nature, Kinds, Causes, and Prevention of Insanity* (1782–86), 2nd ed. (London, 1806), 1:135–36.

19. Johnson, *History of Rasselas*, 88–89.

20. *Some Remarks on the Royal Navy: To Which Are Annexed Some Short but Interesting Reflections on a Future Peace* (London, [1760]), 40. Patriot in Retirement, *A Letter from a Patriot in Retirement, to the Right Honourable Mr. William Pitt, upon Resigning His Employment* (London, 1741 [i.e., 1761]), 64. Rousseau's 1761 reflections were translated in *The Miscellaneous Works of J. J. Rousseau* (Edinburgh, 1774), 4:186–90.

21. Henry Home, Lord Kames, *Sketches of the History of Man*, 2nd ed. (Edinburgh, 1778), 2:289 ("Sketch VI. War and Peace Compared"). Robertson, *History of the Reign of the Emperor Charles V*, 3:432–33.

22. Victor de Riquetti, marquis de Mirabeau, *Philosophie rurale, ou économie générale et politique de l'agriculture, réduite à l'ordre immuable des loix physiques & morales, qui assurent la prospérité des empires* (Amsterdam, 1763), 1:138. Ronald L. Meek, *The Economics of Physiocracy: Essays and Translations* (Cambridge, MA: Harvard University Press, 1963), 70. Our discussion of the Physiocrats also benefited from John Shovlin, *The Political Economy of Virtue: Luxury, Patriotism and the Origins of the French Revolution* (Ithaca, NY: Cornell University Press, 2006); Henry C. Clark, *Compass of Society: Commerce and Absolutism in Old-Regime France* (Plymouth, UK: Lexington Books, 2007); Michael Sonenscher, *Before the Deluge: Public Debt, Inequality, and the Intellectual Origins of the French Revolution* (Princeton, NJ: Princeton University Press, 2007); and Liana Vardi, *The Physiocrats and the World of the Enlightenment* (Cambridge: Cambridge University Press, 2012) (see 116 for the analogy to meteorology; we are grateful to Liana Vardi for sharing her rich work with us before publication). Bernard E. Harcourt, *The Illusion of Free Markets: Punishment and the Myth of Natural Order* (Cambridge, MA: Harvard University Press, 2011), chap. 3, emphasizes the significance of the notion of spontaneous or natural order for the Physiocrats (although his argument that they largely invented it ex nihilo is of course incomplete).

23. Mirabeau, as quoted in Vardi, *Physiocrats and the World of the Enlightenment*, 49.

24. Quoted in Meek, *Economics of Physiocracy*, 212. For the interconnections with physiology, see Paul P. Christensen, "Fire, Motion, and Productivity: The Proto-Energetics of Nature and Economy in François Quesnay," in *Natural Images in Economic Thought*, ed. Philip Mirowski (Cambridge: Cambridge University Press, 1994), 249–88.

25. Pierre-Paul Le Mercier de la Rivière, *L'ordre naturel et essentiel des sociétés politiques* (London, 1767), 359 (and cf. 404, 410, etc.)

26. Ibid., 447. For the same formulation see also Mirabeau and Quesnay, *Philosophie rurale*, 1:417, and *Ephémérides du citoyen* 9 (1770): 26. Although this phrase is often reputed to have been the continuation of "*laissez faire, laissez passer*," it is unclear whether anyone in the eighteenth century had in fact put them together in this way.

27. François Quesnay, "Maximes générales du gouvernement économique d'un royaume agricole," in Pierre Samuel Du Pont de Nemours, *Physiocratie, ou Constitution naturelle du gouvernement le plus avantageux au genre human* (Leiden, 1768), 105. Cf. Meek, *Economics of Physiocracy*, 231, and Sonenscher, *Before the Deluge*, 214. And compare, for example, Le Mercier de la Rivière, *L'ordre naturel et essentiel des sociétés politiques*, 143.

28. Ferdinando Galiani, *On Money*, trans. Peter R. Toscano (Ann Arbor, MI: University Microfilms, 1977), 36, 47. Cf. Clark, *Compass of Society*, 178–79.

29. *Oeuvres de Turgot*, ed. Gustave Schelle (Paris: Félix Alcan, 1913–23), 1:206–8. *Turgot on Progress, Sociology and Economics*, ed. Ronald L. Meek (Cambridge: Cambridge University Press, 1973), 6.

30. *Turgot on Progress, Sociology and Economics*, 41.

31. Ibid., 44. Compare the very similar formulation by Turgot's friend the young André Morellet in *Réflexions sur les avantages de la libre fabrication et de l'usage des toiles peintes en France* (Paris, 1758), 205–6.

32. *Oeuvres de Turgot* 1:228. *Turgot on Progress, Sociology and Economics*, 53. Jean-Pierre Poirier, *Turgot: Laissez-faire et progrès social* (Paris: Perrin, 1999), 49–50.

33. *On Universal History*, in *Turgot on Progress, Sociology and Economics*, 69–70, 84. Although Turgot asserted that "this theory is not at all derogatory of Providence" (71), his very choice of title, the same as Bishop Bossuet's late-seventeenth-century masterpiece that demonstrated how the special conservation of the Judeo-Christian religion through history was possible only through divine providence, underscored the gulf that separated them. Compare also the view of the financial administrator and political economist François Véron de Forbonnais of jealousy and enmity between nations as favoring nature's goals in the growth of population and production: Forbonnais, *Principes et observations économiques* (Paris, 1767), as discussed in Sonenscher, *Before the Deluge*, 188.

34. Turgot's letter to Hume is quoted in Robert Brown, "Social Sciences," in *The Cambridge History of Eighteenth-Century Philosophy*, ed. Knud Haakonssen (Cambridge: Cambridge University Press, 2006), 2:1098. Anne-Robert-Jacques Turgot, baron de l'Aulne, *Lettres sur les grains* (n.p., 1771?), 81. "Éloge de Vincent de Gournay," in *Oeuvres de Turgot* 1:619.

35. Du Pont de Nemours and Condorcet quoted at the beginning of Peter Groenewegen, "Turgot and Adam Smith," in his *Eighteenth-Century Economics: Turgot, Beccaria and Smith and Their Contemporaries* (London: Routledge, 2002), 363. For Turgot's liberal policies in office, and his tensions with Du Pont de Nemours ("I still insist that you purge the term 'tutelary' from your vocabulary"), see Vardi, *The Physiocrats and the World of the Enlightenment*, 241–60 (and 244 for the quote from Turgot's letter to Du Pont de Nemours, 14 March 1774).

36. *Remontrances du parlement de Paris au XVIIIe siècle*, ed. Jules Flammermont (Paris, 1898), 3:288, 346. Emma Rothschild, *Economic Sentiments: Adam Smith, Condorcet, and the Enlightenment* (Cambridge, MA: Harvard University Press, 2001), 22–23 (quoting Adam Smith), 126.

37. *Edinburgh Review* 9, no. 17 (October 1806): 84, in review of Millar's posthumous work; and compare Jeffrey's similar understanding of Millar in *Edinburgh Review* 3, no. 5 (October 1803): 157.

38. Duncan Forbes, "'Scientific' Whiggism: Adam Smith and John Millar," *Cambridge Journal* 7 (1954): 651. Cf. Ronald L. Meek, *The Rise and Fall of the Concept of the Economic Machine* (Leicester: Leicester University Press, 1965). Friedrich A. Hayek, "The Results of Human Action but Not of Human Design," in *Studies in Philosophy, Politics and Economics*, 96–105. Ronald Hamowy, *The Scottish Enlightenment and the Theory of Spontaneous Order* (Carbondale: Southern Illinois University Press, 1987). Christopher J. Berry, *Social Theory of the Scottish Enlightenment* (Edinburgh: Edinburgh University Press, 1997). Craig Smith, *Adam Smith's Political Philosophy: The Invisible Hand and Spontaneous Order* (Abingdon, UK: Routledge, 2006).

39. David Hume, "Of the Rise and Progress of the Arts and Sciences," in *Essays and Treatises on Several Subjects*, new ed. (London, 1770), 1:128, 130–31. The essay originally appeared in 1742.

40. David Hume, *An Enquiry concerning the Principles of Morals* (1751), ed. Peter H. Nidditch (Oxford: Clarendon Press, 1975), "Appendix III: Some Farther Considerations with Regard to Justice," 304. Compare David Hume, *A Treatise of Human Nature* (1739–40), ed. Ernest C. Mossner (London: Penguin, 1985), 631.

41. Hume, *Treatise of Human Nature*, 580–81.

42. Ibid., 548–49, 630. Knud Haakonssen, *The Science of a Legislator: The Natural Jurisprudence of David Hume and Adam Smith* (Cambridge: Cambridge University Press, 1981), 20–21.

43. David Hume, *Dialogues concerning Natural Religion*, ed. Martin Bell (London: Penguin, 1990), 77.

44. David Hume, "Of Commerce," in *Essays: Moral, Political, and Literary*, ed. Eugene F. Miller (Indianapolis, IN: Liberty Fund, 1987), 260. Compare Gilbert Stuart's denial of "intention and design" in the history of social institutions in his *An Historical Dissertation concerning the Antiquity of the English Constitution* (Edinburgh, 1770), 222–23: "It is, however, by circumstance and accident that rules are discovered for the conduct of men; and society must have subsisted for ages, and its different appearances must have been often unfolded, before the wisdom of individuals could plan or project the arrangements of nations." See also his *Observations concerning the Public Law, and the Constitutional History of Scotland* (Edinburgh, 1779), 4.

45. David Hume, *The History of England, from the Invasion of Julius Cæsar to the Accession of Henry VII* (London, 1762), 2:446.

46. David Hume, "Of the Balance of Trade," in *Essays: Moral, Political, and Literary*, 312.

47. Hume, *Treatise of Human Nature*, 366–67, 642–43. John Robertson, *The Case for the Enlightenment: Scotland and Naples, 1680–1760* (Cambridge: Cambridge University Press, 2005), 294–96, 301, 319. And compare Rev. [John] Logan, *Elements of the Philosophy of History: Part First* (Edinburgh, 1781), 13–14: "The same sympathy and imitation which gives a similarity of character, manners, and sentiments to a circle of companions, spreads by a like contagion over nations, which are no more than a collection of individuals. . . . The arrangements and improvements which take place in human affairs result not from the efforts of individuals, but from a movement of the whole society."

48. Adam Ferguson, *An Essay on the History of Civil Society* (Edinburgh, 1767), 186–87, 279, 364.

49. Rothschild, *Economic Sentiments*, 136–37; countered most recently by Peter Harrison, "Adam Smith and the History of the Invisible Hand," *Journal of the History of Ideas* 72, no. 1 (January 2011): 29–49.

50. Adam Smith, *The Theory of Moral Sentiments*, ed. David D. Raphael and Alec L. Macfie (Indianapolis: Liberty Fund, 1984), 9, 11 (I.i.1), 159, (III.iv.7), 87 (II.ii.3.5). Christopher Herbert, *Culture and Anomie: Ethnographic Imagination in the Nineteenth Century* (Chicago: University of Chicago Press, 1991), 82. Haakonssen, *Science of a Legislator*, 61. Pierre Force, *Self-Interest before Adam Smith: A Genealogy of Economic Science* (Cambridge: Cambridge University Press, 2003), 67–68.

51. Adam Smith, *An Inquiry into the Nature and Causes of the Wealth of Nations*, ed. Roy H. Campbell and Andrew S. Skinner (Indianapolis: Liberty Fund, 1981), 1:22 (I.i.11), 1:116 (I.x.1), and see above, p. 229, for the full quote), 2:630 (IV.vii.c.88), 2:687 (IV.ix.51). Herbert, *Culture and Anomie*, 99.

52. Smith, *Theory of Moral Sentiments* 233–34 (VI.ii.2.17, added to the last edition of 1790). Rothschild, *Economic Sentiments*, 123–24.

53. Smith, *Inquiry into the Nature and Causes of the Wealth of Nations* 1:413, 422 (III.iv.5, 17).

54. Ibid. 1:456 (IV.ii.9).

55. Smith, *Theory of Moral Sentiments*, 183–85 (IV.i.9–10).

56. Adam Smith, "The History of Astronomy," in his *Essays on Philosophical Subjects*, ed. William P. D. Wightman and J. C. Bryce (Indianapolis: Liberty Fund, 1982), 48–50 (III.1–2).

57. Alec Macfie, "The Invisible Hand of Jupiter," *Journal of the History of Ideas* 32 (1971): 595–99, believes that Smith reversed the meaning of the phrase between this essay and the later works. Rothschild, *Economic Sentiments*, 116–17. Force, *Self-Interest before Adam Smith*, 70.

58. Horace Walpole, *The Castle of Otranto* (1764), ed. Wilmarth S. Lewis (Oxford, 1982), ix (in the introduction, quoting Walpole's letter to the Rev. William Cole, 9 March 1765), 24, 100. We are grateful to Yael Shapira for helpful conversations about this novel.

59. Walpole, *Castle of Otranto*, 7. Emma J. Clery, *The Rise of Supernatural Fiction, 1762–1800* (Cambridge: Cambridge University Press, 1995), 66. Katherine Rowe, *Dead Hands: Fictions of Agency, Renaissance to Modern* (Stanford, CA: Stanford University Press 1999), xi.

60. Jesse M. Molesworth, "Against All Odds: The Sway of Chance in Eighteenth-Century Britain" (unpublished PhD diss., Stanford University, 2003), 267; and his *Chance and the Eighteenth-Century Novel*, chap. 6.

61. Horace Walpole to Rev. William Cole, 9 March 1765, in *The Yale Edition of Horace Walpole's Correspondence*, vol. 1, ed. Wilmarth S. Lewis (New Haven, 1937), 88.

## CHAPTER 7

1. Richard Payne Knight, *The Landscape, a Didactic Poem . . . Addressed to Uvedale Price, Esq.* (London, 1794), 2–4, [23], 25. Michael Clarke and Nicholas Penny, eds., *The Arrogant*

*Connoisseur: Richard Payne Knight, 1751–1824* (Manchester: Manchester University Press, 1982), [ix].

2. Richard Payne Knight, *The Progress of Civil Society: A Didactic Poem* (London, 1796), [3]–4.

3. Ibid., [2], 10–11, 20, [26], 78–81. The last verses quoted replicate closely those in Knight, *Landscape*, 2.

4. Knight, *Progress of Civil Society*, 73, 81, 84, 88–90.

5. Thomas James Mathias, *The Pursuits of Literature; or, What You Will . . . Part the Second* (London, 1796), 15.

6. Quoted together with other similar condemnations in Stephen Daniels, "The Political Iconography of Woodland in Later Georgian England," in *The Iconography of Landscape*, ed. Denis Cosgrove and Stephen Daniels (Cambridge: Cambridge University Press, 1989), 66.

7. Knight, *Progress of Civil Society*, 11, 81–84, 90. The free-born soul is from Knight, *Landscape*, 71.

8. *Anti-Jacobin, or Weekly Examiner: In Two Volumes*, 4th ed. (London, 1799), 1:526–27 (no. 15, 19 February 1798).

9. Thomas James Mathias, *The Pursuits of Literature . . . Part the Third*, 3rd ed. (London, 1797), 7–9.

10. Charles Daubeny, *A Sermon Applicable to the Present Times, and Designed as an Antidote to Those Dangerous Doctrines Now in Circulation . . .* (Bath, 1793), 16. D[enis] O'Bryen, *Utrum Horum? The Government, or the Country?* 2nd ed. (London, 1796), 68–69.

11. Charles Lloyd, *Edmund Oliver* (Bristol, 1798), 1:128, 178–79.

12. Ibid., 36, 52, 126, 131.

13. Charles Lloyd, *A Letter to the Anti-Jacobin Reviewers* (Birmingham, 1799), 20–24.

14. Ibid., 33–35. Lloyd repeated the same vision in his even more strongly antirevolutionary *Lines Suggested by the Fast, Appointed on Wednesday, February 27, 1799* (Birmingham, [1799]).

15. Edmund Burke, *Reflections on the Revolution in France* (London, 1790), ed. Conor C. O'Brien (Harmondsworth: Penguin, 1984), 119, 280–82 (emphasis added).

16. Edmund Burke, "Speech on a Motion Made in the House of Commons," 7 May 1782, in *Works and Correspondence*, new ed.. (London, 1852), 6:130; and *Thoughts and Details on Scarcity* (1795), in ibid. 5:196. See also 339 for another strong self-organizing statement about the evolution of European states in Burke's *Letters on the Proposals for Peace with the Regicide Directory of France* (1796).

17. Thomas Paine, *Common Sense* (14 February 1776) and *Rights of Man* (1791), in *Collected Writings*, ed. Eric Foner (New York: Library of America, 1995), 34, 534 (emphasis added).

18. Thomas Paine, *Rights of Man Part the Second* (1792), in *Collected Writings*, 552–54.

19. Adam Ferguson, *An Essay on the History of Civil Society* (Edinburgh, 1767), 188.

20. Ibid., 93, 196, 410, 412–413n.

21. Soame Jenyns, *A Free Inquiry into the Nature and Origin of Evil* (1757), in *The Works of Soame Jenyns* (London, 1790), 3:135–36 (and similarly 2:215).

22. "Essay on Parties," in *Miscellaneous Works of the Late Philip Dormer Stanhope, Earl of Chesterfield . . .* (London, 1777), 1:54; originally published in *Common Sense* 25, 16 July 1737.

*King Charles II: His Declaration to All His Loving Subjects of the Kingdom of England, Dated from his Court at Breda* . . . (Edinburgh, 1660), 1.

23. Alexander Hamilton, James Madison, and John Jay, *The Federalist: With Letters of "Brutus"* (1788), ed. Terence Ball (Cambridge: Cambridge University Press, 2003), 41–43.

24. Ibid., 44–45, 252–54. Compare Madison's reflections on parties in the *National Gazette*, 23 January 1792, http://www.constitution.org/jm/17920123_parties.htm.

25. Louis-Sébastien Mercier, *Fragments of Politics and History . . . Translated from the French* (1793; London, 1795), 1:438.

26. Ibid., 93–94.

27. Jean-Jacques Rousseau, *The Social Contract*, bk. 2, chap. 3, in *Social Contract: Locke, Hume, Rousseau*, ed. Ernest Baker (Oxford: Oxford University Press, 1960), 193–94. Paul Friedland, *Political Actors: Representative Bodies and Theatricality in the Age of the French Revolution* (Ithaca, NY: Cornell University Press, 2002), chap. 2. Mona Ozouf, "'Public Opinion' at the End of the Old Regime," *Journal of Modern History* 60, supp. (1988): S16, singles out Mably as a French thinker who more than others resigned himself to the multiplicity of contradictory interests in public life; see Gabriel Bonnot de Mably, *Doutes proposés aux philosophes économistes sur l'ordre naturel et essentiel des sociétés politiques* (Paris, 1768).

28. Mercier, *Fragments of Politics and History* 1:11, 12, 98, 105–6, 108, 122–23, 390; 2:13.

29. Rousseau, *Social Contract*, bk. 2, chap. 6, p. 204 (with minor corrections to the translation).

30. Mercier, *Fragments of Politics and History* 1:130–32, 139.

31. Joseph de Maistre, *Considerations on France* (1796), in *The Works of Joseph de Maistre*, ed. and trans. Jack Lively (New York: Schocken, 1971), 66, 77. Compare his *Du Pape* (1819): Every important institution is "formed of itself by the concurrence of a thousand agents, who are almost always ignorant of what they are doing. . . . The institution thus vegetates insensibly over the course of age" (quoted in Owen Bradley, *A Modern Maistre: The Social and Political Thought of Joseph de Maistre* [Lincoln: University of Nebraska Press, 1999], 99; and more generally for de Maistre's notions of providence, chap. 7).

32. *Considerations on France*, in *Works of Joseph de Maistre*, 66–67.

33. *Essay on the Generative Principle of Political Constitutions* (1809), in *Works of Joseph de Maistre*, 152.

34. *Considerations on France*, in *Works of Joseph de Maistre*, 47–49, 77.

35. Isaiah Berlin, introduction (orig. 1952) to Joseph de Maistre, *Considerations on France*, trans. and ed. Richard A. Lebrun (Cambridge: Cambridge University Press, 1994), xiv–xv.

36. Louis-Antoine Saint-Just, "Discours sur la constitution de la France, prononcé à la Convention Nationale . . . 24 Avril 1793," in his *Œuvres complètes* (Paris: Lebovici, 1984), 416. Dan Edelstein, *The Terror of Natural Right: Republicanism, the Cult of Nature, and the French Revolution* (Chicago: University of Chicago Press, 2009), 2, 200. David Bates, *Enlightenment Aberrations* (Ithaca, NY: Cornell University Press, 2002), 162–69.

37. *Considerations on France*, in *Works of Joseph de Maistre*, 62. De Maistre continues in an explicitly Malthusian vein and argues against the seeming randomness of war by comparing it to meteorology: "If tables of massacre were available like meteorological tables, who knows if some law might not be discovered after centuries of observation?"

38. Samuel Taylor Coleridge, *Lay Sermons*, in *The Collected Works of Samuel Taylor Coleridge*, vol. 6, ed. Reginald J. White (Princeton, NJ: Princeton University Press, 1972), 204–5, and for J. S. Mill, editor's introduction, xlii.

39. Ibid., 206–7. Compare Samuel Taylor Coleridge, *The Friend*, 12 October 1809, in *The Collected Works of Samuel Taylor Coleridge*, vol. 4, ed. Barbara E. Rooke (Princeton: Princeton University Press, 1969), 130–31.

40. Samuel Taylor Coleridge, *On the Constitution of the Church and State*, in *The Collected Works of Samuel Taylor Coleridge*, vol. 10, ed. John Colmer (Princeton, NJ: Princeton University Press, 1976), 40. Compare also Coleridge's celebration of self-organizing aggregation in the formation of life and of beauty as discussed in Denise Gigante, *Life: Organic Form and Romanticism* (New Haven, CT: Yale University Press, 2009), 23.

41. Samuel Taylor Coleridge, *Essays on His Times in "The Morning Post" and "The Courier,"* in *The Collected Works of Samuel Taylor Coleridge*, vol. 3, pts. 1–3, ed. David V. Erdman (Princeton, NJ: Princeton University Press, 1978), pt. 1, p. 255 (*Morning Post*, 14 October 1800). Compare similarly pt. 2, p. 176 (*Courier*, 30 May 1811). And cf. Patrick Brantlinger, *Fictions of State: Culture and Credit in Britain, 1694–1994* (Ithaca, NY, 1996), 127–32.

42. Coleridge, *Lay Sermons*, 229–30. Coleridge, *The Friend*, 15 February 1809 [1810], 326. Elsewhere Coleridge added a third protection, the very much visible hand of "church and state," that is to say, the benevolent supervision of a national "clerisy," comprising clergy and intellectuals: see Brantlinger, *Fictions of State*, 128–29.

43. Thomas Malthus, *An Essay on the Principle of Population; or, A View of Its Past and Present Effects on Human Happiness*, 3rd ed. (London, 1806), 2:238. Speech of Mr. Tooke, *Proceedings of the First General Meeting of the British and Foreign Philanthropic Society for the Permanent Relief of the Labouring Classes . . . the 1st of June, 1822* (London, 1822), 38, 42. Charles Bray, *An Essay upon the Union of Agriculture and Manufactures and upon the Organization of Industry* (London, 1844), 30.

44. Jean Charles Léonard Sismondi (de Simonde), *Nouveaux principes d'économie politique; ou, De la richesse dans ses rapports avec la population* (Paris, 1819), 2:217 (bk. 6, chap. 6).

## EPILOGUE

1. Immanuel Kant, *Allgemeine Naturgeschichte und Theorie des Himmels*, ed. Fritz Krafft (Munich: Kindler, 1971), 57–58, 80, 115, 124.

2. Immanuel Kant, "Prolegomena to Any Future Metaphysics That Will Be Able to Come Forward as Science," in *Theoretical Philosophy after 1781*, ed. Henry Allison and Peter Heath (Cambridge: Cambridge University Press, 2002), 57. Kant, "Attempt to Introduce the Concept of Negative Magnitudes into Philosophy" (1763), in *Theoretical Philosophy, 1755–1770*, ed. David Walford in collaboration with Ralf Meerbote (Cambridge: Cambridge University Press, 2003), 239, 240 (italics in original).

3. Immanuel Kant, *Critique of Pure Reason*, trans. and ed. Paul Guyer and Allan W. Wood (Cambridge: Cambridge University Press, 1988), 101, 99.

4. Kant, "Negative Magnitudes," 240.

5. Kant, *Critique of Pure Reason*, 305.

6. Kant, "The Only Possible Argument in Support of a Demonstration of the Existence of God," in *Theoretical Philosophy*, 156, 172.

7. Kant, *Critique of Judgment*, trans. Paul Guyer and Eric Matthews (Cambridge: Cambridge University Press, 2000), 221, 225.

8. Ibid., 245, 242, 247, 228, 246–47. There is an immense and important literature that addresses the complexity of Kant's formulations here. See, e.g., Alicia Juarrero Roqué, "Self-Organization: Kant's Concept of Teleology and Modern Chemistry," *Review of Metaphysics* 39 (September 1985): 187–35; Philip R. Sloan, "Preforming the Categories: Eighteenth-Century Generation Theory and the Biological Roots of Kant's A Priori," *Journal of the History of Philosophy* 40, no. 2 (2002): 237; Helmut Müller-Sievers, *Self-Generation: Biology, Philosophy, and Literature around 1800* (Stanford, CA: Stanford University Press, 1997), chap. 2. But more generally, see the recent collection edited by Philippe Huneman, *Understanding Purpose: Kant and the Philosophy of Biology* (Rochester, NY: University of Rochester Press, 2007), and especially the essays by Huneman (on the complexities of regulative concepts) and John H. Zammito (on Kant's complex relation to Blumenbach and Herder and the concept of epigenesis).

9. There is still a lively debate among Kant scholars about just these topics, and whether indeed Kant was *ever* able to make living organisms fit either his general philosophical system or his approach to the sciences more specifically. See, e.g., Paul Guyer, "Organisms and the Unity of Science," in *Kant and the Sciences*, ed. Eric Watkins (Oxford: Oxford University Press, 2001), 259–80, and John H. Zammito, "'This Inscrutable Principle of an Original Organization': Epigenesis and 'Looseness of Fit' in Kant's Philosophy of Science," *Studies in the History and Philosophy of Science* 34 (2003): 73–109. Also see Eckart Förster, *Kant's Final Synthesis* (Cambridge, MA: Harvard University Press, 2000), and the review by Michael Friedman, "Eckart Förster and Kant's *Opus postumum*," *Inquiry: An Interdisciplinary Journal of Philosophy* 46, no. 2 (2003): 215–27.

10. Immanuel Kant, *Opus Postumum*, ed. Eckart Förster, trans. Eckart Förster and Michael Rosen (Cambridge: Cambridge University Press, 1993), 66, 68, 86.

11. *Kant: Political Writings*, ed. Hans S. Reiss, 2nd ed. (Cambridge: Cambridge University Press, 1991), 41.

12. Immanuel Kant, *Observations on the Feeling of the Beautiful and Sublime*, ed. John T. Goldthwait, 2nd ed. (Berkeley and Los Angeles: University of California Press, 2003), 74.

13. *Kant: Political Writings*, 108.

14. See above, p. 126.

15. In general, see Müller-Sievers, *Self-Generation*; also Robert J. Richards, *The Romantic Conception of Life: Science and Philosophy in the Age of Goethe* (Chicago: University of Chicago Press, 2002).

16. For the new social sciences of the nineteenth century as efforts to rein in the contingency of the previous period, see Peter Wagner, "Certainty and Order, Liberty and Contingency: The Birth of Social Science as Empirical Political Philosophy," in *Rise of the Social Sciences and the Formation of Modernity: Conceptual Change in Context, 1750–1850*, ed. Johan Heilbron, Lars Magnusson, and Björn Wittrock (Dordrecht: Kluwer, 1998), 241–63.

17. For an extended discussion of just this paradox, in the context of English "friendly societies," see Penelope Gwynn Ismay, "Trust among Strangers: Securing British Modernity 'by

way of friendly society,' 1780s–1870s" (PhD diss., University of California, Berkeley, 2010), esp. chaps. 2–3.

18. John Stuart Mill, *On Liberty*, ed. John Gray and G. W. Smith (London: Routledge, 1991), 26–27, 34–35, 119. Mill is, incidentally, often considered the father of the modern doctrine of emergence. See esp. John Stuart Mill, *A System of Logic*, in *Collected Works of John Stuart Mill*, ed. J. M. Robson (Toronto: University of Toronto Press, 1974), 7:371. On emergence, see, e.g., Ernest Nagel, *The Structure of Science: Problems in the Logic of Scientific Explanation* (New York: Harcourt, Brace, and World, 1961), 372. See also the article "Emergent Properties" in the *Stanford Encyclopedia of Philosophy*, http://plato.stanford.edu/entries/properties-emergent.

19. Marshall Berman, *All That Is Solid Melts into Air: The Experience of Modernity* (New York: Simon and Schuster, 1982), 95. Karl Marx and Friedrich Engels, *The Communist Manifesto*, ed. Gareth Stedman Jones (London: Penguin, 2002), 223. Karl Marx, *Grundrisse: Foundations of the Critique of Political Economy*, ed. Martin Nicolaus (Harmondsworth: Penguin, 1993), 196–97. Stedman Jones, preface to *Communist Manifesto*, 13.

20. Paul H. Barrett et al., eds., *Charles Darwin's Notebooks, 1836–1844: Geology, Transmutation of Species, Metaphysical Enquiries* (Ithaca, NY: Cornell University Press, 1987), 342–43.

21. Charles Darwin, *On the Origin of Species* (Cambridge, MA: Harvard University Press, 1964), 490.

22. Ibid., 43. Barrett et al., *Charles Darwin's Notebooks*, 390 (emphasis added).

23. Frederick Burkhardt et al., eds., *The Correspondence of Charles Darwin* (Cambridge: Cambridge University Press, 1984–), 8:328.

# INDEX